T0190037

Probability Theory and Stochastic Modelling

Volume 89

The **Probability Theory and Stochastic Modelling** series is a merger and continuation of Springer's two well established series Stochastic Modelling and Applied Probability and Probability and Its Applications series. It publishes research monographs that make a significant contribution to probability theory or an applications domain in which advanced probability methods are fundamental. Books in this series are expected to follow rigorous mathematical standards, while also displaying the expository quality necessary to make them useful and accessible to advanced students as well as researchers. The series covers all aspects of modern probability theory including

- Gaussian processes
- Markov processes
- Random fields, point processes and random sets
- Random matrices
- Statistical mechanics and random media
- Stochastic analysis

as well as applications that include (but are not restricted to):

- Branching processes and other models of population growth
- Communications and processing networks
- Computational methods in probability and stochastic processes, including simulation
- Genetics and other stochastic models in biology and the life sciences
- Information theory, signal processing, and image synthesis
- Mathematical economics and finance
- Statistical methods (e.g. empirical processes, MCMC)
- Statistics for stochastic processes
- Stochastic control
- Stochastic models in operations research and stochastic optimization
- Stochastic models in the physical sciences

More information about this series at http://www.springer.com/series/13205

Boris L. Rozovsky • Sergey V. Lototsky

Stochastic Evolution Systems

Linear Theory and Applications to Non-Linear Filtering

Second Edition

Boris L. Rozovsky
Division of Applied Mathematics
Brown University
Providence
Rhode Island, USA

Sergey V. Lototsky
Department of Mathematics
University of Southern California
Los Angeles
California, USA

1st edition (1990) translated from the Russian by A. Yarkho and published under Rozovskii, B.L. as volume 35 in the series "Mathematics and Its Applications" by Kluwer Academic Publishers.
Original Russian language edition: ЭВОЛЮЦИОННЫЕ СТОХАСТИЧЕСКИЕ СИСТЕМЫ, published by Nauka Publishers, Moscow, 1983

ISSN 2199-3130 ISSN 2199-3149 (electronic)
Probability Theory and Stochastic Modelling
ISBN 978-3-030-06933-9 ISBN 978-3-319-94893-5 (eBook)
https://doi.org/10.1007/978-3-319-94893-5

Mathematics Subject Classification (2010): 60H15 (primary), 35R60 (secondary)

This Springer imprint is published by the registered company Springer Nature Switzerland AG
The registered company address is: Gewerbestrasse 11, 6330 Cham, Switzerland

To our parents

Preface to the Second English Edition

The second edition benefits from the addition of a new co-author, S.V. Lototsky, who contributed substantially to this book. Due to his hard work, the quality of the second edition has improved drastically. In particular, all seven chapters were carefully revised, and we believe that the book is now much more reader-friendly. Compared to the first edition [136], labeling of statements and equations has been changed to comply with the current standards of the publisher. Beside these editorial changes, a new chapter was added, *Chaos Expansion for Linear Stochastic Evolution Systems*. Still, a number of interesting and important results related to stochastic evolution systems, stochastic partial differential equations, and filtering of random process did not make it into the book. An interested reader can find some of these topics in references [20, 33, 34, 38, 41, 67, 101, 105, 112, 115, 116, 125, 128, 145, 146, 158, 159, 159, 160].

The first author [BR] acknowledges the support of several grants during the period the second edition of the book was written: AFOSR (5-21024 (inter), FA9550-09-1-0613) ARO (DAAD19-02-1-0374, W911NF-07-1-0044, W911NF-13-1-0012, W911N-16-1-0103), NSF (DMS 0604863, DMS 1148284), ONR (N0014-03-1-0027, N0014-07-1-0044, OSD/AFOSR 9550-05-1-0613, and SD Grant 5-21024 (inter).

The second author [SL] gratefully acknowledges hospitality of the Division of Applied Mathematics at Brown University on several occasions that were crucial to the success of the project.

Providence, RI, USA
Los Angeles, CA, USA
March 2018

Boris L. Rozovsky
Sergey V. Lototsky

Preface to the First English Edition

The subject of this book is linear stochastic partial differential equations and their applications to the theory of diffusion processes and non-linear filtering.

Until recently, the term "stochastic differential equation" did not need any specifications: in 99 cases out of 100 it was applied to ordinary stochastic differential equations. Their theory started to develop at the beginning of the 1940s, based on Itô's stochastic calculus [52, 53], and now forms one of the most beautiful and fruitful branches of the theory of stochastic processes, [37, 51, 55, 65, 93, 149].

In the middle of the 1970s, however, the situation changed: in various branches of knowledge (primarily, in physics, biology, and control theory) a vast number of models were found that could be described by stochastic evolution partial differential equations. Such models were used, for example to describe a free (boson) field in relativistic quantum mechanics, a hydromagnetic dynamo-process in cosmology, diffraction in random-heterogeneous media in statistical physics, the dynamics of populations for models with a geographical structure in population genetics, etc.

The emergence of this new type of equation was simultaneously stimulated by the inner needs of the theory of differential equations. Such equations were effectively used to study parabolic and elliptic second-order equations in infinite-dimensional spaces.

An especially powerful impetus to the development of the theory of evolution stochastic partial differential equations was given by the problem of non-linear filtering of diffusion processes.

The filtering problem (estimation of the "signal" by observing it when it is mixed with a "noise") is one of classical problem in the statistics of stochastic processes. It also belongs to a rare type of purely engineering problems that have a precise mathematical formulation and allows for a mathematically rigorous solution.

The first remarkable results in connection with filtering of stationary processes were obtained by A.N. Kolmogorov [62] and N. Wiener [163]. After the paper by R. Kalman and R. Bucy [57] was published, the 1960s and 1970s witnessed a rapid development of filtering theory for systems whose dynamics could be described by

Itô's stochastic differential equations. The results were first summed up by R.Sh. Liptser and A.N. Shiryayev [93] and G. Kallianpur [56].

One of the key results of the modern non-linear filtering theory states that the solution of the filtering problem for the processes described by Itô's ordinary stochastic equations is equivalent to the solution of an equation commonly called the filtering equation. The filtering equation is a typical example of an evolution stochastic partial differential equation. An equation of this type can be regarded as an "ordinary" Itô equation

$$du(t) = A(t, u(t))dt + B(t, u(t))dW(t)$$

for the process $u(t)$ taking values in a function space $\mathbb{X}$. The coefficients A and B of "drift" and "diffusion" in this equation are operators (unbounded, as a rule), and $\dot{W}(t)$ is a "white noise" taking values in a function space. Such an equation may be regarded as a system (an infinite one, if the space $\mathbb{X}$ is infinite) of one-dimensional Itô equations. Below we shall call equations (systems of equations) of this type stochastic evolution systems.

The theory of stochastic evolution systems is quite a young branch of science but it has nevertheless generated many interesting and important results, much more than it would be reasonable to include in one book. The references [3, 6, 7, 16–18, 84, 121, 158, 161, etc.] contain sections devoted to the theory.

The present monograph has the following objectives:

(1) to cover the general theory of linear stochastic evolution systems (LSESs) with unbounded drift and diffusion operators;
(2) to construct the theory of Itô's second-order parabolic equations;
(3) to investigate, on the basis of the latter, the filtering problem and related issues (interpolation, extrapolation) for processes whose trajectories can be described by Itô's ordinary equations.

The first item is the subject of Chaps. 2 and 3, the second is the subject of Chaps. 4 and 5, while the third item is the subject of Chap. 6. Chapter 1 contains examples and auxiliary results.

Since the time the present book was finished (the Russian edition of this book was published in 1983), very important discoveries have been made in the theory of stochastic differential equations: namely the development of the Malliavin's calculus and its elaborations, which provided the basis for the stochastic interpretation of Hörmander's results on the hypoellipticity of elliptic-parabolic second-order equations.

This made it necessary to include Chap. 7 in the English version of the book, devoted to hypoellipticity of second-order stochastic partial differential equations and, in particular, to the filtering equations. Necessary extensions of other parts of the book have been done as well.

Here our brief chapter-to-chapter summary of what is covered ends, but we would like to remark that each chapter has its own introduction describing its contents sufficiently thoroughly.

Throughout the book the author has tried to adhere to the universal language of functional analysis and has given preference to functional-analytical methods of proof, the rationale for this is that the book was written to be understood by researchers of different interests and educational backgrounds.

The necessary prerequisite for the reader is a familiarity with functional analysis and the theory of stochastic processes within the framework of standard graduate-level university courses. No preliminary reading on partial differential equations is needed.

Even though the book has a strictly hierarchical structure (each chapter to follow is based on the result of the preceding one), the exposition allows the reader interested only in some of the chapters to begin reading them directly and to use the preceding chapters only for reference.

Each section of the book is divided into paragraphs that are enumerated but not titled. Theorems, lemmas, propositions, definitions, notes, and warnings are numbered according to the paragraph they belong to (each paragraph contains no more then one theorem, lemma etc.). The formulas are numbered independently within each section. The formula number includes no less than two numbers: the section and the formula within the section. When formulas are referred to in a subsequent chapter, the number of the chapter is added. Thus (2.3.14) means Formula 14 from Sect. 3 of Chap. 2. Paragraphs (and therefore theorems, lemmas etc.) when referred to, are indicated in a similar way. When references are made within a section, only the number of the paragraph is indicated. A section-paragraph reference is used within chapters but different sections. For reference to a paragraph from another chapter, the number of the chapter is added. Thus, Theorem 3.2.4 belongs to Paragraph 4, Sect. 2, and Chap. 3.

The author owes his heart felt gratitude to N.V. Krylov for numerous important discussions and valuable suggestions on the subject of this book. His thanks are also due to Mrs. A. Yarkho the translator for the English edition and to R.F. Anderson for thoughtful editorial work on it.

The author is much indebted to all the participants of the seminar on the theory of martingales and control at Steklov Mathematical Institute of the USSR Academy of Sciences and of the seminar on stochastic differential equations with partial derivatives at Moscow State University who read and discussed various parts of the book.

Author's thanks are due to Benita Boyd for the patience and efficiency with which she did the word processing.

Moscow, Russia
Charlotte, NC, USA

Boris L. Rozovsky

[illegible]

Moscow, Russia

Cincinnati, OH, USA

Contents

Standard Notations

$\mathbb{E}$	Expectation
$\mathbb{E}[\cdot\vert\mathscr{F}]$	Conditional expectation with respect to the σ-algebra $\mathscr{F}$
$:=$	Equals by definition
$\wedge$	Min
$\vee$	Max (also, see Warning 1.8)
a.s.	Almost sure
a.a.	Almost all
$1_{\{A\}}$	The indicator function of a set A
$\mathbb{I}$	The identity operator
A^*	Transpose of a matrix or dual of an operator
A'	Adjoint of an operator
$\mathbb{R}_+$	The positive half-line $[0, +\infty)$
$\mathbb{N}$	The set of positive integers $1, 2, 3, \ldots$
I	An open interval in $\mathbb{R}_+$
$\mathbb{R}^n$	The Euclidean space of dimension n
f_i	Partial derivative $\partial f/\partial x^i$, either classical or generalized
$\boldsymbol{\Delta}$	The Laplace operator
$\mathcal{T}_\varepsilon,\ \varepsilon > 0$	The (Sobolev) averaging operator
G	A domain in $\mathbb{R}^{\mathrm{d}}$ with sufficiently regular boundary
$\vert\cdot\vert$	The absolute value of a real number, length of a multi-index, or the Euclidean norm of a vector in a finite-dimensional Euclidean space $\mathbb{R}^n$
$(\cdot, \cdot)$	The standard inner product in $\mathbb{R}^{\mathrm{d}}$ or in the default Hilbert space $\mathbb{H}$ (Chaps. 2 and 3)
ℓ_{d}	The Lebesgue measure on $\mathbb{R}^{\mathrm{d}}$ $(\ell := \ell_1)$
$\mathscr{P}$	The σ-algebra of predictable sets
$\mathscr{B}(\mathbb{X})$	The Borel σ-algebra of subsets of a topological space $\mathbb{X}$
$\overline{\mathscr{B}(\mathbb{X})}$	The completion of $\mathscr{B}(\mathbb{X})$ with respect to the underlying measure
$\mathbb{L}_p(G)$	$\mathbb{L}_p\big(G, \overline{\mathscr{B}(G)}, \ell_d; \mathbb{R}^1\big)$, $1 \le p \le \infty$; similarly for $\mathbb{L}_p(\mathbb{R}^{\mathrm{d}})$
$\mathbb{C}(I; \mathbb{X})$	The space of strongly continuous mappings from I to $\mathbb{X}$

$\mathbb{C}^n(G)$	The space of continuous real-valued functions on G with n continuous derivatives, $n \in \mathbb{N}\bigcup\{0\}\bigcup\{\infty\}$
$\mathbb{C}_0^n(G)$	The subset of $\mathbb{C}^n(G)$ containing functions with compact support in G
$\mathbb{C}_b^n(G)$	The space of bounded continuous real-valued functions on G having n bounded continuous derivatives
$\mathbb{C}^{m,n}(I \times G)$	The space of real-valued functions $f = f(t, x)$ on $I \times G$ having m continuous derivatives in t and n continuous derivatives in x
$\mathbb{C}_b^{m,n}(I \times G)$	The subspace of $\mathbb{C}^{m,n}(I \times G)$ consisting of bounded continuous real-valued functions having m bounded continuous derivatives in t and n bounded continuous derivatives in x
$\mathbb{W}_p^m$, $\mathbb{W}_p^m(r, \mathbb{R}^d)$, $\mathbb{H}^m$, $\mathbb{H}^m(r, \mathbb{R}^d)$	Various Sobolev spaces
$\mathbb{H}^0$	$\mathbb{L}_2(\mathbb{R}^d)$
$(\cdot, \cdot)_m$	The inner product in $\mathbb{H}^m$
$[\cdot, \cdot]$	The canonical bilinear functional or the Lie bracket (Chap. 7)
$\mathfrak{L}(\mathbb{X}; \mathbb{Y})$	The space of continuous linear operators from $\mathbb{X}$ to $\mathbb{Y}$
$\mathfrak{L}_1(\mathbb{X}; \mathbb{Y})$	The space of nuclear operators from $\mathbb{X}$ to $\mathbb{Y}$
$\mathfrak{L}_2(\mathbb{X}; \mathbb{Y})$	The space of Hilbert–Schmidt operators from $\mathbb{X}$ to $\mathbb{Y}$
$\|\!\|\!\| \cdot \|\!\|\!\|$	The Hilbert–Schmidt norm of an operator
$\square$	The end of a statement or a proof

Chapter 1
Examples and Auxiliary Results

1.1 Introduction

The first of the four sections in this chapter presents examples of linear stochastic evolution systems (LSESs) arising in various applications. The following three sections collect a number of auxiliary results which are used systematically throughout the book. In particular, Sect. 1.5 surveys the theory of stochastic ordinary differential equations.

1.2 Examples of Stochastic Evolution Systems

1.2.1. In this section we give examples of LSESs. The main example, which will motivate much of the development in the book, is filtering of diffusion processes. Other examples come from physics, chemistry, and biology.

1.2.2 The Filtering Equation

Suppose that we observe the sum of a "signal" $x = x(t)$ and "noise" $w = w(t)$. The problem of the estimation of the unobservable signal $x(t)$ or a function of $x(t)$ on the basis of the observations $y(s) = x(s) + w(s),\ s \geq t$, is usually referred to as the filtering problem.

This model arises in many applications. For example, it may be that the process $y = y(t)$ describes the position of a moving object computed on the basis of radar observations, $w = w(t)$ is the measurement error, and the signal $x(t)$ represents the true coordinates of the object.

B. L. Rozovsky, S. V. Lototsky, *Stochastic Evolution Systems*, Probability Theory and Stochastic Modelling 89, https://doi.org/10.1007/978-3-319-94893-5_1

For a more precise formulation, suppose that the signal $x = x(t)$ is the solution of the ordinary differential equation $dx(t)/dt = b(x(t))$, $x(0) = x_0$, and the observed process y satisfies

$$dy(t) = b\big(x(t)\big)\,dt + dw(t),\ y(0) = x_0,$$

where w is a Wiener process representing the measurement error.

The variance of $w(t)$ is equal to t, whereas in applications the observation error can depend on the value of the observation. Thus, the process $y = y(t)$ satisfying the Itô equation

$$dy(t) = b\big(x(t)\big)\,dt + \sigma_0\big(y(t)\big)dw(t),\ y(0) = x_0,$$

with variance equal to $\int\limits_{[0,t]} \mathbb{E}\sigma_0^2\,(y(s))ds$, is a more realistic model for the observation process. In various problems of practical importance, the evolution equation of the signal process also includes random perturbations. In these cases the signal can be described by the Itô equations

$$dx(t) = b\big(x(t)\big)\,dt + \sigma\big(x(t)\big)dv(t),\ x(0) = x_0,$$

where $v = v(t)$ is another Wiener process and x_0 is a random variable with a known distribution.

Let us specify the filtering problem as follows. Given a function $f = f(x)$ with $\mathbb{E}\big|f\big(x(t)\big)\big|^2$ finite, find the function $\hat{f}(y_t^0)$ that minimizes the functional $\mathbb{E}\big|f\big(x(t)\big) - g(y_t^0)\big|^2$ in the class of square integrable functions $g(y_t^0)$ of the trajectories y_t^0 of the observed process. As will be shown in Chap. 6, if the Wiener processes w and v are independent, then

$$\hat{f}(y_t^0) = \frac{\displaystyle\int_{\mathbb{R}^1} f(x)\varphi(t,x)dx}{\displaystyle\int_{\mathbb{R}^1} \varphi(t,x)dx},$$

and the function $\varphi = \varphi(t,s)$ is the solution of the LSES

$$\begin{aligned} d\varphi(t,x) &= \left(\frac{1}{2}\frac{\partial^2}{\partial x^2}\big(\sigma^2(x)\varphi(t,x)\big) - \frac{\partial}{\partial x}\Big(b(x)\varphi(t,x)\Big)\right) dt \\ &\quad + \frac{b(x)}{\sigma_0\big(y(t),t\big)}\,\varphi(t,x)dy(t),\ t > 0,\ x \in \mathbb{R}^1, \\ \varphi(0,x) &= p(x), \end{aligned}$$

where $p(x)$ is the probability density function of x_0.

As a matter of fact, a more general situation is considered in Chap. 6, namely the processes x and y are multidimensional, the processes v and w are possibly correlated, and the functions b and σ depend on t and y_t^0.

1.2.3 The Krylov Equation (Backward Diffusion Equation)

The Itô equation

$$du(t,s,x) = b\big(u(t,s,x)\big)\,dt + \sigma\big(u(t,s,x)\big)dw(t), \quad t > s,\ u(s,s,x) = x \in \mathbb{R}^1,$$

simulates the dynamics of a particle (in the process of diffusion) starting at the moment s from the point x. Thus, the solution of this equation also depends on s, x. In 1975 it was pointed out by N.V. Krylov that, for fixed t, the function $u = u(t,s,x)$ as a function of s and x is a solution of the equation

$$-du(t,s,x) = \left(\frac{1}{2}\sigma^2(x)\,\frac{\partial^2}{\partial x^2}\,u(t,s,x) + b(x)\,\frac{\partial}{\partial x}\,u(t,s,x)\right)ds$$

$$+\sigma(x)\,\frac{\partial}{\partial x}\,u(t,s,x) * dw(s), \ s < t, \ u(t,t,x) = x.$$

The symbol $*$ before $dw(s)$ indicates that the stochastic integral is interpreted as the backward Itô integral (see Sect. 1.5.13). We study this equation in Chap. 5.

1.2.4 The Helmholtz Parabolic Equation

In statistical radiophysics, LSESs are useful in describing diffraction in a random non-uniform medium [60, 139]. For example, the following version of the Helmholtz equation is proposed in [59] to describe the propagation of a monochromatic light in a medium with large-scale non-uniformity:

$$\frac{\partial u(t,x)}{\partial t} = \frac{\mathrm{i}}{2k}\sum_{j=1}^{2}\frac{\partial^2 u(t,x)}{\partial x_j^2} - \frac{k^2}{8}A_0 u(t,x)$$

$$+\frac{ik}{2}\,u(t,x)\,\dot{W}(t,x), \ t > 0, \ x \in \mathbb{R}^2, \ u(0,x) = u_0(x); \ \mathrm{i} = \sqrt{-1}.$$

In this equation the axis t is chosen in the direction of original propagation of the light, k is the mean wave number, $W(t,x)$ is a “white noise” with respect t and A_0 is the trace of the correlation operator of $W(t,x)$ in x. The white noise describes the relative magnitude of permittivity.

1.2.5 A Continuous Branching Model with Geographical Structure

Models connected with the theory of branching processes are extremely useful in various applications. Let us take a look at one such model.

Suppose that some region consists of a series of subregions and the jth subregion at the moment t contains $p(j,t)$ particles of a certain type. The function $p(j,t)$ can be either the total number or the fraction of the particles. Every particle in this subregion within a small time interval $\Delta(t)$ can die with probability $\lambda(j)\Delta t+o(\Delta t)$, or give birth to a new particle with probability $\mu(j)\Delta t+o(\Delta t)$. Apart from that, a migration (diffusion through the boundary of the subregion) is possible. If the diameter of the partition of the region tends to zero, then a suitably normalized $p(j,t)$ converges to a random field $u=u(t,x)$ representing the density of particles at the time t and at the place x.

This model is successfully used in population biology and chemistry (e.g. [2, 21, 42]). In [2], such a model was used to describe a chemical reaction. It was shown that the function $u=u(t,x)$, representing the proportion of the reacting particles at the time t at the point x of an infinitely long rector, satisfies the equation

$$\frac{\partial u(t,x)}{\partial t}=\frac{\partial}{\partial x}\Big((\lambda-\mu)(\psi(t,x))\,u(t,x)\Big)+D\frac{\partial^2 u(t,x)}{\partial x^2}+G\dot{W}(t,x),$$
$$t>0,\ x\in\mathbb{R},\ u(0,x)=u_0(x),$$

where D is the diffusion coefficient, $\psi=\psi(t,x)$ is the average concentration of the reacting particles, and G is an operator such that

$$GG^*f=-D\frac{\partial}{\partial x}\left(\psi(t,x)\frac{\partial}{\partial x}f\right)+(\lambda+\mu)\big(\psi(t,x)\big)f.$$

The (generalized) function $\dot{W}=\dot{W}(t,x)$ is a random field that is Gaussian white noise with respect to t.

Similar analysis of models in population genetics leads to many other linear and non-linear stochastic evolution equations; cf. [21, 42].

1.2.6 Equation of the Free Field

Let $\mathscr{S}$ be the space of all rapidly decreasing real functions on $\mathbb{R}^{d+1}$. Denote by $\mathscr{S}^*$ the dual space, known as the Schwartz space of tempered distributions on $\mathbb{R}^{d+1}$; see, for example, [165]. Let $\mathfrak{S}$ be the σ-algebra generated by the cylinder subsets of $\mathscr{S}^*$. On the measurable space $(\mathscr{S}^*,\mathfrak{S})$ we can construct a Gaussian probability measure ν associated with the characteristic functional

$$C_\nu(\eta) = \int_{\mathscr{S}^*} e^{\mathrm{i}\eta(\omega)}\, \nu(d\omega) = \exp\left\{-\left(\eta, \frac{1}{2}(-\Delta_{t,x} + m^2)^{-1}\eta\right)_{\mathbb{L}_2(\mathbb{R}^{\mathrm{d}+1})}\right\},$$

$$\eta \in \mathscr{S}, \ \Delta_{t,x} = \frac{\partial^2}{\partial t^2} + \sum_{i=1}^{\mathrm{d}} \frac{\partial^2}{\partial x_i^2},$$

where $m > 0$ is a constant and $\eta(\omega)$ is the result of the application of the functional $\omega \in \mathscr{S}^*$ to $\eta \in \mathscr{S}$. The free (boson) field is one of the simplest objects of relativistic quantum theory. In Euclidean field theory (see, for example, the monograph of B. Simon [144]), this field is interpreted as a canonical generalized field (i.e. $\xi(\omega) = \omega$ for every $\omega \in \mathscr{S}^*$) on the probability space $(\mathscr{S}^*, \mathfrak{S}, \nu)$.

Let μ be the Gaussian measure on $(\mathscr{S}^*, \mathfrak{S})$ with the characteristic functional

$$C_\mu(\eta) = \exp\left\{-\frac{1}{2}\|\eta\|^2_{\mathbb{L}_2(\mathbb{R}^{\mathrm{d}+1})}\right\}.$$

The canonical generalized field $\dot{W}$ on the probability space $(\mathscr{S}^*, \mathfrak{S}, \mu)$ is usually called the generalized white noise. According to T. Hida and L. Strait [47], the free field $\xi = \xi(t,x)$ is a stationary solution of the equation

$$\frac{\partial \xi(t,x)}{\partial t} = -\sqrt{(m^2 - \Delta_x)}\, \xi(t,x) + \dot{W}(t,x), \ \ \Delta_x = \sum_{i=1}^{\mathrm{d}} \frac{\partial^2}{\partial x_i^2},$$

where the equation is interpreted in the sense of distributions (generalized functions).

1.3 Measurability and Integrability in Banach Spaces

1.3.1. We present some classical results concerning measurability and integrability, and introduce a number of special spaces which are used systematically in the main part of the book.

Warning 1.1 *Throughout this book, if the contrary is not stated, we consider only real spaces. Metric spaces are assumed separable wherever possible.* □

1.3.2. Let $\mathbb{X}$ be a Banach space with topological dual $\mathbb{X}^*$. The value of the functional x^* at $x \in \mathbb{X}$ will be denoted by x^*x and the norm in $\mathbb{X}$ by $\|\cdot\|_{\mathbb{X}}$. If $\mathbb{X}$ is a Hilbert space, the inner product, corresponding to the norm $\|\cdot\|_{\mathbb{X}}$, will be denoted by $(\cdot,\cdot)_{\mathbb{X}}$. The Borel σ-algebra of $\mathbb{X}$ (i.e. the σ-algebra generated by open subsets of $\mathbb{X}$) will be denoted by $\mathscr{B}(\mathbb{X})$.

1.3.3. Let (S, Σ) be a measurable space (S is a set and Σ is a σ-algebra of subsets of S) with a countably additive, σ-finite and positive measure μ.

Definition 1.1 The completion of Σ with respect to μ will be called the Lebesgue extension of Σ and denoted by $\overline{\Sigma}$. □

For more details related to this definition, see [26, Theorem III.5.17, Definition III.5.18].

1.3.4.

Definition 1.2 The triple (S, Σ, μ) is called a measure space and the triple $(S, \overline{\Sigma}, \mu)$ is called a complete measure space. If $\mu(S) = 1$, then each of the spaces will be referred to as a **probability space**. □

1.3.5.

Definition 1.3 A function $f : S \to \mathbb{X}$ is called $(\overline{\Sigma}, \mu)$-measurable if

$$f^{-1}(\mathcal{A}) \in \overline{\Sigma}, \ \mathcal{A} \in \mathscr{B}(\mathbb{X}).$$ □

1.3.6.

Definition 1.4 A function $f : S \to \mathbb{X}$ is called Σ-measurable if

$$f^{-1}(\mathcal{A}) \in \Sigma, \ \mathcal{A} \in \mathscr{B}(\mathbb{X}).$$ □

In the future, when there is no danger of confusion, we will refer to the $(\overline{\Sigma}, \mu)$-measurable function as $\overline{\Sigma}$-measurable and call the Σ-measurable function Borel.

Remark 1.1 In Definitions 1.3 and 1.4, the metric structure of the space $\mathbb{X}$ is not important and $(\mathbb{X}, \mathscr{B}(\mathbb{X}))$ can be an arbitrary measurable space. In the case when $\mathbb{X}$ is a Banach space, it is enough to require the inclusions $f^{-1}(\mathcal{A}) \in \overline{\Sigma}$ or $f^{-1}(\mathcal{A}) \in \Sigma$ for all open sets $\mathcal{A}$. □

In particular, it follows that if S is a topological space and Σ contains all the open subsets of S, then every continuous mapping $f : S \to \mathbb{X}$ is Borel.

1.3.7.

Theorem 1.1 ([26, Theorem III.6.10]) *The function $f : S \to \mathbb{X}$ is $(\overline{\Sigma}, \mu)$-measurable (respectively, Σ-measurable) if and only if there exists a sequence of $(\overline{\Sigma}, \mu)$-measurable (Σ-measurable) functions $f_n : S \to \mathbb{X}$ such that $\|f_n(s) - f(s)\|_{\mathbb{X}} \to 0$ μ-a.s. (respectively, for every $s \in S$).* □

Remark 1.2

(a) If f is a $(\overline{\Sigma}, \mu)$-measurable function, then there exists a sequence of $(\overline{\Sigma}, \mu)$-measurable functions $\{f_k,\ k \geq 1\}$ converging to f in the norm of $\mathbb{X}$ for μ- a.a. s, and each function f_k takes at most countably many values.

(b) If f is a Σ-measurable function, then there exists a sequence of Σ-measurable functions $\{f_k,\ k \geq 1\}$ converging to f in the norm of $\mathbb{X}$ for every $s \in S$, and each function f_k takes at most countably many values.

Indeed, let $\{x_n,\ n \in \mathbb{N}\}$ be a dense subset of $\mathbb{X}$, and let $B(x_n, 1/k)$, $k \in \mathbb{N}$, be open balls in $\mathbb{X}$ with the radius equal to $1/k$ and the center at the point x_n. For every k, define disjoint sets

$$B_{n,k} := \left\{ x \in \mathbb{X} : x \in B(x_n, 1/k),\ x \notin \bigcup_{m=1}^{n-1} B(x_m, 1/k) \right\}.$$

Then

$$f_k(s) := \sum_{n=1}^{\infty} f(x_n) 1_{\{f \in B_{n,k}\}}(s),\ k \in \mathbb{N},$$

is the required sequence for both (a) and (b) parts. □

1.3.8. The following is an immediate consequence of the results from the previous paragraph.

Proposition 1.1 *If the mapping $f : S \to \mathbb{X}$ is $(\overline{\Sigma}, \mu)$-measurable, then there exists a Σ-measurable version of f. That is, f can be changed on a set $S_0 \in \overline{\Sigma}$, where $\mu(S_0) = 0$, in such a way that the resulting function will be Σ-measurable.* □

Indeed, with $B_{n,k}$ from the previous paragraph, each $1_{\{f \in B_{n,k}\}}(s)$ can be changed on a subset of S with μ-measure zero to become Σ-measurable.

1.3.9.

Theorem 1.2 (Pettis, [26, Theorem III.6.11]) *The function $f : S \to \mathbb{X}$ is $(\overline{\Sigma}, \mu)$-measurable (Σ-measurable) if and only if, for every $x^* \in X^*$, the mapping $x^* f : S \to \mathbb{R}^1$ is $(\overline{\Sigma}, \mu)$-measurable (Σ-measurable).* □

Theorems 1.1 and 1.2 imply the following result.

Corollary 1.1 *If the sequence of $(\overline{\Sigma}, \mu)$-measurable (or Σ-measurable) functions $f_n : S \to \mathbb{X}$ converges weakly to f μ-a.s. (or for every $s \in S$), then the limit function f is $(\overline{\Sigma}, \mu)$-measurable (or Σ-measurable).* □

1.3.10. Let $\mathbb{Y}$ be a Banach space **continuously embedded** in $\mathbb{X}$, that is, $\mathbb{Y} \subset \mathbb{X}$ and there exists a constant N such that $\|x\|_{\mathbb{X}} \leq N\|x\|_{\mathbb{Y}}$ for every $x \in \mathbb{Y}$.

Proposition 1.2

(i) *If $f : S \to \mathbb{Y}$ is a $(\overline{\Sigma}, \mu)$-measurable (or Σ-measurable) function, then f is also $(\overline{\Sigma}, \mu)$-measurable (or Σ-measurable) as a function from S to $\mathbb{X}$.*

(ii). *If $f : S \to \mathbb{X}$ is a $(\overline{\Sigma}, \mu)$-measurable (or Σ-measurable) function, then $f^{-1}(\Gamma) \in \overline{\Sigma}$ (or $f^{-1}(\Gamma) \in \Sigma$) for every Γ in the Borel σ-algebra of $\mathbb{Y}$.* □

Indeed, the embedding operator $\imath : \mathbb{Y} \to \mathbb{X}$ is continuous and one-to-one. The results of [85, Section IV.39] then imply that, for a one-to-one continuous mapping of a complete separable metric space into a metric space, the image of the first is a Borel subset of the second. In other words, $\mathscr{B}(\mathbb{Y}) \subset \mathscr{B}(\mathbb{X})$ and $\imath(\mathbb{Y}) \in \mathscr{B}(\mathbb{X})$. It remains to apply Remark 1.1.

1.3.11. We now define the **Bochner integral**. Call $g : S \mapsto \mathbb{X}$ an elementary function if g has finitely many values $\{x_1, x_2, \dots, x_n\} \subset \mathbb{X}$. Let $B_i := g^{-1}(x_i) \in \overline{\Sigma}$, $i = 1, 2, \dots, n$, and

$$\int_S g(s)d\mu(s) := \sum_{i=1}^{n} x_i \mu(B_i).$$

Definition 1.5 The mapping $f : S \to \mathbb{X}$ is called μ-integrable if there exists a sequence of elementary functions $\{f_n\}$, $n \in \mathbb{N}$, from S into $\mathbb{X}$ such that

$$\lim_{n\to\infty} \|f_n(s) - f(s)\|_{\mathbb{X}} = 0 \ (\mu\text{- a.s.}) \ and \ \lim_{n\to\infty} \int_S \|f(s) - f_n(s)\|_{\mathbb{X}}\, d\mu(s) = 0.$$

For an arbitrary set $B \in \overline{\Sigma}$ the Bochner integral of f over this set is defined by the equality $\int_B f(s)d\mu(s) := \lim_{n\to\infty} \int_S 1_{\{B\}}(s) f_n(s)d\mu(s)$. □

1.3.12. We present, without proofs, some of the main properties of the Bochner integral. An interested reader can find all the details in [165, Section V.5].

I. The value of the integral does not depend on the approximating sequence $\{f_n\}$ and the mapping $f \mapsto \int_S f d\mu$ is lincar.
II. A $(\overline{\Sigma}, \mu)$-measurable mapping f is μ-integrable if and only if the norm $\|f\|_{\mathbb{X}}$ is μ-integrable.
III. $\|\int_B f(s)d\mu(s)\|_{\mathbb{X}} \le \int_B \|f(s)\|_{\mathbb{X}} d\mu(s), \ B \in \overline{\Sigma}$.
IV. If $f : S \to \mathbb{X}$ is μ-integrable, $\mathbb{Y}$ is a Banach space, and $\mathcal{L} : \mathbb{X} \to \mathbb{Y}$ is a bounded linear operator, then $\mathcal{L}f : S \to \mathbb{Y}$ is μ-integrable and

$$\int_B \mathcal{L}f(s)d\mu(s) = \mathcal{L}\int_B f(s)d\mu(s), \ B \in \overline{\Sigma}.$$

1.3.13. Let (S_1, Σ_1, μ_1) and (S_2, Σ_2, μ_2) be measure spaces. Denote by Σ_1 the sets of the form $\Gamma_1 \times \Gamma_2$, where $\Gamma_1 \in \Sigma_1$ and $\Gamma_2 \in \Sigma_2$. Let $\mu_1 \times \mu_2$ be a measure on $\Sigma_1 \otimes \Sigma_2$ with the property

$$\mu_1 \times \mu_2(\Gamma) = \mu_1(\Gamma_1)\mu_2(\Gamma_2), \ \Gamma = \Gamma_1 \times \Gamma_2, \ \Gamma_1 \in \Sigma_1, \ \Gamma_2 \in \Sigma_2.$$

It is known [26, Theorem III.11.2 and Corollary III.11.6] that the measure $\mu_1 \times \mu_2$ is uniquely determined by this equality, is σ-finite, and countably additive.

The measure space $(S_1 \times S_2, \Sigma_1 \otimes \Sigma_2, \mu_1 \times \mu_2)$ will be called the direct product of the measure spaces. The completion of $\Sigma_1 \otimes \Sigma_2$ with respect to the measure $\mu_1 \times \mu_2$ will be denoted by $\overline{\Sigma_1 \otimes \Sigma_2}$. Note that, in general, completeness of Σ_1 and Σ_2 does not imply completeness of $\Sigma_1 \otimes \Sigma_2$.

1.3.14. In this paragraph we introduce some standard function spaces and list their main properties. The proofs and more details are in [26, Chapters II and IV].

Let $(S, \overline{\Sigma}, \mu)$ be a complete measure space and $1 \le p \le \infty$. Denote by $\mathbb{L}_p(S, \overline{\Sigma}, \mu; \mathbb{X})$ the space (of equivalence classes) of $(\overline{\Sigma}, \mu)$-measurable functions $f : S \to \mathbb{X}$ such that

$$\|f\|_{\mathbb{L}_p(S,\overline{\Sigma},\mu;\mathbb{X})} := \begin{cases} \left(\int\limits_S \|f(s)\|^p{}_{\mathbb{X}} \, d\mu(s) \right)^{1/p} < \infty, & \text{if } 1 \le p < \infty, \\ \operatorname{ess}_\mu \sup\limits_{s \in S} \|f(s)\|_{\mathbb{X}}, & \text{if } p = \infty. \end{cases} \tag{1.3.1}$$

Every function in an equivalence class of functions from $\mathbb{L}_p(S, \overline{\Sigma}, \mu; \mathbb{X})$ is called a representative of the class. The space $\mathbb{L}_p(S, \overline{\Sigma}, \mu; \mathbb{X})$ is a Banach space with respect to the norm defined by (1.3.1). For $1 \le p < \infty$, the topological dual of $\mathbb{L}_p(S, \overline{\Sigma}, \mu; \mathbb{X})$ is isometrically isomorphic to, and will be identified with, the space $\mathbb{L}_q(S, \overline{\Sigma}, \mu; \mathbb{X}^*)$, where $1/p + 1/q = 1$. In particular, the dual of $\mathbb{L}_1(S, \overline{\Sigma}, \mu; \mathbb{X})$ is $\mathbb{L}_\infty(S, \overline{\Sigma}, \mu; \mathbb{X})$.

If $\mathbb{X}$ is a Hilbert space, then $\mathbb{L}_2(S, \overline{\Sigma}, \mu; \mathbb{X})$ is also a Hilbert space with inner product

$$(f, g)_{\mathbb{L}_2(S,\overline{\Sigma},\mu;\mathbb{X})} := \int \big(f(s), g(s)\big)_{\mathbb{X}} \, d\mu(s).$$

1.4 Martingales in $\mathbb{R}^1$

1.4.1. In this section we state some basic results of the general theory of continuous-parameter martingales and related stochastic processes. Detailed proofs of the statements presented below in this section are in [23, 24, 92, 108].

1.4.2. Let $(\Omega, \mathscr{F}, \mathbb{P})$ be a complete probability space and let $\{\mathscr{F}_t\}$, $t \in \mathbb{R}_+$, be a family of σ-algebras with the following properties:

1. $\mathscr{F}_t \subset \mathscr{F}$, $t \in \mathbb{R}_+$.
2. $\mathscr{F}_s \subset \mathscr{F}_t$, $s < t$, $s, t \in \mathbb{R}_+$ (increasing).

3. $\bigcap_{\varepsilon>0} \mathscr{F}_{t+\varepsilon} = \mathscr{F}_t$, $t \in \mathbb{R}_+$ (right-continuous).
4. The σ-algebra $\mathscr{F}_0$ contains all $\mathbb{P}$-null sets in $\mathscr{F}$ ($\mathbb{P}$-complete).

The quadruple $\mathbb{F} := (\Omega, \mathscr{F}, \{\mathscr{F}_t\}, \mathbb{P})$ is called a **stochastic basis with the usual assumptions** and will be fixed throughout this section.

Let $(\mathbb{X}, \mathscr{X})$ be a measurable space. The space $\mathbb{X}$ will be referred to as the state space.

1.4.3.

Definition 1.6 An $\mathscr{F}$-measurable mapping $\xi : \Omega \to \mathbb{X}$ is called an $\mathbb{X}$-valued **random variable**. A mapping $x : \mathbb{R}_+ \times \Omega \to \mathbb{X}$, which is a random variable for every $t \in \mathbb{R}_+$, is called a **stochastic process taking values in** $\mathbb{X}$, or an **$\mathbb{X}$-valued process**, or just an $\mathbb{X}$-process. If this mapping is $\mathscr{B}(\mathbb{R}_+) \otimes \mathscr{F}$-measurable, then it is called a measurable stochastic process. □

When there is no danger of confusion, we will not mention the space where the process takes its values. The parameter $t \in \mathbb{R}_+$ usually stands for "time" and the parameter ω, for "random event" (fluctuation); usually, the argument ω will be omitted. In the scope of this definition we can also consider processes defined on an interval in $\mathbb{R}_+$.

Warning 1.2 *All stochastic processes considered in this book will be assumed to be measurable.* □

If the state space $\mathbb{X}$ of a random variable (or stochastic process) is the extended real line $\overline{\mathbb{R}}^1$ and $\mathscr{X}$ is the Borel σ-algebra on $\overline{\mathbb{R}}^1$, then we call the corresponding random variable (or stochastic process) **real-valued**. The attribute "real-valued" will usually be omitted.

1.4.4.

Definition 1.7 Let $x = x(t)$ and $y = y(t)$ be stochastic processes. If $x(t) = y(t)$ $\mathbb{P}$- a.s. for every $t \in \mathbb{R}_+$, then y will be called a **version** of x. □

More generally, if $(\mathbb{X}, \mathscr{X})$ is a measurable space, $f : S_1 \times S_2 \to \mathbb{X}$, and for every $s_2 \in S_2$, $f(\cdot, s_2)$ is a $\overline{\Sigma}_1$-measurable mapping of the complete measure space $(S_1, \overline{\Sigma}_1, \mu_1)$ into $\mathbb{X}$, then every function $\tilde{f}$ with the same measurability properties and such that $f(\cdot, s_2) = \tilde{f}(\cdot, s_2)$ μ_1-a.s., for every $s_2 \in S_2$, is called a version of f (with respect to s_1).

1.4.5.

Definition 1.8 For a fixed $\omega \in \Omega$, the function $x(\cdot, \omega) : \mathbb{R}_+ \to \mathbb{X}$ is called a sample path or a trajectory of the process x. A stochastic process will be referred to as continuous (left-continuous, right-continuous, etc.) if its trajectories have the corresponding property for $\mathbb{P}$-*a.a.* ω. □

1.4.6. The stochastic process $x = x(t, \omega)$ is called **$\mathscr{F}_t$-adapted** if, for every fixed $t \in \mathbb{R}_+$, the random variable $x(t, \cdot)$ is $\mathscr{F}_t$-measurable. The process is called

progressively measurable if, for each $t \in \mathbb{R}_+$, the mapping $x : [0, t] \times \Omega \to \mathbb{X}$ is $\mathscr{B}([0, t]) \otimes \mathscr{F}_t$-measurable. It follows that every progressively measurable process is $\mathscr{F}_t$-adapted.

A measurable set $\mathcal{A} \subset \mathbb{R}_+ \times \Omega$ is said to be progressively measurable if its indicator $1_{\{\mathcal{A}\}} = 1_{\{\mathcal{A}\}}(t, \omega)$ is a progressively measurable stochastic process. For example, if the process $x = x(t, \omega)$ is progressively measurable and $A \in \mathscr{X}$, then the set $\mathcal{A} = \{(t, \omega) : x(t, \omega) \in A\}$ is progressively measurable.

1.4.7. Denote by $\mathscr{P}$ the σ-algebra on $\mathbb{R}_+ \times \Omega$ generated by all $\mathscr{F}_t$-adapted, left-continuous, real-valued stochastic processes. It is called the **σ-algebra of predictable sets**.

Definition 1.9 A $\mathscr{P}$-measurable stochastic process is called **predictable.** □

Warning 1.3 *If, in addition to $\{\mathscr{F}_t\}$, some other family of σ-algebras on $(\Omega, \mathscr{F})$ is given we will modify the attribute "predictable" by reference to the corresponding family of σ-algebras.* □

Example 1.1 Every process $B = B(t)$ of the form

$$B(t) := B_0(\omega)1_{\{0\}}(t) + \sum_{i=0}^{\infty} B_i(\omega)1_{\{(t_i,t_{i+1}]\}}(t),$$

where $0 = t_0 < t_1 < \ldots < t_n < \ldots$, and B_i is an $\mathscr{F}_{t_i}$-measurable random variable, for every i, is predictable. Processes of this type are called simple predictable processes.

Approximation of left-continuous and $\mathscr{F}_t$-adapted (hence predictable) stochastic processes by simple predictable processes shows that the following statement holds.

Proposition 1.3 *Sets of the form $(a, b] \times \Gamma$, and $\{0\} \times A$, where $a, b \in \mathbb{R}_+$, $\Gamma \in \mathscr{F}_a$, and $A \in \mathscr{F}_0$, generate the σ-algebra $\mathscr{P}$.* □

The σ-algebra generated by the sets mentioned in the proposition, where $a, b \in [T_0, T]$, $T_0, T \in \mathbb{R}_+$, will be denoted by $\mathscr{P}_{[T_0,T]}$, and by $\mathscr{P}_T$ if $T_0 = 0$.

It is clear that $\mathscr{P}_{[T_0,T]}$ is generated by left-continuous, $\mathscr{F}_t$-adapted processes defined on the interval $[T_0, T]$. A $\mathscr{P}_{[T_0,T]}$-measurable stochastic process on $[T_0, T]$ will also be called predictable.

Remark 1.3 By definition, every predictable process is progressively measurable and consequently $\mathscr{F}_t$-adapted.

Sometimes instead of a predictable (or $\mathscr{F}_t$-adapted or progressively measurable) stochastic process we speak of a predictable (or $\mathscr{F}_t$-adapted or progressively measurable) function of (t, ω). □

1.4.8.

Definition 1.10 A random variable τ taking values in $\overline{\mathbb{R}}_+$ is called a **stopping time** with respect to (or relative to) an increasing family of σ-algebras $\{\mathscr{Y}_t\}$, $t \in \mathbb{R}_+$, if $\{\omega : \tau(\omega) \leq t\} \in \mathscr{Y}_t$ for every $t \geq 0$. □

Warning 1.4 *All stopping times are considered with respect to the family* $\{\mathscr{F}_t\}$ *unless stated otherwise.* □

For a measurable set $\mathcal{A} \subset \mathbb{R}_+ \times \Omega$, define

$$\mathscr{D}_{\mathcal{A}}(\omega) = \begin{cases} \inf\{t : (t, \omega) \in \mathcal{A}\}, & \text{if } \{t : \ (t, \omega) \in \mathcal{A}\} \neq \emptyset; \\ +\infty, & \text{if } \{t : \ (t, \omega) \in \mathcal{A}\} = \emptyset. \end{cases}$$

The random variable $\mathscr{D}_{\mathcal{A}}$ will be called the début of the set $\mathcal{A}$.

The following result [108, Chapter IV, Theorem 48] establishes a connection between progressively measurable sets and stopping times.

Theorem 1.3 *If* $\mathcal{A}$ *is a progressively measurable set, then the début of* $\mathcal{A}$ *is a stopping time.* □

Let $x = x(t)$ be a progressively measurable $\mathbb{X}$-process, $A \in \mathscr{X}$, and $\mathcal{A} := \{(t, \omega)\colon x(t, \omega) \in A\}$. By the above theorem, the début of $\mathcal{A}$ is a stopping time; this stopping time is called the hitting time of A. The following relation provides the reason for this terminology:

$$\mathscr{D}_{\mathcal{A}}(\omega) = \begin{cases} \inf\{t : x(t, \omega) \in A\}, & \text{if } \{t : \ x(t, \omega) \in A\} \neq \emptyset; \\ +\infty, & \text{if } \{t : \ x(t, \omega) \in A\} = \emptyset. \end{cases}$$

1.4.9.

Definition 1.11 The real-valued stochastic process $M = M(t)$ is called a real-valued **supermartingale** (**submartingale**) with respect to, or relative to, an increasing family of σ-algebras $\{\mathscr{Y}_t\}$, $t \in \mathbb{R}_+$, if it is $\mathscr{Y}_t$-adapted, $\mathbb{P}$-integrable for every $t \in \mathbb{R}_+$, and $\mathbb{E}[M(t)|\mathscr{Y}_s] \leq M(s)$ $\big($respectively, $\mathbb{E}[M(t)|\mathscr{Y}_s] \geq M(s)\big)$ for every $s, t \in \mathbb{R}_+$, $s \leq t$.

If $M = M(t)$ is both a submartingale and a supermartingale relative to the family $\{\mathscr{Y}_t\}$, then$M$ is called a **martingale** relative to this family. □

Warning 1.5 *Throughout this section, unless otherwise stated, we will consider real-valued supermartingales, submartingales, and martingales relative to the same family* $\{\mathscr{F}_t\}$*. Therefore, we will not explicitly refer to this family. The adjective "real-valued" will usually be dropped as well.* □

It follows from the definition that (a) if $M = M(t)$ is a martingale, then its expected value is constant: $\mathbb{E}M(t) = \mathbb{E}M(s)$ for all s and $t \in \mathbb{R}_+$; (b) if $M = M(t)$

is a square-integrable martingale, then its increments are uncorrelated: for every $t_1 \le t_2 \le t_3 \le t_4$,

$$\mathbb{E}\Big(\big(M(t_4) - M(t_3)\big)\big(M(t_2) - M(t_1)\big)\Big) = 0.$$

1.4.10. An important property of martingales is the invariance of the martingale property with respect to **optional sampling transformation**, namely: under fairly general conditions if $M = M(t)$ is a martingale then, for each stopping time τ, the process $t \mapsto M(t \wedge \tau)$ is also a martingale.

Let $M = M(t)$ be a right-continuous supermartingale. Suppose that there exists a $\mathbb{P}$-integrable, real-valued random variable Y such that

$$M(t) \ge \mathbb{E}[Y|\mathscr{F}_t] \quad \text{for every} \quad t \in \mathbb{R}_+. \tag{1.4.1}$$

Let τ and σ be stopping times such that $\tau \le \sigma$ ($\mathbb{P}$-a.s.), and define

$$M_\tau = \begin{cases} M(\tau), & \text{if } \tau < +\infty, \\ Y, & \text{if } \tau = +\infty; \end{cases} \quad M_\sigma = \begin{cases} M(\sigma), & \text{if } \sigma < +\infty, \\ Y, & \text{if } \sigma = +\infty. \end{cases}$$

Then the following result holds [108, Chapter VI, Theorem 13].

Theorem 1.4 *The random variables M_τ and M_σ are $\mathbb{P}$-integrable and satisfy*

$$M_\tau \ge \mathbb{E}[M_\sigma|\mathscr{F}_\tau] \quad (\mathbb{P}\text{- a.s.}). \tag{1.4.2}$$

□

Corollary 1.2 *If $M = M(t)$ is a martingale and the family of random variables $\{M(t),\ t \ge 0\}$ is uniformly $\mathbb{P}$-integrable, then the inequality sign in* (1.4.2) *can be replaced by equality.* □

1.4.11. We recall some inequalities for real-valued submartingales and supermartingales; the details are in [108, Chapter VI, §1].

Let $I := [a, b]$ be a bounded interval in $\mathbb{R}_+$. If $M = M(t)$ is a right-continuous supermartingale and $\lambda > 0$, then

$$\lambda \mathbb{P}\big(\inf_{t \in I} M(t) \le -\lambda\big) \le \mathbb{E}|M(b)|. \tag{1.4.3}$$

Let $M = M(t)$ be a positive right-continuous submartingale. Let p and q be real numbers such that $1 < p < \infty$ and $1/p + 1/q = 1$. Then

$$\mathbb{E}\big(\sup_{t \in I} M(t)\big)^p \le q^p \mathbb{E} M^p(b). \tag{1.4.4}$$

Inequality (1.4.4) implies, in particular, the following result due to J. Doob: if $M = M(t)$ is a right-continuous martingale, then

$$\mathbb{E} \sup_{t \in I} |M(t)|^p \leq q^p \, \mathbb{E}|M(b)|^p. \tag{1.4.5}$$

1.4.12.

Definition 1.12 A right-continuous supermartingale is said to belong to class $(\mathcal{D})$ if the family of random variables $\{M(\tau),\ \tau \in \mathscr{T}\}$, where $\mathscr{T}$ is the collection of all stopping times relative to $\{\mathscr{F}_t,\ t \geq 0\}$, is uniformly $\mathbb{P}$-integrable. □

The following statement [23, Chapter V, i. 50] is the starting point in the development of differential and integral calculus for martingales.

Theorem 1.5 *Let $M = M(t)$ be a supermartingale of class $(\mathcal{D})$. Then there exists a unique (up to a version) predictable real-valued stochastic process $A = A(t, \omega)$ such that*

(a) *A is increasing and right-continuous;*
(b) $A(0) = 0$ ($\mathbb{P}$- a.s.) *and* $\sup \mathbb{E}A(t) < \infty$*;*
(c) *the process $t \mapsto M(t) + A(t)$ is a martingale.*

If the supermartingale M is continuous, then the corresponding process A is also continuous. □

Remark 1.4 The process A can be continuous even when the process M is not [23, Chapter V, i. 52]. □

The decomposition

$$M(t) = N(t) - A(t), \tag{1.4.6}$$

where $N = N(t)$ is a martingale, is usually referred to as the **Doob–Meyer decomposition** of M. An $\mathscr{F}_t$-adapted real-valued stochastic process $A = A(t)$ with property (a) and such that $A(0) = 0$ ($\mathbb{P}$-a.s.) and $\mathbb{E}A(t) < \infty$ for every $t > 0$ is called an **increasing process**.

Let $\mathfrak{M}_2^c(\mathbb{R}_+, \mathbb{R}^1)$ be the set of all continuous real-valued martingales $M = M(t)$ such that $M(0) = 0$ ($\mathbb{P}$-a.s.) and, for each $t \in \mathbb{R}_+$, $\mathbb{E}|M(t)|^2 < \infty$. The following result is an easy consequence of Theorem 1.5; see [23, Chapter II, §3].

Corollary 1.3 *For every $M \in \mathfrak{M}_2^c(\mathbb{R}_+, \mathbb{R}^1)$, there exists a stochastic process $\langle M \rangle = \langle M \rangle_t,\ t \geq 0$, such that*

(a) *$\langle M \rangle$ is continuous;*
(b) *$\langle M \rangle_0 = 0$ ($\mathbb{P}$-a.s.);*
(c) *The process $t \mapsto |M(t)|^2 - \langle M \rangle_t$ is a martingale.*

The process $\langle M \rangle$ is the only predictable process (up to a version) with properties (a), (b), (c). □

We call the process $\langle M\rangle$ the **quadratic variation** process or simply quadratic variation of M. This term is motivated by the following well-known result [24, Chapter 2, §3]).

Proposition 1.4 *If* $M \in \mathfrak{M}_2^c(\mathbb{R}_+, \mathbb{R}^1)$, $T \in \mathbb{R}_+$, *and* $\{0 = t_0^n < \ldots < t_{k(n)+1}^n = T\}$ *is a sequence of partitions of the interval* $[0, T]$ *such that* $\lim_{n\to\infty} \max_i |t_i^n - t_{i-1}^n| = 0$, *then*

$$\lim_{n\to\infty} \sum_{i=0}^{k(n)} |M(t_{i+1}^n) - M(t_i^n)|^2 = \langle M\rangle_T$$

in probability. □

1.4.13.

Definition 1.13 A real-valued stochastic process $M = M(t)$ is called a real-valued **local martingale** relative to an increasing family of σ-algebras $\{\mathscr{Y}_t\}$ if there exists an increasing sequence of stopping times $\{\tau_n,\ n \in \mathbb{N}\}$ (relative to the same family) such that $\lim_{n\to\infty} \tau_n = +\infty$ ($\mathbb{P}$- a.s.) and, for each n, the process $t \mapsto M(t \wedge \tau_n)$, $t \geq 0$, is a martingale relative to $\{\mathscr{Y}_t\}$. □

Local submartingales and supermartingales are defined in a similar way.

Remark 1.5 Let τ be a stopping time relative to $\{\mathscr{Y}_t\}$ and let $M = M(t)$ be a $\mathscr{Y}_t$-adapted stochastic process. Then the process $M(t \wedge \tau)$, $t \geq 0$, is a martingale relative to $\{\mathscr{Y}_{t\wedge\tau}\}$ if and only if it is a martingale relative to $\{\mathscr{Y}_t\}$.

Thus, in Definition *1.13*, we can require $M(t \wedge \tau_n)$ to be a martingale relative to $\{\mathscr{Y}_{t\wedge\tau_n}\}$ rather than $\{\mathscr{Y}_t\}$. □

The set of all continuous local martingales with $M(0) = 0$ ($\mathbb{P}$-a.s.) will be denoted by $\mathfrak{M}_{loc}^c(\mathbb{R}_+, \mathbb{R}^1)$.

Lemma 1.1 *If* $M = M(t)$ *is a continuous local martingale, then there exists a sequence of stopping times* $\{\sigma_n,\ n \in \mathbb{N}\}$ *such that, for every* $n \in \mathbb{N}$,

$$\sup_{t>0} |M(t \wedge \sigma_n)| \leq n \ (\mathbb{P}\text{- a.s.})$$ □

The proof is left to the reader as an exercise.

Let $M = M(t)$ and $\{\sigma_n\}$ be the local submartingale and the sequence of stopping times from Lemma 1.1. Then, for each $n \in \mathbb{N}$, the process $t \mapsto -M(t \wedge \sigma_n)$, $t \geq 0$, is a continuous supermartingale of class $(\mathcal{D})$, and therefore, by Theorem 1.5, there exists a unique, up to a version, increasing predictable process $A_n = A_n(t)$ such that the process $A_n(t) - M(t \wedge \sigma_n)$ is a martingale.

Define the process $A = A(t)$ by

$$A(t) := A_n(t) \text{ for } t < \sigma_n. \tag{1.4.7}$$

It is easy to see that the process A is well defined by (1.4.7) and the following theorem holds.

Theorem 1.6 *Let $M = M(t)$ be a continuous local submartingale. Then there exists a predictable real-valued stochastic process $A = A(t)$ such that*

(a) $A(0) = 0$ ($\mathbb{P}$-*a.s.*);
(b) $A(t) \geq A(s)$ ($\mathbb{P}$-*a.s.*) *for every* $t, s \in \mathbb{R}_+$, $t \geq s$;
(c) *the process* $t \mapsto M(t) - A(t)$ *is a local martingale.*

The process A is continuous. It is the only (up to a version) predictable, continuous stochastic process with properties (a), (b), (c). □

Corollary 1.4 *If $M \in \mathfrak{M}^c_{loc}(\mathbb{R}_+, \mathbb{R}^1)$, then there exists a unique (up to a version) continuous, increasing process $\langle M \rangle = \langle M \rangle_t$ such that $|M|^2 - \langle M \rangle \in \mathfrak{M}^c_{loc}(\mathbb{R}_+, \mathbb{R}^1)$.* □

The process $\langle M \rangle$ is called the quadratic variation process or simply **quadratic variation** of the local martingale M.

1.4.14. Assume that $M, N \in \mathfrak{M}^c_{loc}(\mathbb{R}_+, \mathbb{R}^1)$. Define $\langle M, N \rangle_t := 1/4\big(\langle M + N \rangle_t - \langle M - N \rangle_t\big)$. The definition implies that the process $\langle M, N \rangle$ is continuous and has bounded variation (being a difference of two increasing processes). This process is called **quadratic covariation** of M and N.

Theorem 1.7 *If $M, N \in \mathfrak{M}^c_{loc}(\mathbb{R}_+, \mathbb{R}^1)$, then $MN - \langle M, N \rangle \in \mathfrak{M}^c_{loc}(\mathbb{R}_+, \mathbb{R}^1)$. Moreover, suppose that $A = A(t)$ is a real-valued process such that*

(a) *A has bounded variation;*
(b) $A(0) = 0$ ($\mathbb{P}$- a.s.);
(c) $MN - A \in \mathfrak{M}^c_{loc}(\mathbb{R}_+, \mathbb{R}^1)$.

Then A is a version of $\langle M, N \rangle$. □

Note that if $M, N \in \mathfrak{M}^c_2(\mathbb{R}_+, \mathbb{R}^1)$, then $MN - \langle M, N \rangle$ is a continuous martingale. More generally, a process is called a semimartingale if it is a sum of a martingale and a process of bounded variation. By Theorem 1.7, the product of two processes from $\mathfrak{M}^c_2(\mathbb{R}_+, \mathbb{R}^1)$ is a semimartingale.

Using Proposition 1.4 and the equality $4ab = (a + b)^2 - (a - b)^2$, we conclude that if $\{0 = t^n_0 < \ldots < t^n_{k(n)+1} = T\}$ is a sequence of partitions of the interval $[0, T]$ such that $\lim\limits_{n \to \infty} \max\limits_i |t^n_i - t^n_{i-1}| = 0$, then

$$\lim_{n \to \infty} \sum_{i=0}^{k(n)} \big(M(t^n_{i+1}) - M(t^n_i)\big)\big(N(t^n_{i+1}) - N(t^n_i)\big) = \langle M, N \rangle_T$$

in probability.

1.4.15. In this paragraph we construct the stochastic Itô integral with respect to a martingale $M \in \mathfrak{M}^c_2(\mathbb{R}_+, \mathbb{R}^1)$.

Given $t > 0$, let $0 = t_0 < t_1 < \ldots < t_n = t$ be a partition of $[0, t]$. Recall that a process $B = B(s)$, $0 \leq s \leq t$, of the form

$$B(s) = B_0 1_{\{0\}}(s) + \sum_{i=0}^{n-1} B_i 1_{\{(t_i, t_{i+1}]\}}(s),$$

where each B_i is an $\mathscr{F}_{t_i}$-measurable random variable, is called a simple predictable process.

Let $B = B(s)$, $0 \leq s \leq t$, be a simple predictable process such that

$$\mathbb{E} \int_{[0,t]} |B(s)|^2 \, d\langle M \rangle_s < \infty. \tag{1.4.8}$$

Define

$$\int_{[0,t]} B(s) dM(s) := \sum_{i=0}^{n-1} B_i \big(M(t_{i+1}) - M(t_i) \big). \tag{1.4.9}$$

By direct computation,

$$\mathbb{E} \int_{[0,t]} B(s) dM(s) = 0 \tag{1.4.10}$$

and

$$\mathbb{E} \left| \int_{[0,t]} B(s) dM(s) \right|^2 = \mathbb{E} \int_{[0,t]} |B(s)|^2 \, d\langle M \rangle_s. \tag{1.4.11}$$

Next, we introduce the measure $\langle M \rangle \circ \mathbb{P}_t$ on the measurable space $\big([0, t] \times \Omega, \mathscr{B}([0, t]) \otimes \mathscr{F}_t\big)$ by the equality

$$\langle M \rangle \circ \mathbb{P}_t(A) := \mathbb{E} \int_{[0,t]} 1_{\{A\}}(s, \omega) \, d\langle M \rangle_s.$$

It follows that $\langle M \rangle \circ \mathbb{P}_t$ is a countably additive positive measure; it is sometimes referred to as the **Dolean measure**. This measure is completely defined by its restriction to the σ-algebra $\mathscr{P}_t$ of predictable sets from $\mathscr{B}([0, t]) \otimes \mathscr{F}_t$; see [28, Chapter V, §3].

Denote by $\mathscr{L}_2(\langle M \rangle \circ \mathbb{P}_t)$ the collection of real-valued, predictable (but not necessarily simple) stochastic processes B on $[0, t]$ satisfying (1.4.8). We will now extend the definition of the stochastic integral with properties (1.4.10) and (1.4.11) to the case of integrands from $\mathscr{L}_2(\langle M\rangle \circ \mathbb{P}_t)$.

To begin, we identify $\mathscr{L}_2(\langle M\rangle \circ \mathbb{P}_t)$ with

$$\mathbb{L}_2\Big([0,t]\times\Omega, \overline{\mathscr{P}}_t, \langle M\rangle \circ \mathbb{P}_t; \mathbb{R}^1\Big)$$

by identifying each element B of $\mathscr{L}_2(\langle M\rangle \circ \mathbb{P}_t)$ with a suitable equivalence class of functions.

Let $\mathcal{A} \in \mathscr{P}$ and let $\mathscr{E}$ be the ring of subsets of $[0,t]\times\Omega$, consisting of finite disjoint unions of the sets of the form $(a,b]\times\Gamma$, where $\Gamma \in \mathscr{F}_a$, and $\{0\}\times A$, where $A\in\mathscr{F}_0$. It follows from Proposition 1.3 that there exists a sequence of sets $\{\mathcal{A}_n\}$ belonging to $\mathscr{E}$ such that

$$\lim_{n\to\infty} \mathbb{E}\int\limits_{[0,t]} \Big|1_{\{\mathcal{A}\}}(s) - 1_{\{\mathcal{A}_n\}}(s)\Big|^2 d\langle M\rangle_s = 0.$$

It therefore follows that the subspace $\mathscr{M}^2$ generated by simple predictable processes with property (1.4.8) is dense in $\mathscr{L}_2(\langle M\rangle \circ \mathbb{P}_t)$.

By (1.4.11), the mapping $B \mapsto \int\limits_{[0,\cdot]} B(s)dM(s)$ of $\mathscr{M}^2$ to $\mathbb{L}_2(\Omega,\mathscr{F}_t,\mathbb{P};\mathbb{R}^1)$ given by (1.4.9) is an isometry and therefore has a unique extension to an isometry from $\mathscr{L}_2(\langle M\rangle \circ \mathbb{P}_t)$ to $\mathbb{L}_2(\Omega,\mathscr{F}_t,\mathbb{P};\mathbb{R}^1)$. This isometry defines the **stochastic integral** of a predictable stochastic process B with respect to M. We denote this integral by $\int\limits_{[0,t]} B(s)dM(s)$.

By construction, the stochastic integral has properties (1.4.10) and (1.4.11). As a function of t it is a real-valued stochastic process, and, as we will see below, has an $\mathscr{F}_t$-measurable version.

1.4.16. Let us list the basic properties of the stochastic integral constructed in Sect. 1.4.15.

(i) If $B_1, B_2 \in \mathscr{L}_2(\langle M\rangle \circ \mathbb{P}_t)$ and $\alpha,\beta\in\mathbb{R}^1$, then

$$\int\limits_{[0,t]} \big(\alpha B_1(s)+\beta B_2(s)\big)dM(s)$$

$$= \alpha\int\limits_{[0,t]} B_1(s)dM(s) + \beta\int\limits_{[0,t]} B_2(s)dM(s) \quad (\mathbb{P}\text{- a.s.}).$$

(ii) If $B \in \mathscr{L}_2(\langle M\rangle \circ \mathbb{P}_t)$ and $0 \le s \le u \le t$, then

$$\int_{[0,t]} 1_{\{[s,u]\}}(r)B(r)dM(r) = \int_{[0,u]} B(r)dM(r) - \int_{[0,s]} B(r)dM(r) \quad (\mathbb{P}\text{- a.s.})$$

$$:= \int_{[s,u]} B(r)dM(r).$$

(iii) If $B_1, B_2 \in \mathscr{L}_2(\langle M\rangle \circ \mathbb{P}_t)$ for every $t \in \mathbb{R}^1$, then the stochastic processes $t \mapsto \int_{[0,t]} B_1(r)dM(r)$ and $t \mapsto \int_{[0,t]} B_2(r)dM(r)$ have versions $B_1 \circ M = B_1 \circ M(t)$, $B_2 \circ M = B_2 \circ M(t)$ in $\mathfrak{M}^c_{loc}(\mathbb{R}_+, \mathbb{R}_1)$, and

$$\langle B_1 \circ M, B_2 \circ M\rangle_t = \int_{[0,t]} B_1(s)B_2(s)\, d\langle M\rangle_s.$$

(iv) If $B, B^n \in \mathscr{L}_2(\langle M\rangle \circ \mathbb{P}_t)$, $n \in \mathbb{N}$, and

$$\lim_{n\to\infty} \mathbb{E} \int_{[0,t]} |B^n(s) - B(s)|^2\, d\langle M\rangle_s = 0,$$

then

$$\lim_{n\to\infty} \mathbb{E} \sup_{s\le t} |B^n \circ M(s) - B \circ M(s)|^2 = 0.$$

(v) If $B \in \mathscr{L}_2(\langle M\rangle \circ \mathbb{P}_t)$ for every $t \in \mathbb{R}_+$ and $N \in \mathfrak{M}^c_{loc}(\mathbb{R}_+, \mathbb{R}^1)$, then

$$\langle B \circ M, N\rangle_t = \int_{[0,t]} B(s)\, d\langle M, N\rangle_s \quad (\mathbb{P}\text{- a.s.}).$$

Remark 1.6 The stochastic integral defined above extends to continuous local martingales M. This will be done in the next chapter in a more general situation.

1.4.17. To conclude this section, we state several extensions of the Cauchy–Schwarz inequality to stochastic integrals.

Theorem 1.8 (The Kunita–Watanabe Inequalities)

(a) *If B is a predictable bounded process, $M, N \in \mathfrak{M}^c_{loc}(\mathbb{R}_+, \mathbb{R}^1)$, and $T \in \mathbb{R}_+$, then*

$$\int_{[0,T]} B(s)\, d\langle M, N\rangle_s \le \left(\int_{[0,T]} |B(s)|^2\, d\langle M\rangle_s \right)^{\frac{1}{2}} \langle N\rangle_T^{\frac{1}{2}} \quad (\mathbb{P}\text{- a.s.}).$$

(b) *If B_1 and B_2 are predictable bounded processes, $M, N \in \mathfrak{M}^c_{loc}(\mathbb{R}_+, \mathbb{R}^1)$, and $T \in \mathbb{R}_+$, then*

$$\left| \int_{[0,T]} B_1(s) B_2(s)\, d\langle M, N\rangle_s \right|$$

$$\leq \left(\int_{[0,T]} |B_1(s)|^2\, d\langle M\rangle_s \right)^{\frac{1}{2}} \left(\int_{[0,T]} |B_2(s)|^2\, d\langle N\rangle_s \right)^{\frac{1}{2}} \quad (\mathbb{P}\text{- a.s.}).$$

(c) *If, in addition, $M, N \in \mathfrak{M}^c_2(\mathbb{R}_+, \mathbb{R}^1)$, then*

$$\mathbb{E} \left| \int_{[0,T]} B_1(s) B_2(s)\, d\langle M, N\rangle_s \right|$$

$$\leq \left(\mathbb{E} \int_{[0,T]} |B_1(s)|^2\, d\langle M\rangle_s \right)^{\frac{1}{2}} \left(\mathbb{E} \int_{[0,T]} |B_2(s)|^2\, d\langle N\rangle_s \right)^{\frac{1}{2}}. \qquad \square$$

1.5 Diffusion Processes

1.5.1. Diffusion processes are one of the most important and thoroughly investigated classes of stochastic processes. The term "diffusion process" came into use because this type of process is a good model of physical diffusion phenomena.

The mathematical theory of diffusion processes was developed in the works of Wiener [162], Kolmogorov [61], Lévy [90], Gikhman [31, 32] and many others.

The modern theory of diffusion processes relies on the theory of Itô stochastic differential equations [53].

In this section we present some of the main definitions and results of the theory of diffusion processes which we will need later.

Warning 1.6 *Throughout the section, we fix the stochastic basis $\mathbb{F} = (\Omega, \mathscr{F}, \{\mathscr{F}_t\}_{t\in\mathbb{R}_+}, \mathbb{P})$ with the usual assumptions, as well as the numbers* $\mathrm{d}, \mathrm{d}_1 \in \mathbb{N}$ *and* $T > 0$.

Given $x \in \mathbb{R}^{\mathrm{d}}$, we denote the ith coordinate of x by x^i, that is, $x = (x^1, x^2, \ldots, x^{\mathrm{d}})$. We will also use the following notation:

$$dx = dx^1 \cdots dx^{\mathrm{d}}, \; x \in \mathbb{R}^{\mathrm{d}}, \quad f_i(x) = \frac{\partial f(x)}{\partial x^i}, \; f_{ij} = \frac{\partial^2 f(x)}{\partial x^i \partial x^j}.$$

We use the Einstein summation convention (summation over repeated indices). For example,

$$a^{ij}h^ih^j + \sigma^{il}\sigma^{jl} + \sigma^{il}f_i(x) = \sum_{i,j=1}^{\mathrm{d}} a^{ij}h^ih^j + \sum_{l=1}^{\mathrm{d}_1}\sigma^{il}\sigma^{jl} + \sum_{i=1}^{\mathrm{d}}\sigma^{il}\frac{\partial f(x)}{\partial x^i}.$$

Note that the repeated indices can be either subscripts or superscripts. □

1.5.2.

Definition 1.14 A real-valued stochastic process $w = w(t)$ is called a **standard Wiener process** with respect to $\{\mathscr{F}_t\}$ if

(i) w is a continuous square-integrable martingale with respect to $\{\mathscr{F}_t\}$;
(ii) $\langle w\rangle_t = t,\ t \in \mathbb{R}_+$.

The process $W = W(t)$ with values in $\mathbb{R}^{\mathrm{d}_1}$ is called a d_1-dimensional **standard Wiener process** if the components $w^1, \dots, w^{\mathrm{d}_1}$ of W are independent standard Wiener processes. □

It is well known that there exists a stochastic basis $\mathbb{F}$ with the usual assumptions on which a d_1-dimensional standard Wiener process can be constructed for every d_1.

Theorem 1.9 *Let $w = w(t)$ be a standard Wiener process in $\mathbb{R}^1$. Then, for every $s, t \in \mathbb{R}_+$ such that $s \le t$, the random variable $w(t) - w(s)$ is independent of $\mathscr{F}_s$ and is Gaussian with mean zero and variance $t - s$.* □

This result goes back to P. Lévy. For a proof, see [93, Chapter 4, §1].

Given a standard Wiener process w in $\mathbb{R}^1$, it follows from Sects. 1.4.15 and 1.4.16 that the stochastic integral $\int_{[0,t]} b(s)\,dw(s)$ is defined for every predictable function $b : \mathbb{R}_+ \times \Omega \to \mathbb{R}^1$ such that

$$\mathbb{E}\int_{[0,t]} |b(s)|^2 ds < \infty. \tag{1.5.1}$$

The integral has a continuous in t version that belongs to $\mathfrak{M}_2^c(\mathbb{R}_+, \mathbb{R}^1)$. This version, for which we use the same notation, has quadratic variation

$$\Big\langle \int_{[0,\cdot]} b(s)\,dw^l(s)\Big\rangle_t = \int_{[0,t]} b^2(s)\,ds,\ t \in \mathbb{R}_+\ (\mathbb{P}\text{-a.s.}),$$

and possesses all other properties of the stochastic integral listed in Sect. 1.4.16.

In general, a stochastic integral with respect to a one-dimensional Wiener process can be defined for every predictable process $b : \mathbb{R}_+ \times \Omega \to \mathbb{R}^1$ satisfying

$$\int_{[0,t]} b^2(s)\,ds < \infty, \quad t \in \mathbb{R}_+ \quad (\mathbb{P}\text{- a.s.}). \tag{1.5.2}$$

Such an integral will be constructed below in a much more general situation (see Theorem 2.11 and Remark 2.4 in the following chapter).

A **stochastic integral** with respect to a Wiener process for a predictable function satisfying condition (1.5.2) has properties similar to properties (i), (ii) from Sect. 1.4.16 (see Sect. 2.4.4 below). In particular,

(a) The integral $\int_{[0,t]} b(s)\,dw(s)$ has a version belonging to $\mathfrak{M}^c_{\text{loc}}(\mathbb{R}_+, \mathbb{R}^1)$ and this version, for which we keep the same notation, has quadratic variation

$$\left\langle \int_{[0,\cdot]} b(s)\,dw(s) \right\rangle_t = \int_{[0,t]} b^2(s)\,ds \quad (\mathbb{P}\text{- a.s.}).$$

(b) If $\{b_n,\ n \in \mathbb{N}\}$ is a sequence of predictable real-valued processes such that

$$\lim_{n\to\infty} \int_{[0,T]} |b_n(t) - b(t)|^2 dt = 0$$

in probability, then

$$\lim_{n\to\infty} \int_{[0,T]} b_n(t)\,dw(t) = \int_{[0,T]} b(t)\,dw(t)$$

in probability.

Warning 1.7 *Throughout this section, when speaking about a stochastic integral with respect to a one-dimensional Wiener process for a predictable function satisfying condition* (1.5.2), *we have in mind the version of the integral belonging to* $\mathfrak{M}^c_{loc}(\mathbb{R}_+, \mathbb{R}^1)$. □

1.5.3. From now on, we fix a d_1-dimensional standard Wiener process $W = W(t) = \big(w^1(t), \dots, w^{d_1}(t)\big),\ t \geq 0$.

Suppose that $\sigma^{il}(t, \omega),\ b^i(t, \omega)$ are predictable real-valued stochastic processes on $\mathbb{R}_+$, $i = 1, 2, \dots, d$, $l = 1, 2, \dots d_1$, such that $\sigma^{ij}(\cdot, \omega) \in \mathbb{L}_2([0, T])$ and $b^i(\cdot, \omega) \in \mathbb{L}_1([0, T])$ ($\mathbb{P}$-a.s.), and fix $T \in \mathbb{R}_+$.

Definition 1.15 Let τ be a stopping time (relative to $\{\mathscr{F}_t\}$) and $\xi = \xi(t)$ be a continuous predictable stochastic process in $\mathbb{R}^{\mathrm{d}}$. We say that ξ has stochastic differential

$$d\xi^i(t) = b^i(t)\,dt + \sigma^{il}(t)\,dw^l(t),\ t < \tau,\ i = 1, \dots, \mathrm{d}, \tag{1.5.3}$$

if, for each $i = 1, 2, \dots, \mathrm{d}$,

$$\mathbb{P}\left\{ \sup_{t<\tau} \left| \xi^i(t) - \xi^i(0) - \int\limits_{[0,t]} b^i(s)\,ds - \int\limits_{[0,t]} \sigma^{il}(s)\,dw^l(s) \right| = 0 \right\} = 1. \quad \square$$

One of the fundamental results of the Itô stochastic integro-differential calculus is the following change of variable formula; cf. [93, Chapter 4, §3].

Theorem 1.10 (The Itô Formula) *If $f \in \mathbb{C}^{1,2}(\mathbb{R}_+, \mathbb{R}^{\mathrm{d}})$ and $\xi = \xi(t)$ is a stochastic process with stochastic differential* (1.5.3)*, then the process*

$$t \mapsto f\big(t,\ \xi(t)\big),\ t \geq 0,$$

has stochastic differential

$$df\big(t, \xi(t)\big) = \Big(\frac{\partial f(t,x)}{\partial t}\Big|_{x=\xi(t)} + \frac{1}{2}\sigma^{il}\sigma^{jl} f_{ij}\big(t, \xi(t)\big) + b^i f_i\big(t, \xi(t)\big) \Big)\,dt$$
$$+ \sigma^{il} f_i\big(t, \xi(t)\big)\,dw^l(t),\ t < \tau. \quad \square$$

1.5.4. In this paragraph, we fix the numbers $T_0,\ K \in \mathbb{R}_+$, where $T_0 < T$.

Let $b^i = b^i(t, x, \omega)$ and $\sigma^{il} = \sigma^{il}(t, x, \omega)$, $i = 1, 2, \dots, \mathrm{d}$, $l = 1, 2, \dots, \mathrm{d}_1$, be functions defined on $[T_0, T] \times \mathbb{R}^{\mathrm{d}} \times \Omega$ and taking values in $\mathbb{R}^1$, and let $X_0 \in \mathbb{R}^{\mathrm{d}}$ be an $\mathscr{F}_{T_0}$-measurable random variable. We assume that, for every $x \in \mathbb{R}^{\mathrm{d}}$, each of the functions $b^i(\cdot, x, \cdot)$ and $\sigma^{il}(\cdot, x, \cdot)$ is predictable

Consider the system of equations

$$\begin{aligned} X^i(t) = X_0^i + \int\limits_{[T_0,t]} b^i(s, X^i(s))\,ds \\ + \int\limits_{[T_0,t]} \sigma^{il}(s, X^i(s))\,dw^l(s),\ t \in [T_0, T],\ i = 1, \dots, \mathrm{d}. \end{aligned} \tag{1.5.4}$$

Equations of this type were considered for the first time by K. Itô and are traditionally called **Itô (ordinary) differential equations**.

Definition 1.16 A solution of *(1.5.4)* is a continuous predictable stochastic process $X = X(t, T_0, X_0)$ taking values in $\mathbb{R}^{\mathrm{d}}$ and satisfying the equality *(1.5.4)* for all $t \in [T_0, T]$, $\omega \in \Omega'$, where $\Omega' \subset \Omega$ and $\mathbb{P}(\Omega') = 1$. □

The parameters T_0 and X_0 in the notation of the solution of system (1.5.4) represent the initial condition and will be dropped when the starting point of the process does not matter.

Here is a basic existence, uniqueness, and regularity result for (1.5.4); cf. [65, Section 2.5].

Theorem 1.11 *Assume that there exists a number $K \in \mathbb{R}_+$ such that, for all $t \in [0, T]$, $\omega \in \Omega$ and $x, y \in \mathbb{R}^{\mathrm{d}}$,*

$$\sum_{i=1}^{\mathrm{d}} \sum_{l=1}^{\mathrm{d}_1} |\sigma^{il}(t, x) - \sigma^{il}(t, y)|^2 \leq K|x - y|^2, \ |b(t, x) - b(t, y)| \leq K|x - y|.$$

Then the solution of (1.5.4) *exists and is unique. If, in addition,*

$$\mathbb{E}\left(|X_0|^{2p} + \int_{[T_0,T]} \left(|b(t, 0)|^{2p} + \sum_{i=1}^{\mathrm{d}} \sum_{l=1}^{\mathrm{d}_1} |\sigma^{il}(t, 0)|^{2p} \right) dt \right) < \infty,$$

for some $p \geq 1$, then there exists a number C depending on p and K such that

$$\begin{aligned} \mathbb{E} \sup_{s \in [T_0,T]} |X(s, T_0, X_0) - X_0|^{2p} &\leq C|T - T_0|^{p-1} e^{C(T-T_0)} \\ &\times \mathbb{E} \int_{[T_0,T]} \left(|X_0|^{2p} + |b(s, 0)|^{2p} + \sum_{i=1}^{\mathrm{d}} \sum_{l=1}^{\mathrm{d}_1} |\sigma^{il}(s, 0)|^{2p} \right) ds \end{aligned} \tag{1.5.5}$$

and, for every $\mathscr{F}_{T_0}$-measurable $\tilde{X}_0$ with $\mathbb{E}|\tilde{X}_0|^{2p} < \infty$,

$$\mathbb{E} \sup_{s \in [T_0,T]} |X(s, T_0, X_0) - X(s, T_0, \tilde{X}_0)|^{2p} \leq Ce^{C(T-T_0)} \mathbb{E}|X_0 - \tilde{X}_0|^{2p}. \tag{1.5.6}$$

□

The term $|T - T_0|^{p-1}$ in (1.5.5) comes from an estimate of the stochastic integral; cf. [65, Corollary 2.5.3].

The process $X = X(t, T_0, X_0)$ is called a **diffusion process** with **drift coefficient** b and **diffusion coefficient** σ. Here is a reason for this terminology. Suppose $X_0(\omega) = x \in \mathbb{R}^{\mathrm{d}}$, $T_0 = s$, and the drift and the diffusion coefficients do not depend on ω and are sufficiently smooth in x. Then, for every

$f \in \mathbb{C}_b^2(\mathbb{R}^d)$, the function

$$u(s,x) := \mathbb{E}f\big(X(t,s,x)\big) \tag{1.5.7}$$

satisfies the backward Kolmogorov equation

$$-\frac{\partial u(s,x)}{\partial s} = \frac{1}{2}\sigma^{il}\sigma^{jl}(s,x)\,u_{ij}(s,x) + b^i(s,x)\,u_i(s,x), \tag{1.5.8}$$

$$u(t,x) = f(x),\ 0 \le s < t,\ x \in \mathbb{R}^d. \tag{1.5.9}$$

Physicists often call a system of this type a diffusion equation and use it to model various physical phenomena with diffusive behavior. As a result, solutions of the Itô equations become a mathematical model of trajectories of diffusing particles.

1.5.5. Representations of the type (1.5.7) and related topics are studied in Chap. 5. At this point, let us only mention the following result.

Theorem 1.12 *Suppose that the conditions of Theorem* 1.11 *are satisfied and* $X_0(\omega) = x \in \mathbb{R}^d$. *Suppose also that the drift and the diffusion coefficients in* (1.5.4) *belong to* $\mathbb{C}_b^{0,m+1}([T_0,T] \times \mathbb{R}^d)$ ($\mathbb{P}$-a.s.), *and the bounds on the coefficients and their derivatives do not depend on* ω. *Then there exists a function* $X = X(t,s,x,\omega)$ *on* $[s,T] \times [T_0,T] \times \mathbb{R}^d \times \Omega$ *such that, for every* $s \in [T_0,T]$, $X(\cdot,s,\cdot,\omega) \in \mathbb{C}^{0,m}([s,T] \times \mathbb{R}^d)$ ($\mathbb{P}$-a.s.) *and* X *is a solution of* (1.5.4) *for* $t \in [s,T]$ *with initial condition* $X(s,s,x) = x$. □

The proof of this statement follows by combining the results of [65, Ch. II, §5] with the following continuity criterion of random function; see also [12].

1.5.6.

Theorem 1.13 (Kolmogorov's Criterion) *Let* G *be a domain in* $\mathbb{R}^d$ *with sufficiently regular boundary, and let* $Y = Y(x,\omega)$ *be a* $\mathscr{B}(G) \otimes \mathscr{F}$*-measurable mapping of* $G \times \Omega$ *to* $\mathbb{R}^m$. *Assume that there exist numbers* $\nu > 0$, $\varepsilon > 0$, *and* $N \subset \mathbb{R}_+$ *such that*

$$E|Y(x) - Y(y)|^\nu \le N|x-y|^{d+\varepsilon},\quad x,y \in G.$$

Then Y *has a continuous in* x *version.* □

Note that the dimension m of Y does not appear in any of the conditions of the theorem.

One of the earlier proofs of this theorem when $d = 1$ can be found in the translator's notes (page 576) in the Russian translation of [25]. The theorem itself goes back to at least 1937 and, by now, has been proved in various forms by many authors.

We now show how to obtain the existence of a version of $X = X(t,s,x)$ that is jointly continuous in t, s, x.

Proposition 1.5 *Suppose that conditions of Theorem* 1.11 *are satisfied with* $p > \mathrm{d}+3$. *Then there exists a function* $X = X(t,s,x,\omega)$ *on* $[s,T]\times[T_0,T]\times\mathbb{R}^{\mathrm{d}}\times\Omega$ *such that* $X(\cdot,\cdot,\cdot,\omega)$ *is continuous and* X *is a solution of* (1.5.4) *for* $t\in[s,T]$ *with initial condition* $X(s,s,x)=x$. □

Proof Let $x,\ y\in\mathbb{R}^{\mathrm{d}}$ and $T\ge t_2\ge t_1\ge s_2\ge s_1\ge T_0$. Let $X=X(t,s,x)$ be the solution of (1.5.4) for $t\ge s$ with initial condition $X(s,s,x)=x$. Using

$$(a+b+c)^2\le 9(a^2+b^2+c^2),\ a,b,c\in\mathbb{R}^1,$$

we find

$$\begin{aligned}|X(t_2,s_2,x)-X(t_1,s_1,y)|^{2p}&\le 9^p\big(|X(t_2,s_2,x)-X(t_1,s_2,x)|^{2p}\\&\quad+|X(t_1,s_2,x)-X(t_1,s_2,y)|^{2p}+|X(t_1,s_2,y)-X(t_1,s_1,y)|^{2p}\big)\\&:=9^p(I_1+I_2+I_3).\end{aligned}$$

It follows from (1.5.6) that there exists a constant N_2 depending on $p,\ K,\ T_0,\ T_1$ such that $I_2\le N_2|x-y|^{2p}$.

By uniqueness, $X(t_1,s_1,y)=X\big(t_1,s_2,X(s_1,s_2,y)\big)$ ($\mathbb{P}$-a.s.). Thus it follows from (1.5.5) and (1.5.6) that there exists a constant N_3 depending on p,K,T and T_0 such that

$$\begin{aligned}\mathbb{E}I_3&=\mathbb{E}|X(t_1,s_2,y)-X(t_1,s_1,y)|^{2p}\\&=\mathbb{E}|X(t_1,s_2,y)-X\big(t_1,s_2,X(s_1,s_2,y)\big)|^{2p}\le N_3\mathbb{E}|X(s_1,s_2,y)-y|^{2p}\\&\le N_3|s_1-s_2|^{p-1}.\end{aligned}$$

Similarly, we can show that there exists a constant N_1 depending only on p,K,T_0,T such that $\mathbb{E}I_1\le N_1|t_1-t_2|^p$. Then continuity follows by the Kolmogorov criterion. □

1.5.7. Under the assumptions of Theorem 3, the process X generates a measure μ_X on the measurable space $\big(\mathbb{C}((T_0,T);\mathbb{R}^{\mathrm{d}}),\mathscr{B}(\mathbb{C}(T_0,T);\mathbb{R}^{\mathrm{d}})\big)$ by

$$\mu_X(\Gamma):=\mathbb{P}\big(X(\cdot,T_0,X_0)\in\Gamma\big),\quad \Gamma\in\mathscr{B}(\mathbb{C}(T_0,T);\mathbb{R}^{\mathrm{d}}).$$

Let $\tilde{X}=\tilde{X}(t,T_0,X_0)$ be the solution of (1.5.4) with the same initial condition, but with the drift coefficient b changed to $\tilde{b}$, where $\tilde{b}^i:=b^i+h^l\sigma^{il}$ and $h^l=h^l(t,x,\omega)$, $l=1,\dots,\mathrm{d}_1$, are uniformly bounded and predictable functions. The following result establishes a connection between the measures μ_X and $\mu_{\tilde{X}}$; cf. [93, Chapter 7].

Theorem 1.14 (Girsanov)

(i) The measures μ_X and $\mu_{\tilde{X}}$ are mutually absolutely continuous,

$$\frac{d\mu_{\tilde{X}}}{d\mu_X}(X) = \mathbb{E}\left(\exp\left\{ \int\limits_{[T_0,T]} h^l\big(t, X(t, T_0, X_0)\big)\, dw^l(t) \right.\right.$$

$$\left.\left. - \frac{1}{2} \int\limits_{[T_0,T]} \left|h\big(t,\; X(t, T_0, X_0)\big)\right|^2 dt \right\} \middle| \mathscr{F}^X_{T_0,T} \right) \quad (\mathbb{P}\text{- a.s.})$$

and

$$\mathbb{E}\, \frac{d\mu_{\tilde{X}}}{d\mu_X}(X) = 1,$$

where X is a trajectory of the process $X(t, X_0, T_0)$, $t \in [T_0, T]$, and $\mathscr{F}^X_{T_0,T}$ is the σ-algebra generated by this process.

(ii) If $\Phi : \mathbb{C}([T_0, T]; \mathbb{R}^d) \to \mathbb{R}^1$ is a Borel-measurable functional and $\mathbb{E}|\Phi(\tilde{X})| < \infty$, then $\mathbb{E}\Phi(\tilde{X}) = \mathbb{E}\left(\Phi(X)\, \frac{d\mu_{\tilde{X}}}{d\mu_X}(X)\right)$. □

1.5.8. In this and the following paragraphs we discuss stochastic versions of the Fubini theorem.

Let $w = w(t)$ be a standard Wiener process. Denote by $\tilde{\mathscr{F}}^s_t$ the σ-algebra generated by the increments $w(u_1) - w(u_2)$, $u_1, u_2 \in [s, t]$. Let $\mathfrak{F}_0$ be a sub-σ-algebra of $\mathscr{F}_0$ completed with respect to $\mathbb{P}$, and define

$$\mathfrak{F}_t := \mathfrak{F}_0 \vee \tilde{\mathscr{F}}^0_t.$$

Warning 1.8 *Here and in the sequel, if $\mathscr{F}$ and $\mathscr{G}$ are two σ-algebras, then $\mathscr{F} \vee \mathscr{G}$ denotes the smallest σ-algebra containing both of them.* □

The next result shows that, in a sense,

$$\mathbb{E}\left(\int\limits_{[0,T]} b(t)\, dw(t) \middle| \mathfrak{F}_T \right) = \int\limits_{[0,T]} \mathbb{E}\big[b(t)\big|\mathfrak{F}_t\big] dw(t).$$

Theorem 1.15 *Let $b = b(t)$ be a predictable real-valued process satisfying*

$$\mathbb{E} \int_{[0,T]} |b(t)|^2 dt < \infty.$$

Then there exists a predictable function $b_{\mathfrak{F}}$ on $\Omega \times [0, T]$ such that

$$b_{\mathfrak{F}}(t) = \mathbb{E}[b(t)|\mathfrak{F}_t]$$

for $\langle w \rangle \circ \mathbb{P}_T$- a.a. (t, ω) and

$$\mathbb{E}\left(\int_{[0,T]} b(t)\, dw(t) | \mathfrak{F}_T \right) = \int_{[0,T]} b_{\mathfrak{F}}(t)\, dw(t) \quad (\mathbb{P}\text{- a.s.}). \qquad \square$$

Proof Similar to Sect. 1.4.15, we can approximate the process b, as an element of $\mathbb{L}_2([0, T]\ \Omega, \mathscr{P}_T, \langle w \rangle \circ \mathbb{P}_T; \mathbb{R}^1)$, by simple predictable processes $b^n = b^n(t)$, and then it suffices to prove the theorem when b is of the form

$$b(t) = 1_{\{0\}}(t)\, b_0 + \sum_{i=1}^{n-1} b_i 1_{\{(t_i, t_{i+1}]\}}(t),$$

where $0 = t_0 < t_1 < \ldots < t_n = T$, and b_i are $\mathbb{P}$-integrable, $\mathscr{F}_{t_i}$-measurable random variables, $i = 0, \ldots, n-1$.

Then

$$\mathbb{E}\big[b(t)|\mathfrak{F}_t\big] = \mathbb{E}\big[b_0|\mathfrak{F}_t\big] 1_{\{0\}}(t) + \sum_{i=1}^{n-1} \mathbb{E}\big[b_i\big|\mathfrak{F}_t\big]\ 1_{\{(t_i, t_{i+1}]\}}(t).$$

The $\mathscr{F}_{t_i}$-measurability of b_i and independence of the σ-algebras $\tilde{\mathscr{F}}_t^{t_i}$ and $\mathscr{F}_{t_i}$ for $t_i < t$ imply

$$\mathbb{E}\big[b_i\big|\mathfrak{F}_{t_i} \vee \tilde{\mathscr{F}}_t^{t_i}\big] = \mathbb{E}\big[b_i\big|\mathfrak{F}_{t_i}\big] \quad (\mathbb{P}\text{- a.s.}).$$

Therefore, $\mathbb{E}\big[b(t)\big|\mathfrak{F}_t\big]$ has a predictable, relative to $\{\mathfrak{F}_t\}$, version

$$b_{\mathfrak{F}}(t) := \mathbb{E}\big[b_0\big|\mathfrak{F}_t\big]\ 1_{\{0\}}(t) + \sum_{i=0}^{n-1} \mathbb{E}\big[b_i\big|\mathfrak{F}_{t_i}\big]\ 1_{\{(t_i, t_{i+t}]\}}(t).$$

Similarly, $\mathfrak{F}_T$-measurability of the increments $w(u_1)-w(u_2)$ for $u_1, u_2 \in [0, T]$ and independence between b_i and $\tilde{\mathscr{F}}_T^{t_i}$ for every i imply

$$\mathbb{E}\left(\int\limits_{[0,T]} b(t)\,dw(t)\big|\mathfrak{F}_T\right) = \mathbb{E}\left(\sum_{i=0}^{n-1} b_i\big(w(t_{i+1}) - w(t_i)\big)\big|\mathfrak{F}_T\right)$$

$$= \sum_{i=0}^{n-1} \mathbb{E}\big[b_i|\mathfrak{F}_{t_i}\big]\big(w(t_{i+1}) - w(t_i)\big) = \int\limits_{[0,T]} b_{\mathfrak{F}}(t)dw(t) \ (\mathbb{P}\text{- a.s.}),$$

completing the proof of the theorem. □

1.5.9. The following statement is a different version of the Fubini theorem in the stochastic setting; cf. [28].

Theorem 1.16 *Let $f = f(t, x, \omega)$ be a $\mathscr{B}([0, T] \times \mathbb{R}^{\mathrm{d}}) \otimes \mathscr{F}$-measurable real-valued function belonging to*

$$\mathbb{L}_2\big([0, T] \times \mathbb{R}^{\mathrm{d}}\big) \bigcap \mathbb{L}_2\big([0, T]; \mathbb{L}_1(\mathbb{R}^{\mathrm{d}})\big) \ (\mathbb{P}\text{- a.s.})$$

and predictable for every x. Then the function $t \mapsto \int\limits_{\mathbb{R}^{\mathrm{d}}} f(t, x, \omega)dx$ has a predictable version, denoted by $F = F(t, \omega)$, such that

$$\int\limits_{\mathbb{R}^{\mathrm{d}}}\left(\int\limits_{[0,T]} f(t, x, \omega)dw(t)\right)dx = \int\limits_{[0,T]} F(t, \omega)dw(t) \ (\mathbb{P}\text{- a.s.}). \qquad \square$$

1.5.10. In this paragraph, we extend the Itô formula from Theorem 1.10 to random functions $f = f(t, x)$. Let $\xi = \xi(t)$ be a stochastic process with stochastic differential (1.5.3), and let F be a $\mathscr{P} \otimes \mathscr{B}(\mathbb{R}^{\mathrm{d}})$-measurable real-valued function belonging to $\mathbb{C}^{0,2}(\mathbb{R}_+, \mathbb{R}^{\mathrm{d}})$ for $\mathbb{P}$-a.a. ω. Assume that, for every $x \in \mathbb{R}$, the function F has stochastic differential

$$dF(t, x) = J(t, x)\,dt + H^l(t, x)\,dw^l(t), \ t < \tau,$$

where $J = J(t, x)$, $H^l = H^l(t, x)$, $l = 1, 2, \ldots \mathrm{d}_1$, are $\mathscr{P} \otimes \mathscr{B}(\mathbb{R}^{\mathrm{d}})$-measurable real-valued functions and τ is the same stopping time as in (1.5.3). We also assumed that $H^l \in \mathbb{C}^{0,1}(\mathbb{R}_+ \times \mathbb{R}^{\mathrm{d}})$ for all $l = 1, 2, \ldots, \mathrm{d}_1$ and $\mathbb{P}$- a.a. ω.

Theorem 1.17 (The Itô–Ventcel Formula) *The process* $t \mapsto F\big(t, \xi(t)\big)$, $t \geq 0$, *has stochastic differential*

$$
\begin{aligned}
dF\big(t,\xi(t)\big) &= J\big(t,\xi(t)\big)\,dt + H^l(t,\xi(t))\,dw^l(t)\\
&+ \Big(b^i F_i\big(t,\xi(t)\big) + \frac{1}{2}\sigma^{il}\sigma^{il}F_{ij}\big(t,\xi(t)\big)\Big)\,dt + \sigma^{il}F_i\big(t,\xi(t)\big)\,dw^l(t) \qquad (1.5.10)\\
&+ H_i^l\sigma^{il}\big(t,\xi(t)\big)\,dt,\ t < \tau.
\end{aligned}
$$

□

1.5.11. Before proving Theorem 1.17, we consider a regularization (or smoothing) method which will be used in the proof and in many other places below.

Set

$$
\zeta(x) := \begin{cases} \exp\left(-\dfrac{|x|^2}{1-|x|^2}\right), & \text{if } |x| < 1;\\[2ex] 0, & \text{if } |x| \geq 1. \end{cases}
$$

By definition, $\zeta \in \mathbb{C}_0^\infty(\mathbb{R}^{\mathrm{d}})$, $\zeta(0) = 1$, and $0 \leq \zeta(x) \leq 1$. Next, we normalize ζ:

$$
\bar{\zeta}(x) = \frac{\zeta(x)}{\int_{\mathbb{R}^{\mathrm{d}}}\zeta(y)dy}.
$$

Let f be a locally integrable function on $\mathbb{R}^{\mathrm{d}}$. Define

$$
\mathcal{T}_\varepsilon f(x) := \varepsilon^{-\mathrm{d}}\int\limits_{\mathbb{R}^{\mathrm{d}}}\bar{\zeta}\left(\frac{x-y}{\varepsilon}\right)f(y)dy.
$$

Similarly, for a function $f = f(t,x)$, we define

$$
\mathcal{T}_\varepsilon f(t,x) := \varepsilon^{-\mathrm{d}}\int\limits_{\mathbb{R}^{\mathrm{d}}}\bar{\zeta}\left(\frac{x-y}{\varepsilon}\right)f(t,y)dy.
$$

The function $\mathcal{T}_\varepsilon f$ is called the Sobolev average of f. Let us establish some properties of this average.

Lemma 1.2 *If* $f \in \mathbb{C}^{0,0}(\mathbb{R}_+ \times \mathbb{R}^{\mathrm{d}})$, *then* $\mathcal{T}_\varepsilon f \in \mathbb{C}^{0,\infty}(\mathbb{R}_+ \times \mathbb{R}^{\mathrm{d}})$ *and, for every* $r > 0$ *and a compact interval* $I \in \mathbb{R}_+$,

$$
\lim_{\varepsilon\to 0}\sup_{\substack{t\in I\\|x|\leq r}}|f(t,x) - \mathcal{T}_\varepsilon f(t,x)| = 0. \qquad (1.5.11)
$$

If $f \in \mathbb{C}^{0,1}(\mathbb{R}_+ \times \mathbb{R}^d)$, *then*

$$\frac{\partial \mathcal{T}_\varepsilon f}{\partial x^i} = \mathcal{T}_\varepsilon \frac{\partial f}{dx^i},\ i = 1, 2, \ldots, \mathrm{d}. \qquad \square$$

Proof Similar statements for functions from $\mathbb{L}_p$ are well known; see below (Lemma 4.1.3 and the reference in Sect. 4.2.4). We will only prove (1.5.11).

By direct computation,

$$\begin{aligned}
&\sup_{\substack{t\in I\\|x|\le r}} |f(t,x) - \mathcal{T}_\varepsilon f(t,x)| \\
&= \sup_{\substack{t\in I\\|x|\le r}} \left| \int_{|z|\le 1} \bar{\zeta}(z)[f(t,x) - f(t,x+\varepsilon z)]dz \right| \\
&\le \left(\int_{|z|\le 1} |\bar{\zeta}(z)|^2 dz \right)^{\frac{1}{2}} \left(\sup_{\substack{t\in I\\|x|\le r}} \int_{|z|\le 1} |f(t,x) - f(t,x+\varepsilon z)|^2 dz \right)^{\frac{1}{2}}.
\end{aligned}$$

Then uniform continuity of f implies (1.5.11). $\square$

1.5.12.

Proof of Theorem 1.17 For the sake of simplicity we assume that $\tau = \infty$. To begin, we establish an auxiliary result.

Lemma 1.3 *Let the conditions of Theorem 1.17 be satisfied. Suppose also that, for every* $x \in \mathbb{R}^d$ *and* $i = 1, 2, \ldots, \mathrm{d}$, *the function* $F_i = \partial F/\partial x^i$ *has stochastic differential*

$$dF_i(t,x) = J_i(t,x)\,dt + H_i^l(t,x)\,dw^l(t),\ t \in \mathbb{R}_+.$$

Then the assertion of Theorem 1.17 holds. $\square$

Proof We will prove the result when the functions b^i, σ^{il} do not depend on t and ω. The general case follows in the same way as for the original Itô formula.

Fix $s, t \in \mathbb{R}_t$, $s \le t$, and let $\{s = t_0^n < t_1^n < \ldots < t_{k(n)+1}^n = t\}$ be a sequence of partitions of the interval $[s, t]$ for $n \in \mathbb{N}$. We use the following notation:

$$\delta_n = \max_m |t_{m+1}^n - t_m^n|,\ \ \Delta_m^n \xi = \xi(t_{m+1}^n) - \xi(t_m^n).$$

By direct computation,

$$\begin{aligned}F\big(t,\xi(t)\big)-F\big(s,\xi(s)\big)&=\sum_{m=0}^{k(n)}\big[F\big(t^n_{m+1},\xi(t^n_{m+1})\big)-F\big(t^n_m,\xi(t^n_m)\big)\big]\\&=\sum_{m=0}^{k(n)}\big[F\big(t^n_{m+1},\xi(t^n_{m+1})\big)-F\big(t^n_m,\xi(t^n_{m+1})\big)\big]\\&+\sum_{m=0}^{k(n)}\big[F\big(t^n_m,\xi(t^n_{m+1})\big)-F\big(t^n_m,\xi(t^n_m)\big)\big]\\&:=I_1^{(n)}+I_2^{(n)}.\end{aligned}$$

First, consider $I_2^{(n)}$. By the Taylor formula

$$\begin{aligned}\sum_{m=0}^{k(n)}\big[F\big(t^n_m,\xi(t^n_{m+1})\big)-F\big(t^n_m,\xi(t^n_m)\big)\big]&=\sum_{m=0}^{k(n)}F_i\big(t^n_m,\xi(t^n_m)\big)\Delta^n_m\xi^i\\&+\frac{1}{2}\sum_{m=0}^{k(n)}F_{ij}\big(t^n_m,\xi(t^n_m)+\theta\Delta^n_m\xi\big)\Delta^n_m\xi^i\Delta^n_m\xi^j,\end{aligned}$$

where $\theta=\theta(\omega)\in[0,1]$ for every ω. Because $d\xi^i=b^i dt+\sigma^{ik}dw^l$ (cf. (1.5.3)),

$$\lim_{\delta_n\to 0}I_2^{(n)}=\int_{[s,t]}F_i\big(u,\xi(u)\big)d\xi^i(u)+\frac{1}{2}\int_{[s,t]}F_{ij}\big(u,\xi(u)\big)\sigma^{il}(u)\sigma^{jl}(u)du$$

in probability.

Next, consider $I_1^{(n)}$. This time, we only go to the first-order term in the Taylor expansion:

$$\begin{aligned}F\big(t^n_{m+1},\xi(t^n_{m+1})\big)&=F\big(t^n_{m+1},\xi(t^n_m)+\Delta^n_m\xi\big)\\&=F\big(t^n_{m+1},\xi(t^n_m)\big)+F_i\big(t^n_{m+1},\xi(t^n_m)+\theta_1\Delta^n_m\xi\big)\Delta^n_m\xi^i,\\F\big(t^n_m,\xi(t^n_{m+1})\big)&=F\big(t^n_m,\xi(t^n_m)\big)+F_i\big(t^n_m,\xi(t^n_m)+\theta_2\Delta^n_m\xi\big)\Delta^n_m\xi^i,\end{aligned}$$

with $\theta_1(\omega), \theta_2(\omega) \in [0, 1]$ for every ω. Because $dF = J dt + H^l dw^l$,

$$\lim_{\delta_n \to 0} \sum_{m=0}^{k(n)} \left[F\left(t_{m+1}^n, \xi(t_m^n)\right) - F\left(t_m^n, \xi(t_m^n)\right) \right]$$

$$= \int_{[s,t]} J\left(u, \xi(u)\right) du + \int_{[s,t]} H^l\left(u, \xi(u)\right) dw^l(u)$$

in probability.

Similarly, because $dF_i = J_i dt + H_i^l dw^l$,

$$\lim_{\delta_n \to 0} \sum_{m=0}^{k(n)} \left[F_i\left(t_{m+1}^n, \xi(t_m^n)\right) - F_i\left(t_m^n, \xi(t_m^n)\right) \right] \Delta_m^n \xi^i = \int_{[s,t]} \sigma^{il}(u) H_i^l(u, \xi(u)) du$$

in probability. That is,

$$\lim_{\delta_n \to 0} I_1^{(n)} = \int_{[s,t]} J\left(u, \xi(u)\right) du + \int_{[s,t]} H^l\left(u, \xi(u)\right) dw^l(u) + \int_{[s,t]} \sigma^{il}(u) H_i^l(u, \xi(u)) du$$

in probability.

Combining the limits for $I_1^{(n)}$ and $I_2^{(n)}$, we get formula (1.5.10) and complete the proof of Lemma 1.3. □

To complete the proof of Theorem 1.17, fix $\varepsilon > 0$ and define $F_\varepsilon(t, x) = \mathcal{T}_\varepsilon F(t, x)$, $J_\varepsilon(t, x) = \mathcal{T}_\varepsilon J(t, x)$ and $H_\varepsilon^l(t, x) = \mathcal{T}_\varepsilon H^l(t, x)$. We now show that the function $(t, x) \mapsto F_\varepsilon(t, x)$ satisfies the conditions of Lemma 1.3.

The assumptions about F imply that the stochastic integral

$$\int_{[0,t]} H^l(s, x)\, dw^l(s)$$

has a continuous in (t, x) version, because the functions $(t, x) \mapsto F(t, x)$ and $(tmx) \mapsto \int_{[0,t]} J(s, x)\, ds$ are continuous in (t, x). As a result, for every $t > 0$, the function $x \mapsto \int_{[0,t]} H^l(s, x)\, dw^l(s)$ is locally integrable in x ($\mathbb{P}$- a.s.), and we can apply the operator $\mathcal{T}_\varepsilon$ to this function.

Using Fubini theorems, both the classical one and the stochastic version from Theorem 1.16, we conclude that

$$dF_\varepsilon(t, x) = J_\varepsilon(t, x)\, dt + H_\varepsilon^l(t, x)\, dw^l(t).$$

By Lemma 1.2, the functions F_ε, H_ε^l, and J_ε belong to $\mathbb{C}^{0,\infty}(\mathbb{R}_+ \times \mathbb{R}^{\mathrm{d}})$ for $\mathbb{P}$- a.a. ω.

Then Theorem 1.16 and Lemma 1.2 imply that, for every $t \in \mathbb{R}_+$ and $i = 1, 2, \dots, \mathrm{d}$,

$$\frac{\partial}{\partial x^i}\left(\int\limits_{[0,t]} H^l_\varepsilon(s,x)\,dw^l(s)\right) = \frac{\partial}{\partial x^i}\left(\mathcal{T}_\varepsilon\left(\int\limits_{[0,t]} H^l(s,\cdot)\,dw^l(s)\right)(x)\right)$$

$$= \varepsilon^{-\mathrm{d}}\int\limits_{\mathbb{R}^\mathrm{d}} \bar{\zeta}_i\left(\frac{x-y}{\varepsilon}\right)\left(\int\limits_{[0,t]} H^l(s,y)w^l(s)\right)dy$$

$$= \varepsilon^{-\mathrm{d}}\int\limits_{[0,t]}\left(\int\limits_{\mathbb{R}^\mathrm{d}} \bar{\zeta}_i\left(\frac{x-y}{\varepsilon}\right)H^l(s,y)dy\right)dw^l(s)$$

$$= \int\limits_{[0,t]} \big(H^l_\varepsilon(s,x)\big)_i\,dw^l(s)\ \ (\mathbb{P}\text{- a.s.})$$

Similarly,

$$\frac{\partial}{\partial x^i}\left(\int\limits_{[0,t]} J_\varepsilon(s,x)\,ds\right) = \int\limits_{[0,t]} \big(J_\varepsilon(s,x)\big)_i\,ds.$$

Consequently, for every $i = 1, 2, \dots, \mathrm{d}$, the function $(F_\varepsilon(t,x))_i$ has stochastic differential

$$d\big(F_\varepsilon(t,x)\big)_i = \big(J_\varepsilon(t,x)\big)_i\,dt + \big(H^l_\varepsilon(t,x)\big)_i\,dw^l(t),\ t \in \mathbb{R}_+,$$

and therefore satisfies the conditions of Lemma 11.

Passing to the limit $\varepsilon \to 0$ in the corresponding equation for $F_\varepsilon\big(t,\xi(t)\big)$ and using Lemma 10 and property (b) of the stochastic integral from Sect. 1.2, we complete the proof of Theorem 9. □

1.5.13. In this paragraph, we define a **backward stochastic integral** with respect to a Wiener process.

Suppose that a family of σ-algebras $\{\mathscr{F}^t_T\}_{0\le t\le T}$ is given on $(\Omega, \mathscr{F})$ such that, for every $s,\ t \in [0,T]$,

1. $\mathscr{F}^t_T \subset \mathscr{F},\ t \in \mathbb{R}_+$;
2. $\mathscr{F}^t_T \subset \mathscr{F}^s_T,\ \ s < t$ (decreasing);
3. $\bigcap\limits_{\varepsilon>0} \mathscr{F}^{t-\varepsilon}_T = \mathscr{F}^t_T$ (left-continuous);
4. The σ-algebra $\mathscr{F}^T_T$ contains all $\mathbb{P}$-null sets from $\mathscr{F}$ ($\mathbb{P}$-complete).

Definition 1.17 The function $f : [0,T] \times \Omega \to \mathbb{R}^1$ is called **backward predictable** (or $\overleftarrow{\mathscr{P}}_{[0,T]}$-measurable) on $[0,T]$ relative to $\{\mathscr{F}_T^t\}$, if the function $(t,\omega) \mapsto f(T-t,\omega)$ is predictable relative to $\{\mathscr{F}_T^{T-t}\}_{0 \le t \le T}$. □

Given a standard Wiener process $w = w(t)$, define

$$w_T(t) = w(T) - w(T-t), \quad t \in [0,T].$$

It follows that w_T is a standard Wiener process relative to $\{\mathscr{F}_T^{T-t}\}$, $t \in [0,T]$.

1.5.14. Let f be a backward predictable function on $[0,T]$ relative to $\{\mathscr{F}_T^t\}$. Suppose that

$$\int_{[0,T]} |f(r)|^2 dr < \infty \quad (\mathbb{P}\text{- a.s.}).$$

Definition 1.18 A **backward stochastic integral** with respect to a one-dimensional Wiener process w is

$$\int_{[s,t]} f(r) * dw(r) := \int_{[T-t,T-s]} f(T-r)\, dw_T(r). \qquad \square$$

It can be shown that the definition does not depend on T by first considering simple functions f and then passing to the limit. The first step is the proposition below. The details are left to the interested reader.

Proposition 1.6 *If*

$$f(t) = \sum_j f_{j+1} 1_{\{[t_j, t_{j+1})\}}$$

is a simple backward predictable function on $[0,T]$, *then*

$$\int_{[0,T]} f(r) * dw\,(r) = \sum_j f_{j+1}\big(w(t_{j+1}) - w(t_j)\big) \quad (\mathbb{P}\text{- a.s.}). \qquad \square$$

1.5.15. Suppose that b^i, σ^{il}, $i = 1,2,\ldots,\mathrm{d}$, $l = 1,2,\ldots,\mathrm{d}_1$, are backward predictable functions on $[0,T]$. Suppose also that $b^i \in \mathbb{L}_1([0,T])$ and $\sigma^{il} \in \mathbb{L}_2([0,T])$ ($\mathbb{P}$- a.s.) for all i, l. Let w^l, $l = 1,\ldots,\mathrm{d}_1$, be independent standard Wiener processes.

Definition 1.19

(a) A stopping time relative to $\{\mathscr{F}_T^t\}$ is a random variable $\tau \in [0,T]$ such that $T - \tau$ is a stopping time relative to $\{\mathscr{F}_T^{T-t}\}$.

(b) A continuous backward predictable process $\xi = \xi(t) \in \mathbb{R}^{\mathrm{d}}$ has **backward stochastic differential**

$$-d\xi^i(t) = b^i(t)\,dt + \sigma^{il}(t) * dw^l(t),\ t \in [\tau, T),\ i = 1, 2, \ldots, \mathrm{d}, \tag{1.5.12}$$

if, for every $i = 1, 2, \ldots, \mathrm{d}$,

$$\mathbb{P}\left(\sup_{t\in[\tau,T]}\left|\xi^i(t) - \xi^i(T) - \int\limits_{[t,T]} b^i(s)\,ds - \int\limits_{[t,T]} \sigma^{il}(s)\,dw^l(s)\right| = 0\right) = 1.$$

□

The following result is an obvious modification of the Itô formula.

Proposition 1.7 *Let $f \in \mathbb{C}^{1,2}(\mathbb{R}_+ \times \mathbb{R}^{\mathrm{d}})$ and let $\xi = \xi(t)$ be a stochastic process with the backward stochastic differential (1.5.12). Then*

$$-df\big(t, \xi(t)\big) = \Big(-\frac{\partial}{\partial t} f(t, x)\big|_{x=\xi(t)} + b^i(t) f_i\big(t, \xi(t)\big)$$
$$+\frac{1}{2}\sigma^{il}(t)\sigma^{jl}(t) f_{ij}\big(t, \xi(t)\big)\Big)\,dt + \sigma^{il}(t) f_i\big(t, \xi(t)\big) * dw^l(t),\ t \in [\tau, T).$$

□

We can study a backward Itô equation in the same way as we studied the "forward" one in Sect. 1.5. By reversing the time, every result for a forward equation leads to the corresponding analog for a backward one.

Warning 1.9 *In the future, when considering backward Itô equations, we will use the corresponding results for forward equations without special reference. A knowledgeable reader should also keep in mind that the backward equations considered in this book are very different from BSDEs, which are also solved backward in time, but require the solution to be adapted to the original (forward) system $\{\mathscr{F}_t\}$.* □

1.5.16. To conclude this section, we present a well-known result used in the study of differential equations.

Lemma 1.4 (Gronwall–Bellman) *Let $u = u(t)$ and $v = v(t)$ be non-negative, $\mathscr{B}(\mathbb{R}_+)$-measurable functions on $[T, +\infty)$. Assume that there exists a number $C > 0$ such that, for every $t > T_0$,*

$$u(t) \le C + \int\limits_{[T_0,t]} u(s)v(s)\,ds.$$

Then

$$u(t) \leq C \exp\left\{ \int\limits_{[T_0,t]} v(s)\, ds \right\}. \qquad \square$$

For a proof, see [5].

Chapter 2
Stochastic Integration in a Hilbert Space

2.1 Introduction

This chapter is about stochastic calculus for continuous martingales and local martingales in a Hilbert space. The topics include definitions and investigations of martingales, local martingales and a Wiener process in a Hilbert space, construction of stochastic integrals with respect to these processes, and a detailed proof of the Itô formula for the square of a norm of a continuous semimartingale.

2.2 Martingales and Local Martingales

2.2.1. In this section we introduce and investigate martingales and local martingales with values in a Hilbert space. Given the subject of this book, we consider only continuous martingales and local martingales.

2.2.2. Let $\mathbb{F} := (\Omega, \mathscr{F}, \{\mathscr{F}_t\}_{t\in\mathbb{R}_+}, \mathbb{P})$ be a stochastic basis with the usual assumptions, and let $\mathbb{H}$ be a separable Hilbert space with topological dual $\mathbb{H}^*$. At this point, we do not identify $\mathbb{H}$ and $\mathbb{H}^*$.

We use the following notation:

- $\|\cdot\|$, the norm in $\mathbb{H}$;
- $(\cdot,\cdot)$ the inner product in $\mathbb{H}$;
- yx, the value of the functional $y \in \mathbb{H}^*$ on $x \in \mathbb{H}$.

Let $x = x(\omega)$ be an $\mathbb{H}$-valued random variable. Recall that x is $\mathbb{P}$-integrable if and only if $\mathbb{E}\|x\| < \infty$.

For a $\mathbb{P}$-integrable random variable $x \in \mathbb{H}$, the integral $\int_\Omega x(\omega)\, d\mathbb{P}(\omega)$ is denoted by $\mathbb{E}x$ and called the **expectation** of x.

B. L. Rozovsky, S. V. Lototsky, *Stochastic Evolution Systems*, Probability Theory and Stochastic Modelling 89, https://doi.org/10.1007/978-3-319-94893-5_2

2.2.3. Let $\mathscr{G}$ be a sub-σ-algebra of $\mathscr{F}$ and let x be a $\mathbb{P}$-integrable $\mathbb{H}$-valued random variable.

Definition 2.1 The random variable $\mathbb{E}[x|\mathscr{G}]$ with values in $\mathbb{H}$ is called the conditional expectation of x with respect to $\mathscr{G}$ if, for every $y \in \mathbb{H}^*$,

$$y\mathbb{E}[x|\mathscr{G}] = \mathbb{E}[yx|\mathscr{G}] \quad (\mathbb{P}\text{- a.s.}).$$

□

Conditional expectation for an $\mathbb{H}$-valued random variable has the same properties as the conditional expectation for a scalar random variable. The reader is encouraged to confirm that the random variable $\mathbb{E}(x|\mathscr{G})$ is well-defined up to a $\mathbb{P}$-null set [cf. the first property of the Bochner integral in Sect. 1.3.12] and is $\mathscr{G}$-measurable [cf. the Pettis Theorem 1.2].

Warning 2.1 *A random process taking values in $\mathbb{H}$ will usually be referred to as an $\mathbb{H}$-processes.* □

2.2.4. Let $\{\mathscr{G}_t\}_{t\in\mathbb{R}_+}$ be an increasing family of sub-σ-algebras of $\mathscr{F}$ that is right-continuous and $\mathbb{P}$-complete.

Definition 2.2 An $\mathbb{H}$-process $M = M(t)$ is called a **martingale** relative to $\{\mathscr{G}_t\}$ if

(i) $M(t)$ is $\mathscr{G}_t$-measurable for every $t \in \mathbb{R}_+$;
(ii) $\mathbb{E}\|M(t)\| < \infty$ for every $t \in \mathbb{R}_+$;
(iii) $\mathbb{E}[M(t)|\mathscr{G}_s] = M(s)$, $\mathbb{P}$- a.s., for every $s, t \in \mathbb{R}_+$ such that $s \le t$. □

2.2.5.

Definition 2.3 An $\mathbb{H}$-process $M = M(t)$ is called a **local martingale** relative to the family $\{\mathscr{G}_t\}$ if there exists a sequence of stopping times $\{\tau_n\}$, relative to the same family, such that the $\tau_n \uparrow \infty$, ($\mathbb{P}$- a.s.) and, for every $n \ge 1$, the $\mathbb{H}$-process $M(t \wedge \tau_n)$, $t \ge 0$, is a martingale relative to $\{\mathscr{G}_t\}$. The sequence $\{\tau_n\}$ is called a localizing sequence for M. □

Warning 2.2 *In this section, we typically consider martingales and local martingales relative to $\{\mathscr{F}_t\}$. As a result, we will usually omit references to the family of σ-algebras with respect to which a particular $\mathbb{H}$-process is a martingale or a local martingale.* □

With no loss of generality, we assume that all martingales and local martingales have the property $M(0) = 0$.

The following result summarizes many of the constructions in this paragraph; the proof is a good exercise for the reader.

Lemma 2.1

(a) *An $\mathbb{H}$-process $M = M(t)$ is a martingale if and only if $\mathbb{E}\|M(t)\| < \infty$ for all $t \in \mathbb{R}_+$ and, for every $y \in \mathbb{H}^*$, the real-valued process yM is a martingale.*

(b) An $\mathbb{H}$-process $M = M(t)$ is a local martingale if and only if there exists a localizing sequence $\{\tau_n\}$ such that $\mathbb{E}\|M(t \wedge \tau_n)\| < \infty$ for all $t \in \mathbb{R}_+$, $n \geq 1$, and, for every $y \in \mathbb{H}^*$, the real-valued process yM is a local martingale. □

2.2.6. Let us fix a complete orthonormal system (CONS) $\{h_i,\ i \in \mathbb{N}\}$ in $\mathbb{H}$ and denote by $\mathcal{P}_n$ the orthogonal projection on the sub-space generated by the first n elements of the system $\{h_i\}$:

$$\mathcal{P}_n(x) = \sum_{i=1}^{n}(x, h_i)h_i.$$

Warning 2.3 *In what follows, for $M \in \mathbb{H}$, $M_i := (h_i, M)$.* □

Theorem 2.1 *If $M = M(t)$ is a martingale in $\mathbb{H}$, then $\|M\|$ is a submartingale. If, in addition, $\mathbb{E}\|M(t)\|^2 < \infty,\ t \in \mathbb{R}_+$, then $\|M\|^2$ is also a submartingale.* □

Proof As a first step, we prove that $\|\mathcal{P}_n(M)\|$ is a submartingale for every $n \geq 1$. Let us fix s and $t \in \mathbb{R}_+$ such that $s < t$. Denote by $P_{n,s}$ the regular conditional distribution of the random vector $\big(M_1(t), \dots, M_n(t)\big)$ relative to $\{\mathscr{F}_s\}$; cf. [36, Chapter 1, §3]. By Minkowski's inequality,

$$\mathbb{E}\big[\|\mathcal{P}_n(M(t))\| \,|\, \mathscr{F}_s\big] = \int_{\mathbb{R}^n} \left(\sum_{i=1}^{n} x_i^2\right)^{\frac{1}{2}} dP_{n,s}(x_1, \dots, x_n)$$

$$\geq \left(\sum_{i=1}^{n}\left(\int_{\mathbb{R}^n} x_i dP_{n,s}(x_1, \dots, x_n)\right)^2\right)^{\frac{1}{2}} = \left(\sum_{i=1}^{n}[\mathbb{E}M_i(t)|\mathscr{F}_s]^2\right)^{\frac{1}{2}}$$

$$= \left(\sum_{i=1}^{n} M_i^2(s)\right)^{\frac{1}{2}} = \|\mathcal{P}_n(M(s))\| \ \ (\mathbb{P}\text{- a.s.}).$$

Passing to the limit $n \to \infty$ on both sides of this inequality and using a version of the dominated convergence theorem for conditional expectations (cf. [25, Chapter 1, §8]), we get the first statement of the theorem. The second assertion follows immediately from the first one and a version of Jensen's inequality (cf. [25, Chapter 1, §9]). □

2.2.7. Recall that an $\mathbb{H}$-valued function $f = f(t)$ is called strongly continuous if $\lim_{t\to s}\|f(t) - f(s)\| = 0$; it is called weakly continuous if, for every $y \in \mathbb{H}^*$, the real-valued function $t \mapsto yf(t)$ is continuous. For martingales, the two notions coincide.

Theorem 2.2 *If $M = M(t)$ is a weakly continuous martingale in $\mathbb{H}$, then it is strongly continuous.* □

Proof Weak continuity of M implies strong continuity of $\mathcal{P}_n(M)$ for every n. Then, because

$$\|M(t) - \mathcal{P}_n(M(t))\| = \sup\Big\{y \in \mathbb{H}^*,\ \|y\|_{\mathbb{H}^*} = 1 :\ y\Big(M(t) - \mathcal{P}_n(M(t))\Big)\Big\},$$

we conclude that the function $\|M - \mathcal{P}_n M\|$ is lower semi-continuous, which, in turn, implies

$$\sup_{t \le T} \|M(t) - \mathcal{P}_n(M(t))\| = \sup_{I_r} \|M(t) - \mathcal{P}_n(M(t))\|,$$

where I_r is the set of all rational points of the interval $[0, T]$.

On the other hand, the process $\|M - \mathcal{P}(M)\|$ is a submartingale by Theorem 2.1, and we can apply inequality (1.4.3):

$$\mathbb{P}(\sup_{I_r} \|M(t) - \mathcal{P}_n(M(t))\| \ge \varepsilon) \le \varepsilon^{-1}\, \mathbb{E}\|M(T) - \mathcal{P}_n(M(T))\| \to 0,\ n \to \infty.$$

Hence, there exists a subsequence $\{n'\}$ such that

$$\lim_{n' \to \infty} \sup_{t \le T} \big\|M(t) - \mathcal{P}_{n'}(M(t))\big\| = 0$$

with probability one, which implies that M is a strongly continuous process in $\mathbb{H}$. □

Corollary 2.1 *A weakly continuous local martingale M in $\mathbb{H}$ is strongly continuous.* □

Proof Let $\{\tau_n\}$ be a localizing sequence for M. Then, for every $n \ge 1$, the process $M(t \wedge \tau_n),\ t \ge 0$, is a weakly, hence strongly, continuous martingale. On the other hand, for every $\varepsilon > 0$ and $T \in \mathbb{R}_+$,

$$\mathbb{P}(\sup_{t \le T} \|M(t) - M(t \wedge \tau_n)\| > \varepsilon) \le \mathbb{P}(\tau_n < T) \to 0,\ n \to \infty,$$

and therefore there exists a subsequence $\{n'\} \subset \mathbb{N}$ such that

$$\lim_{n' \to \infty} \sup_{t \le T} \|M(t) - M(t \wedge \tau_{n'})\| = 0$$

with probability one, which concludes the proof. □

It is now clear that, for martingales and local martingales in $\mathbb{H}$, there is no need to distinguish between weak and strong continuity, and we will refer to such processes simply as continuous.

2.2.8. Denote by $\mathfrak{M}_2^c(\mathbb{R}_+, \mathbb{H})$ the set of continuous martingales in $\mathbb{H}$ such that $\mathbb{E}\|M(t)\|^2 < \infty$ for every $t \in \mathbb{R}_+$. We call such processes continuous **square**

integrable martingales in $\mathbb{H}$. Similarly, $\mathfrak{M}^c_{loc}(\mathbb{R}_+, \mathbb{H})$ denotes the set of continuous local martingales with values in $\mathbb{H}$. Sometimes, it is necessary to restrict the values of t to a bounded interval $[0, T]$, leading to the corresponding notation $\mathfrak{M}^c_2([0, T], \mathbb{H})$ and $\mathfrak{M}^c_{loc}([0, T], \mathbb{H})$.

Remark 2.1 Similar to Lemma 1.1, if $M \in \mathfrak{M}^c_{loc}(\mathbb{R}_+, \mathbb{H})$, then there exists a localizing sequence $\{\sigma_n,\ n \in \mathbb{N}\}$ and a set $\Omega' \subset \Omega$ with $\mathbb{P}(\Omega') = 1$, such that, for all $t \in \mathbb{R}_+$, $n \geq 1$, and $\omega \in \Omega'$,

$$\|M(t \wedge \sigma_n)\| \leq n. \qquad \square$$

From this remark and Theorem 2.1 it follows that if $M \in \mathfrak{M}^c_{loc}(\mathbb{R}_+, \mathbb{H})$ then $\|M(t)\|$ is a continuous local submartingale in $\mathbb{R}^1$.

Given $M \in \mathfrak{M}^c_{loc}(\mathbb{R}_+, \mathbb{H})$ denote by $\langle M \rangle$ the increasing process such that $\|M\|^2 - \langle M \rangle \in \mathfrak{M}^c_{loc}(\mathbb{R}_+, \mathbb{R}^1)$ (cf. Corollary 1.4). We call this process the **quadratic variation** of M.

The following result is very important and very well-known; cf. [106, 122, etc.].

Theorem 2.3 (Burkholder–Davis–Gundy Inequality) *Let τ be a stopping time and $M \in M^c_{loc}(\mathbb{R}_+, \mathbb{H})$. Then, for every $p \geq 1$, there exists a number $B = B(p)$ such that*

$$\mathbb{E} \sup_{t \leq \tau} \|M(t)\|^p \leq B(p)\mathbb{E}\langle M \rangle_\tau^{p/2}.$$

In particular, $B(1) = 3$ and $B(2) = 4$. $\square$

For future use we need the following auxiliary result.

Proposition 2.1 *Let $M \in \mathfrak{M}^c_{loc}(\mathbb{R}_+, \mathbb{H})$ and let τ be a stopping time. Define $\langle M \rangle_\tau$ on the set $\{\omega : \tau = \infty\}$ by $\langle M \rangle_\infty := \lim_{t\to\infty} \langle M \rangle_t$.*

(i) *If $\mathbb{E} \sup_{t \leq \tau} \|M(t)\| < \infty$, then the process $t \mapsto M(t \wedge \tau)$, $t \geq 0$, is a continuous martingale.*
(ii) *If $\mathbb{E}\langle M \rangle_\tau^{1/2} < \infty$, then $\mathbb{E} \sup_{t \leq \tau} \|M(t)\| < \infty$.*
(iii) *If $\mathbb{E}\langle M \rangle_\tau < \infty$, then $\mathbb{E} \sup_{t \leq \tau} \|M(t)\|^2 < \infty$.* $\square$

Proof Let $\{\sigma_n\}$ be the sequence of stopping times from Remark 2.1. By the optional sampling theorem (see Sect. 1.4.10) and Lemma 2.1, the process $t \mapsto M(t \wedge \sigma_n \wedge \tau)$, $t \geq 0$, is in $\mathfrak{M}^c(\mathbb{R}_+, \mathbb{H})$. On the other hand, because $\mathbb{E} \sup_{n \in \mathbb{N}} \|M(t \wedge \sigma_n \wedge \tau)\| < \infty$, the family $\{\|M(t \wedge \sigma_n \wedge \tau)\|\}$, $n \in \mathbb{N}$, is uniformly integrable and consequently

$$\mathbb{E}\Big| \|M(t \wedge \sigma_n \wedge \tau) - \|M(t \wedge \tau)\| \Big| = 0$$

for every $t \geq 0$.

Similarly,

$$\mathbb{E}[M(t \wedge \tau)|\mathscr{F}_s] = M(s \wedge \tau) \quad (\mathbb{P}\text{- a.s.}),$$

for every $s,\ t \in \mathbb{R}_+,\ s \leq t$.

The rest follows from Theorem 2.3. □

Corollary 2.2 *Let $M(t) \in \mathfrak{M}^c_{loc}(\mathbb{R}_+, \mathbb{H})$, let τ be a stopping time, and $a, b \in \mathbb{R}_+$. Then*

$$\mathbb{P}\big(\sup_{t\leq\tau} \|M(t)\| \geq a\big) \leq 3a^{-1}\,\mathbb{E}\big(\langle M\rangle_\tau^{1/2} \wedge b\big) + \mathbb{P}\big(\langle M\rangle_\tau^{1/2} \geq b\big). \qquad \square$$

2.2.9. Let $M, N \in \mathfrak{M}^c_{loc}(\mathbb{R}_+, \mathbb{H})$. Define

$$\langle M, N\rangle_t := \frac{1}{4}\Big(\langle M+N\rangle_t - \langle M-N\rangle_t\Big).$$

The following result is similar to the finite-dimensional case.

Theorem 2.4 *If $M, L \in \mathfrak{M}^c_{loc}(\mathbb{R}_+, \mathbb{H})$, then $(M, N) - \langle M, N\rangle \in \mathfrak{M}^c_{loc}(\mathbb{R}_+, \mathbb{R}^1)$. Moreover, if $A = A(t)$ is a continuous predictable real-valued process with bounded variation on bounded intervals and $(M, L) - A \in \mathfrak{M}^c_{loc}(\mathbb{R}_+, \mathbb{R}^1)$, then A is a version of $\langle M, N\rangle$.* □

In the future, given an orthonormal basis $\{h_i,\ i \geq 1\}$ in $\mathbb{H}$, we use the short-hand notation

$$\langle M_i\rangle := \langle (M, h_i)\rangle.$$

Proposition 2.2 *For every $h \in \mathbb{H}$, there exists a set $\Omega_h \subset \Omega$ with $\mathbb{P}(\Omega_h) = 1$ such that, for all $\omega \in \Omega_h$ and every $0 \leq s \leq t < \infty$,*

$$\langle (M, h)\rangle_t - \langle (M, h)\rangle_s \leq \|h\|^2\big(\langle M\rangle_t - \langle M\rangle_s\big). \tag{2.2.1}$$

□

Proof Without loss of generality, we assume that $\|h\| = 1$, and then let $h = h_1$ be the first element of the CONS $\{h_n\}$ in $\mathbb{H}$. Consider process

$$N(t) := \|M(t)\|^2 - \big(M(t), h\big)^2, \quad t \in \mathbb{R}_+.$$

It follows that

$$N(t) = \sum_{i\geq 2} \big(M(t), h_i\big)^2,$$

and hence

$$N - \sum_{i \geq 2} \langle M_i \rangle$$

is a continuous local martingale.

On the other hand, by uniqueness of the Doob–Meyer decomposition for N,

$$\sum_{i \geq 2} \langle M_i \rangle_t = \langle M \rangle_t - \langle (M, h \rangle_t \quad (\mathbb{P}\text{- a.s.}),$$

which means that the process $t \mapsto \langle M \rangle_t - \langle (M, h) \rangle_t$, $t \in \mathbb{R}_+$, is increasing, completing the proof. □

Lemma 2.2 *Let $M \in \mathfrak{M}^c_{loc}(\mathbb{R}_+, \mathbb{H})$ and let $\{h_i,\ i \geq 1\}$ be an orthonormal basis in $\mathbb{H}$. There exists a set $\Omega' \subset \Omega$ with $\mathbb{P}(\Omega') = 1$ such that, for all $t \in \mathbb{R}_+$ and $\omega \in \Omega'$,*

$$\langle M \rangle_t = \sum_{i \geq 1} \langle M_i \rangle_t. \tag{2.2.2}$$

□

Proof If $\mathbb{H}$ is finite-dimensional, then the result follows directly from the definition of $\langle M \rangle$.

If $\mathbb{H}$ is infinite-dimensional, then it is enough to consider the case $M \in \mathfrak{M}^c_2(\mathbb{R}_+, \mathbb{H})$. As in the proof of Proposition 2.2, it follows that, for every $n \in \mathbb{N}$, the process $\|M\|^2 - \sum_{i=1}^{n} M_i^2$ is a continuous local submartingale and $\langle M \rangle - \sum_{i=1}^{n} \langle M_i \rangle$ is the increasing process in the Doob–Meyer decomposition of $\|M\|^2 - \sum_{i=1}^n M_i^2$. Using monotonicity of the function $t \mapsto \langle M \rangle_t - \sum_{i=1}^{n} \langle M_i(t) \rangle$ for fixed n, we get the following convergence in probability:

$$\begin{aligned}
\lim_{n \to \infty} \mathbb{P}\left(\sup_{t \leq T} \left(\langle M \rangle_t - \sum_{i=1}^{n} \langle M_i \rangle_t \right) > \varepsilon \right) &\leq \lim_{n \to \infty} \varepsilon^{-1} \mathbb{E}\left(\langle M \rangle_T - \sum_{i=1}^{n} \langle M_i \rangle_T \right) \\
&= \lim_{n \to \infty} \varepsilon^{-1} \mathbb{E}\Big(\|M(T)\|^2 - \|\mathcal{P}_n(M(T))\|^2 \Big) = 0,
\end{aligned}$$

$\varepsilon > 0$, $t > 0$. Extracting a subsequence along which the convergence is with probability one concludes the proof. □

This lemma and Proposition 1.4 imply the following result.

Corollary 2.3 *If $M \in \mathfrak{M}^c_2(\mathbb{R}_+, \mathbb{H})$, $T \in \mathbb{R}_+$, and*

$$\{0 = t_0^n < t_1^n < \ldots < t_{k(n)+1}^n = T,\ n \in \mathbb{N}\}$$

is a sequence of partitions of the interval $[0, T]$ *with*

$$\lim_{n\to\infty} \max_i (t^n_{i+1} - t^n_i) = 0,$$

then

$$\lim_{n\to\infty} \sum_{i=0}^{k(n)} \|M(t^n_{i+1}) - M(t^n_i)\|^2 = \langle M\rangle_T$$

in probability. □

2.2.10. The following result is easy but useful; the proof is left as an exercise to the interested reader.

Proposition 2.3 *If* $M^n \in \mathfrak{M}^c_2(\mathbb{R}_+, \mathbb{H})$, $n \in \mathbb{N}$, *and* $M = M(t)$ *is an* $\mathbb{H}$*-process such that, for every* $T \in \mathbb{R}_+$,

$$\lim_{n\to\infty} \mathbb{E} \sup_{t\le T} \|M^n(t) - M(t)\|^2 = 0,$$

then $M \in \mathfrak{M}^c_2(\mathbb{R}_+, \mathbb{H})$ *and, for every* $T \in \mathbb{R}_+$,

$$\lim_{n\to\infty} \mathbb{E} \sup_{t\le T} |\langle M^n\rangle_t - \langle M\rangle_t| = 0.$$ □

2.2.11. Next, we review some facts from functional analysis and introduce a few additional definitions. As before, $\mathbb{H}$ is a separable Hilbert space with norm $\|\cdot\|$, inner product $(\cdot,\cdot)$, and topological dual $\mathbb{H}^*$. Recall that yx is the notation for the value of $y \in \mathbb{H}^*$ on $x \in \mathbb{H}$.

Let $\mathbb{X}$ be another separable Hilbert space with norm $\|\cdot\|_{\mathbb{X}}$, inner product $(\cdot,\cdot)_{\mathbb{X}}$, and topological dual $\mathbb{X}^*$. The set of continuous linear operators from $\mathbb{H}$ and $\mathbb{X}$ will be denoted by $\mathfrak{L}(\mathbb{H}, \mathbb{X})$; it is a Banach space with norm

$$\|A\| = \sup_{x\in\mathbb{H}, \|x\|=1} \|Ax\|_{\mathbb{X}}.$$

Warning 2.4 *In the future, we write* $\mathfrak{L}(\mathbb{H})$ *instead of* $\mathfrak{L}(\mathbb{H}, \mathbb{H})$ *and use the same convention for various subspaces of* $\mathfrak{L}(\mathbb{H})$ *to be discussed below. In particular,* $\mathfrak{L}_1(\mathbb{H}) \equiv \mathfrak{L}_1(\mathbb{H}, \mathbb{H})$, $\mathfrak{L}_2(\mathbb{H}) \equiv \mathfrak{L}_2(\mathbb{H}, \mathbb{H})$. □

Recall that, as a rule, we do not identify $\mathbb{H}$ and $\mathbb{H}^*$. Instead, by the Riesz representation theorem, there is a isometric isomorphism $\mathcal{J}_{\mathbb{H}} \in \mathfrak{L}(\mathbb{H}, \mathbb{H}^*)$ given by $h \mapsto (\cdot, h)_{\mathbb{H}}$; we call $\mathcal{J}_{\mathbb{H}}$ the **canonical isomorphism**. Given an operator $A \in \mathfrak{L}(\mathbb{H}, \mathbb{X})$, the dual operator $A^* \in \mathfrak{L}(\mathbb{X}^*, \mathbb{H}^*)$ is defined by

$$(A^*y)x = y(Ax), \ y \in \mathbb{X}^*, \ x \in \mathbb{H}.$$

The adjoint operator $A' \in \mathfrak{L}(\mathbb{X}, \mathbb{H})$ is then

$$A' := \mathcal{J}_{\mathbb{H}}^{-1} A^* \mathcal{J}_{\mathbb{X}},$$

so that

$$(Ah, x)_{\mathbb{X}} = (h, A'x)_{\mathbb{H}}.$$

In particular, an operator $A \in \mathfrak{L}(\mathbb{H})$ is called **self-adjoint** if $A = A'$.

2.2.12. Next, we review two particular types of linear operators: Hilbert–Schmidt and nuclear. For details, see [30, 143].

Let $\{h_i,\ i \in \mathbb{N}\}$ be an orthonormal basis, or CONS, in $\mathbb{H}$.

Definition 2.4 An operator $A \in \mathfrak{L}(\mathbb{H}, \mathbb{X})$ is called **Hilbert–Schmidt** if

$$\sum_{i \geq 1} \|Ah_i\|_{\mathbb{X}}^2 < \infty.$$

The collection of all Hilbert–Schmidt operators from $\mathbb{H}$ to $\mathbb{X}$ is denoted by $\mathfrak{L}_2(\mathbb{H}, \mathbb{X})$. □

The following theorem shows that the definition does not depend on the basis and presents additional properties of the Hilbert–Schmidt operators.

Theorem 2.5

(i) *If* $\{h'_i,\ i \in \mathbb{N}\}$ *is another CONS in* $\mathbb{H}$ *and* $A \in \mathfrak{L}_2(\mathbb{H}, \mathbb{X})$, *then*

$$\sum_{i \geq 1} \|Ah_i\|_{\mathbb{X}}^2 = \sum_{i \geq 1} \|Ah'_i\|_{\mathbb{X}}^2.$$

(ii) $\mathfrak{L}_2(\mathbb{H}, \mathbb{X})$ *is a separable Hilbert space with inner product*

$$(A, B) := \sum_{i \geq 1} (Ah_i, Bh_i)_{\mathbb{X}}, \quad A, B \in \mathfrak{L}_2(\mathbb{H}, \mathbb{X}).$$

The corresponding norm is

$$|||A||| := \left(\sum_{i \geq 1} \|Ah_i\|_{\mathbb{X}}^2 \right)^{1/2}$$

and

$$\|A\| \leq |||A|||.$$

(iii) *A Hilbert–Schmidt operator is compact.*
(iv) *If $A \in \mathfrak{L}_2(\mathbb{H})$ is self-adjoint, then*

$$|||A||| = \Big(\sum_{i \geq 1} \lambda_i^2\Big)^{1/2},$$

where $\{\lambda_i\}$ are the eigenvalues of A, counting multiplicity.
(v) *Let $\mathbb{Y}$ be a separable Hilbert space, $A \in \mathfrak{L}_2(\mathbb{H}, \mathbb{X})$, $B \in \mathfrak{L}(\mathbb{X}, \mathbb{Y})$, $C \in \mathfrak{L}(\mathbb{Y}, \mathbb{H})$. Then $BA \in \mathfrak{L}_2(\mathbb{H}, \mathbb{Y})$, $AC \in \mathfrak{L}_2(\mathbb{Y}, \mathbb{X})$, and*

$$|||BA||| \leq |||A||| \cdot \|B\|, \quad |||AC||| \leq |||A||| \cdot \|C\|.$$

□

2.2.13.

Definition 2.5 An operator $A \in \mathfrak{L}(\mathbb{H})$ is called **nuclear** if it can be represented as

$$A = \sum_{i=1}^{n} B_i C_i,$$

where $B_i, C_i \in \mathfrak{L}_2(\mathbb{H})$ and $n < \infty$. The collection of all nuclear operators in $\mathfrak{L}(\mathbb{H})$ is denoted by $\mathfrak{L}_1(\mathbb{H})$. □

Theorem 2.6

(i) *Given $A \in \mathfrak{L}_1(\mathbb{H})$, define the trace of A by*

$$\mathrm{tr}(A) := \sum_{i \geq 1} (Ah_i, h_i).$$

Then $\mathrm{tr}(A)$ is finite and does not depend on the choice of the CONS.
(ii) *If A is a compact self-adjoint operator in $\mathfrak{L}(\mathbb{H})$, then $A \in \mathfrak{L}_1(\mathbb{H})$ if and only if*

$$\sum_i |\lambda_i| < \infty,$$

where $\{\lambda_i\}$ are the eigenvalues of A, counting multiplicity. In that case, $\mathrm{tr}(A) = \sum_i \lambda_i$.
(iii) *If $A \in \mathfrak{L}_1(\mathbb{H})$ and $B \in \mathfrak{L}(\mathbb{H})$, then both AB and BA are nuclear operators.*
(iv) *$\mathfrak{L}_1(\mathbb{H})$ is a separable Banach space with norm*

$$\|A\|_{\mathfrak{L}_1(\mathbb{H})} := \sup\big\{B \in \mathfrak{L}(\mathbb{H}),\ \|B\| \leq 1 :\ |\mathrm{tr}(AB)|\big\},$$

and $|||A||| \leq \|A\|_{\mathfrak{L}_1(\mathbb{H})}$. □

2.2.14. We now present an important result connecting continuous local martingales and nuclear operators; see [107, 109].

Theorem 2.7 *For every $M \in \mathfrak{M}^c_{loc}(\mathbb{R}_+, \mathbb{H})$, there exists a unique, up to a version, predictable process $Q_M = Q_M(t,\omega)$ with the following properties:*

1) *For all $(t,\omega) \in \mathbb{R}_+ \times \Omega$, $Q_M(t,\omega) \in \mathfrak{L}_1(\mathbb{H})$ and $\operatorname{tr}\big(Q_M(t,\omega)\big) = 1$;*
2) *for all $(t,\omega) \in \mathbb{R}_+ \times \Omega$ and $h \in \mathbb{H}$, $(Q_M h, h) \geq 0$;*
3) *for every $g, h \in \mathbb{H}$ and $(t,\omega) \in \mathbb{R}_+ \times \Omega$, $(Q_M g, h) = (g, Q_M h)$ and*

$$\langle (M,g), (M,h) \rangle_t = \int_{[0,t]} \big(Q_M(s) g, h\big)\, d\langle M \rangle_s. \qquad \square$$

Proof To begin, let us assume that $M \in \mathfrak{M}^c_2(\mathbb{R}_+, \mathbb{H})$. Given $g \in \mathbb{H}$, define the real-valued process $y = y(t)$ by

$$y(t) := \|g\|^2 \langle M \rangle_t - \langle (M,g) \rangle_t.$$

It follows from Proposition 2.2 that y is an increasing process, and then, for every interval $[a,b] \subset \mathbb{R}_+$ and every bounded non-negative function f on $[a,b]$,

$$\|g\|^2 \int_{[a,b]} f(t)\, d\langle M \rangle_t = \int_{[a,b]} f(t)\, d\langle (M,g) \rangle_t + \int_{[a,b]} f(t)\, dy(t)$$

$$\geq \int_{[a,b]} f(t)\, d\langle (M,g) \rangle_t$$

with probability one.

By a version of the Radon–Nikodym theorem (cf. [23, Theorem 33]), there exists a unique, up to a version, non-negative predictable process $q = q(t,g)$ such that

$$\mathbb{E} \int_{[0,t]} q(s,g)\, d\langle M \rangle_s < \infty, \; t \in \mathbb{R}_+,$$

and

$$\mathbb{P}\left(\langle (M,g) \rangle_t = \int_{[0,t]} q(s,g)\, d\langle M \rangle_s, \; t \in \mathbb{R}_+ \right) = 1.$$

Taking into account that

$$\langle (M,g), (M,h) \rangle_t = \tfrac{1}{4}\Big(\langle (M, g+h) \rangle_t - \langle (M, g-h) \rangle_t \Big),$$

we define the process $\mathfrak{q} = \mathfrak{q}(t, g, h)$ by

$$\mathfrak{q}(t, g, h) = \frac{1}{4}\Big(q(t, g+h) - q(t, g-h)\Big)$$

and conclude that, for every $g, h \in \mathbb{H}$, the process $\mathfrak{q}$ is predictable,

$$\mathbb{E} \int\limits_{[0,t]} \mathfrak{q}(s, g, h)\, d\langle M \rangle_s < \infty, \quad t \in \mathbb{R}_+,$$

and there exists a set $\Omega_{g,h} \subset \Omega$ such that $\mathbb{P}(\Omega_{g,h}) = 1$ and

$$\langle (M, g), (M, h) \rangle_t = \int\limits_{[0,t]} \mathfrak{q}(s, g, h)\, d\langle M \rangle_s, \ t \in \mathbb{R}_+, \ \omega \in \Omega_{g,h}. \tag{2.2.3}$$

Moreover, by (2.2.2) and (2.2.3),

$$\langle M \rangle_t = \sum_{i \geq 1} \int\limits_{[0,t]} \mathfrak{q}(s, h_i, h_i)\, d\langle M \rangle_s.$$

Denote by $\mathscr{D}$ the collection of finite linear combinations of $\{h_i\}$ with rational coefficients. Then $\mathscr{D}$ is a countable dense set in $\mathbb{H}$. By (2.2.3), the following holds for $\langle M \rangle \circ \mathbb{P}$-a.a. (t, ω):

(a) $\mathfrak{q}(t, \cdot, \cdot)$ is a symmetric bi-linear form on $\mathscr{D} \times \mathscr{D}$;
(b) $\mathfrak{q}(t, g, h) \leq \|g\| \cdot \|h\|, \ t \in \mathbb{R}_+, \ g, h \in \mathscr{D}$;
(c) $\mathfrak{q}(t, g, g) \geq 0, \ g \in \mathscr{D}$;
(d) $\sum_{i \geq 1} \mathfrak{q}(t, h_i, h_i) = 1$.

Denote by S_q the corresponding exceptional set, that is, the collection of $(t, \omega) \in \mathbb{R}_+ \times \Omega$ for which at least one of the properties (a)–(d) fails for at least one g or h from $\mathscr{D}$. Since $\mathscr{D}$ is countable, the set S_q has zero $\langle M \rangle \circ \mathbb{P}$-measure.

For (t, ω) in the complement of S_q, we extend $\mathfrak{q}$ by continuity to all of $\mathbb{H} \times \mathbb{H}$ while preserving all the properties (a)–(d), and then deduce the existence of a non-negative, self-adjoint continuous linear operator $Q_M = Q_M(t, \omega)$ on $\mathbb{H}$ such that, for all $g, h \in \mathbb{H}$,

$$\big(Q_M(t, \omega) g, h\big) = \mathfrak{q}(t, g, h).$$

By Theorem 2.6, $Q_M(s, \omega)$ is a nuclear operator with $\mathrm{tr}(Q_M) = 1$. On the exceptional set S_q, we set Q_M equal to an arbitrary but fixed non-negative, self-adjoint, nuclear operator with unit trace, for example, $Q_M h_1 = h_1, \ Q_M h_i = 0, \ i \geq 2$. Predictability of Q_M then follows from predictability of $\mathfrak{q}$ and the Pettis theorem from Sect. 1.2.9.

By (2.2.3),

$$\langle (M, g), (M, h) \rangle_t = \int_{[0,t]} \big(Q_M(s, \omega) g, h \big) \, d\langle M \rangle_s \tag{2.2.4}$$

for all (t, ω); on the exceptional set S_q, we get the trivial equality $0 = 0$.

This concludes the construction of the operator Q_M when M is a continuous square-integrable martingale.

Now let $M \in \mathfrak{M}^c_{loc}(\mathbb{R}_+, \mathbb{H})$ and let $\{\sigma_n\}$ be the localizing sequence of stopping times from Remark 2.1. Denote by Q^n_M the corresponding operator for the martingale $M(\cdot \wedge \sigma_n)$. By construction, $Q^n_M = Q^{n+1}_M$ for $\langle M \rangle \circ \mathbb{P}$-a.a (t, ω) from the set $\{(t, \omega) : t < \sigma_n(\omega)\}$, $n \geq 1$.

Setting $Q_M(t, \omega) := Q^n_M(t, \omega)$ for $t < \sigma_n(\omega)$, we conclude the proof of the theorem. □

Warning 2.5 *In future, we will occasionally refer to Q_M as the correlation operator of the (local) martingale M.* □

Definition 2.6 Let Q be a nuclear symmetric non-negative operator on $\mathbb{H}$. The process $W \in \mathfrak{M}^c_2(\mathbb{R}_+, \mathbb{H})$ with correlation operator $Q_W = \big(\mathrm{tr}(Q)\big)^{-1} Q$ is called the Wiener process (martingale) with the covariance operator Q, or simply a Q-**Wiener process.** □

Remark 2.2 By direct computation, a Wiener process in $\mathbb{H}$ with the covariance operator Q can be written as

$$W(t) = \sum_{i \geq 1} \sqrt{\lambda_i} \; w^i(t) h_i,$$

where $\{h_i, \; i \geq 1\}$ are normalized (that is, $(h_i, h_j) = 0, i \neq j, \; \|h_i\| = 1$) eigenfunctions of Q, corresponding to eigenvalues $\lambda_i > 0$ and $\{w^i = (W, h_i)/\sqrt{\lambda_i}\}$ are independent one-dimensional standard Wiener processes. The series converges $\mathbb{P}$- a.s. and in $\mathbb{L}_2(\Omega; \mathbb{H})$. □

2.3 Stochastic Integral with Respect to a Square Integrable Martingale

2.3.1. Throughout this section, we fix $M \in \mathfrak{M}^c_2(\mathbb{R}_+, \mathbb{H})$. Let $\mathbb{H}$ and $\mathbb{X}$ be separable Hilbert spaces with orthonormal bases $\{h_i\}$ and $\{e_i\}$, respectively. As before, we write $\|\cdot\|_{\mathbb{X}}$ and $(\cdot, \cdot)_{\mathbb{X}}$ for the norm and inner product in $\mathbb{X}$ and omit the subscript in the case of $\mathbb{H}$.

Denote by $\mathbb{L}_2\big(\langle M\rangle \circ \mathbb{P}, \mathfrak{L}(\mathbb{H}, \mathbb{X})\big)$ the set of predictable mappings $B : \Omega \times \mathbb{R}_+ \to \mathfrak{L}(\mathbb{H}, \mathbb{X})$ such that, for all $t \in \mathbb{R}_+$,

$$\int_{[0,t]} \|B(s)\|^2 \, d\langle M\rangle_s < \infty, \quad (\mathbb{P}\text{- a.s.}). \tag{2.3.1}$$

2.3.2. In this section, we define the integral

$$\int_{[0,t]} B(s) dM(s)$$

for $B \in \mathbb{L}_2\big(\langle M\rangle \circ \mathbb{P}, \mathfrak{L}(\mathbb{H}, \mathbb{X})\big)$.

Note that

$$\big|\big(B(s,\omega)h_i, e_i\big)_{\mathbb{X}}\big| \le \|B(s,\omega)\| \tag{2.3.2}$$

for all s and ω.

By the Pettis theorem, the real-valued process $t \mapsto \big(B(t,\omega)h_i, e_j\big)_{\mathbb{X}}$ is predictable. Therefore, the stochastic integral

$$B \circ M^{ij}(t) := \int_{[0,t]} \big(B(s)h_i, e_j\big)_{\mathbb{X}} \, dM_i(s) \tag{2.3.3}$$

is defined and belongs to $\mathfrak{M}_2^c(\mathbb{R}_+, \mathbb{R}^1)$.

Indeed, in view of (2.3.1), (2.3.2), and Theorem 1.7, we have

$$\begin{aligned}\mathbb{E}\int_{[0,t]} \big(B(s)h_i, e_j\big)_{\mathbb{X}}^2 \, d\langle M_i\rangle_s &= \mathbb{E}\int_{[0,t]} \big(B(s)h_i, e_j\big)_{\mathbb{X}}^2 \big(Q_M(s)h_i, h_i\big) \, d\langle M\rangle_s \\ &\le \mathbb{E}\int_{[0,t]} \|B(s)\|^2 \, d\langle M\rangle_s < \infty.\end{aligned}$$

From this inequality and the results of Sects. 1.4.15 and 1.4.16, it follows that the stochastic integral (2.3.3) exists and belongs to $\mathfrak{M}_2^c(\mathbb{R}_+, \mathbb{R}^1)$.

If

$$B \circ M_{m,n}(t) := \sum_{j=1}^{m} \sum_{i=1}^{n} B \circ M^{ij}(t) e_j,$$

then $B \circ M_{m,n} \in \mathfrak{M}_2^c(\mathbb{R}_+, \mathbb{X})$, and, by Lemma 2.2

$$\langle B \circ M_{m,n}\rangle_t = \sum_{j=1}^{m} \left\langle \sum_{i=1}^{n} B \circ M^{ij} \right\rangle_t \quad (\mathbb{P}\text{- a.s.}),$$

where, by (2.3.3),

$$\left\langle \sum_{i=1}^{n} B \circ M^{ij} \right\rangle_t = \sum_{i,k=1}^{n} \int_0^t \left(B(s)h_i, e_j\right)_{\mathbb{X}} \left(B(s)h_k, e_j\right)_{\mathbb{X}} d\langle M_i, M_k\rangle_s$$

$$= \sum_{i,k=1}^{n} \int_0^t \left(h_i, B'(s)e_j\right) \left(h_k, B'(s)e_j\right) d\langle M_i, M_k\rangle_s;$$

recall that $B'(s)$ denotes the operator adjoint to $B(s)$ (cf. Sect. 2.2). Then, by Theorem 2.7,

$$\sum_{i,k=1}^{n} \left(h_i, B'(s)e_j\right) \left(h_k, B'(s)e_j\right) d\langle M_i, M_k\rangle_s$$

$$= \int_{[0,t]} \left\| \mathcal{Q}_M^{1/2}(s)\, \mathcal{P}_n\big(B'(s)\, e_j\big) \right\|^2 d\langle M\rangle_s,$$

where $\mathcal{P}_n$ is the projection operator from Sect. 2.2.6; the square root of the positive-definite self-adjoint linear operator $\mathcal{Q}_M(s)$ can be defined in many ways (see e.g. [43] or Sect. 3.3.2 below).

As a result, for every $t \in \mathbb{R}_+$,

$$\langle B \circ M_{m,n}\rangle_t = \sum_{j=1}^{m} \int_{[0,t]} \left\| \mathcal{Q}_M^{1/2}(s)\mathcal{P}_n\big(B'(s)e_j\big) \right\| d\langle M\rangle_s \ \ (\mathbb{P}\text{- a.s.}). \tag{2.3.4}$$

Remark 2.3 By the Pettis theorem, the process

$$t \mapsto \|\mathcal{Q}_M^{1/2}(t)\mathcal{P}_n\big(B'(t)e_j\big)\|, \ t \in \mathbb{R}_+,$$

is predictable. □

The next step is to show that the sequence $\{B \circ M_{m,n},\ m, n \in \mathbb{N}\}$ is Cauchy in $\mathbb{L}_2\big(\Omega; \mathbb{L}_2((0,T); \mathbb{X})\big)$.

Let m, n and $m', n' \in \mathbb{N}$, with $m' > m,\ n' > n$, and fix $T \in \mathbb{R}_+$. From the Doob inequality (1.4.5), we have

$$U := \mathbb{E} \sup_{t \le T} \|B \circ M_{m',n'}(t) - B \circ M_{m,n}(t)\|_{\mathbb{X}}^2$$

$$\le 4\mathbb{E} \left\| \sum_{j=1}^{m'} \sum_{i=n+1}^{n'} B \circ M^{ij}(T)e_j + \sum_{j=m+1}^{m'} \sum_{i=1}^{n'} B \circ M^{ij}(T)e_j \right\|_{\mathbb{X}}^2 \tag{2.3.5}$$

$$\le 8 \sum_{j=1}^{m'} \mathbb{E} \left| \sum_{i=n+1}^{n'} B \circ M^{ij}(T) \right|^2 + 8 \sum_{j=m+1}^{m'} \mathbb{E} \left| \sum_{i=1}^{n'} B \circ M^{ij}(T) \right|^2 .$$

Similar to (2.3.4),

$$\mathbb{E}\left|\sum_{i=n+1}^{n'} B \circ M^{ij}(T)\right|^2 = \mathbb{E}\int_{[0,t]} \|Q_M^{1/2}(t)\,\mathcal{P}_{n+1,n'}\big(B'(t)e_j\big)\|^2\, d\langle M\rangle_t, \tag{2.3.6}$$

where $\mathcal{P}_{n+1,n'}$ is the orthogonal projection on the linear span of $h_{n+1}, \ldots, h_{n'}$, and

$$\mathbb{E}\left|\sum_{i=1}^{n'} B \circ M^{ij}(T)\right|^2 = \mathbb{E}\int_{[0,T]} \left\|Q_M^{1/2}(t)\mathcal{P}_{n'}\big(B'(t)\big)(t)e_j\right\|^2 d\langle M\rangle_t. \tag{2.3.7}$$

From (2.3.5)–(2.3.7) it follows that

$$\begin{aligned}
U &\le 8\mathbb{E}\int_{[0,T]} \left(\sum_{j=1}^{m} \| Q_M^{1/2}(t)\mathcal{P}_{n+1,n'}B'(t)e_j\|^2 \right.\\
&\qquad \left. + \sum_{j=m+1}^{m'} \left\|Q_M^{1/2}(t)B'(t)e_j\right\|^2\right) d\langle M\rangle_t\\
&\le 8\left(\mathbb{E}\int_{[0,T]} |\!|\!| Q_M^{1/2}(t)\mathcal{P}_{n+1,n'}|\!|\!|^2\, \|B(t)\|^2\, d\langle M\rangle_t \right. \qquad (2.3.8)\\
&\qquad \left. + \sum_{j=m+1}^{m'} \int_{[0,T]} \|Q_M^{1/2}(t)B'(t)e_j\|^2\right) d\langle M\rangle_t \to 0
\end{aligned}$$

as $m' > m \to \infty$, $n' > n \to \infty$ by the dominated convergence theorem. Indeed, $Q_M(t)$ is nuclear for every t, ω, with unit trace, and so $Q_M^{1/2}(t)$ is Hilbert–Schmidt with

$$|\!|\!| Q_M^{1/2}(t)|\!|\!|^2 = \sum_{i\ge 1} \|Q_M^{1/2}(t)h_i\|^2 \le 1$$

for all t and ω. As a result,

$$|\!|\!| Q_M^{1/2}(t)\mathcal{P}_{n+1,n'}|\!|\!|^2 = \sum_{i=n+1}^{n'} \|Q_M^{1/2}(t)h_i\|^2$$

and

$$\sum_{j\geq 1} \|Q_M^{1/2}(t)B'(t)e_j\|^2 = \|\!|Q_M^{1/2}(t)B'(t)|\!\|^2 \leq \|B'(t)\|^2 \cdot \|\!|Q_M^{1/2}(t)|\!\|^2.$$

Consequently, there exists a stochastic process $B \circ M \in \mathfrak{M}_2^c(\mathbb{R}_+, \mathbb{X})$ such that, for every $T \in \mathbb{R}_+$,

$$\lim_{\substack{m\to\infty\\ n\to\infty}} \mathbb{E} \sup_{t\leq T} \|B \circ M(t) - B \circ M_{m,n}(t)\|_{\mathbb{X}}^2 = 0. \tag{2.3.9}$$

Accordingly, we define the stochastic integral by

$$\int_{[0,t]} B(s)dM(s) := B \circ M(t),\ t \in \mathbb{R}_+.$$

This equality defines the stochastic integral uniquely up to a version. By (2.3.4), (2.3.9), and Proposition 2.3,

$$\begin{aligned}\langle B \circ M\rangle_t &= \lim_{\substack{m\to\infty\\ n\to\infty}} \int_{[0,t]} \sum_{j=1}^{m} \|Q_M^{1/2}(s)\mathcal{P}_n\big(B'(s)e_j\big)\|^2\, d\langle M\rangle_s \\ &= \int_{[0,t]} \sum_{j=1}^{\infty} \|Q_M^{1/2}(s)B'(s)e_j\|^2\, d\langle M\rangle_s = \int_{[0,t]} \|\!|B(s)Q_M^{1/2}(s)|\!\|^2\, d\langle M\rangle_s.\end{aligned} \tag{2.3.10}$$

Note that $\|\!|Q_M^{1/2}(s)B(s)|\!\|^2 \leq \|B(s)\|^2$, $s \in \mathbb{R}_+$. Remark 2.3 and the results of Sect. 1.3.7 imply that $\|\!|BQ_M^{1/2}|\!\|^2$ is a predictable process.

2.3.3. The objective of this paragraph is to verify that the above construction of the stochastic integral does not depend on the choice of the bases in $\mathbb{H}$ and $\mathbb{X}$.

Given $y \in \mathbb{X}$, define the operator B_y by the formula

$$B_y h = (B'y, h),\ h \in \mathbb{H},$$

so that $B_y \in \mathbb{L}_2\big(\langle M\rangle \circ \mathbb{P}, \mathfrak{L}_2(\mathbb{H}, \mathbb{R}^1)\big)$. Next, define

$$B_y \circ M^{h_i}(t) := \int_{[0,t]} \big(B'(s)y, h_i\big)\, dM_i(s).$$

By (2.3.9), for every $T \in \mathbb{R}_+$,

$$\lim_{n\to\infty} \mathbb{E} \sup_{t\leq T} \left| B_y \circ M(t) - \sum_{i=1}^{n} B_y \circ M^{h_i}(t) \right|^2 = 0. \tag{2.3.11}$$

We now prove that $B_y \circ M$ does not depend on the choice of the CONS $\{h_i\}$. Let $\{\psi_i\}$ be another CONS in $\mathbb{H}$. Define $B_y \circ M^{\psi_i}(t)$ similarly to $B_y \circ M^{h_i}(t)$:

$$B_y \circ M^{\psi_i}(t) := \int\limits_{[0,t]} \big(B'(s)y, \psi_i\big)\, d\big(M(s), \psi_i\big).$$

We need to show that

$$\lim_{n\to\infty} \mathbb{E} \sup_{t\le T} \left| \sum_{i=1}^{n} \Big(B_y \circ M^{h_i}(t) - B_y \circ M^{\psi_i}(t) \Big) \right|^2 = 0. \tag{2.3.12}$$

From Doob's inequality (1.4.5) and the results of Sect. 1.4.16 it follows that

$$\begin{aligned}
&\mathbb{E} \sup_{t\le T} \left| \sum_{i=1}^{n} \Big(B_y \circ M^{h_i}(t) - B_y \circ M^{\psi_i}(t) \Big) \right|^2 \\
&\le 4\mathbb{E}\Bigg(\sum_{i,j=1}^{n} \int\limits_{[0,T]} \big(B'(s)y, h_i\big)\big(B'(s)y, h_j\big)\, d\langle M_i, M_j\rangle_s \\
&- 2 \sum_{i,j=1}^{n} \int\limits_{[0,T]} \big(B'(s)y, h_i\big)\big(B'(s)y, \psi_i\big)\, d\langle (M, h_i), (M, \psi_i)\rangle_s \\
&+ \sum_{i,j=1}^{n} \int\limits_{[0,T]} \big(B'(s)y, \psi_i\big)\big(B'(s)y, \psi_j\big)\, d\langle (M, \psi_i), (M, \psi_j)\rangle_s \Bigg) \\
&:= 4(I_1(n) - I_2(n) + I_3(n)).
\end{aligned}$$

Theorem 2.7 implies that

$$I_2(n) = 2\mathbb{E} \int\limits_{[0,T]} \left(Q_M(s) \sum_{i=1}^{n} \big(B'(s)y, h_i\big)h_i, \sum_{j=1}^{n} \big(B'(s)y, \psi_j\big)\psi_j \right) d\langle M\rangle_s$$

and consequently

$$\lim_{n\to\infty} I_2(n) = 2\mathbb{E} \int\limits_{[0,T]} \big(Q_M(s)B'(s)y, B'(s)y\big)\, d\langle M\rangle_s.$$

It follows in the same way that

$$\lim_{n\to\infty} I_1(n) = \lim_{n\to\infty} I_3(n) = \mathbb{E} \int_{[0,T]} \big(Q_M(s)B'(s)y, B'(s)y\big)\, d\langle M\rangle_s.$$

Thus we get the equality (2.3.12) and conclude that $B_y \circ M$ does not depend on the CONS in $\mathbb{H}$.

Next, we show that, for every fixed $y \in \mathbb{X}$ and every $t \in \mathbb{R}_+$,

$$\big(y, B \circ M(t)\big)_{\mathbb{X}} = B_y \circ M(t) \quad (\mathbb{P}\text{-a.s.}), \tag{2.3.13}$$

which will imply that the construction of the integral does not depend on the CONS in either $\mathbb{H}$ or $\mathbb{X}$.

Note that

$$\begin{aligned}\mathbb{E}\Big|\big(y, B \circ M(t)\big)_{\mathbb{X}} - B_y \circ M(t)\Big|^2 &\le 3\mathbb{E}\Big|\big(y, B \circ M(t) - B \circ M_{m,n}(t)\big)_{\mathbb{X}}\Big|^2 \\ &+ 3\mathbb{E}\left|\big(y, B \circ M_{m,n}(t)\big)_{\mathbb{X}} - \sum_{i=1}^{n} B_y \circ M^{h_i}(t)\right|^2 \\ &+ 3\mathbb{E}\left|B_y \circ M(t) - \sum_{i=1}^{n} B_y \circ M^{h_i}(t)\right|^2 = 3(V_1 + V_2 + V_3).\end{aligned} \tag{2.3.14}$$

By (2.3.9),

$$\lim_{\substack{m\to\infty \\ n\to\infty}} V_1 = \lim_{n\to\infty} V_3 = 0. \tag{2.3.15}$$

To study V_2, write

$$\big(y, B \circ M_{m,n}(t)\big) = \sum_{i=1}^{n} \int_{[0,t]} \sum_{j=1}^{m} \big(B(s)h_i, e_j\big)_{\mathbb{X}} \, (y, e_j)_{\mathbb{X}} \, dM_i(s),$$

so that

$$\begin{aligned}V_2 &= \mathbb{E}\left|\big(y, B \circ M_{m,n}(t)\big)_{\mathbb{X}} - \sum_{i=1}^{n} B_y \circ M^{h_i}(t)\right|^2 \\ &= \mathbb{E} \sum_{i,k=1}^{n} \int_{[0,t]} \left[\sum_{j=1}^{m} \big(B(s)h_i, e_j\big)_{\mathbb{X}} \, (y, e_j)_{\mathbb{X}} - \big(B'(s)y, h_i\big)\right] \\ &\times \left[\sum_{j=1}^{m} \big(B(s)h_k, e_j\big)_{\mathbb{X}}(y, e_j)_{\mathbb{X}} - \big(B'(s)y, h_k\big)\right] d\langle M_i, M_k\rangle_s\end{aligned}$$

$$= \mathbb{E} \int_{[0,t]} \left(Q_M(s) \sum_{i=1}^{n} \left[\sum_{j=1}^{m} \big(B(s)h_i, e_j\big)_{\mathbb{X}} (y, e_j)_{\mathbb{X}} - \big(B'(s)y, h_i\big) \right] h_i, \right.$$

$$\left. \sum_{k=1}^{n} \left[\sum_{j=1}^{m} \big(B(s)h_k, e_j\big)_{\mathbb{X}} (y, e_j)_{\mathbb{X}} - \big(B'(s)y, h_k\big) \right] h_k \right) d\langle M\rangle_s.$$

We have

$$\lim_{\substack{m\to\infty \\ n\to\infty}} \sum_{i=1}^{n} \sum_{j=1}^{m} \big(B(s)h_i, e_j\big)_{\mathbb{X}} (y, e_j)_{\mathbb{X}} h_i = \sum_{i\geq 1} \big(B(s)h_i, y\big)_{\mathbb{X}} h_i = B'(s)y$$

for all s and ω. Therefore,

$$\lim_{\substack{m\to\infty \\ n\to\infty}} \left\| \sum_{i=1}^{n} \left[\sum_{j=1}^{m} \big(B(s)h_i, e_j\big)_{\mathbb{X}} (y, e_j)_{\mathbb{X}} - \big(B'(s)y, h_i\big) \right] h_i \right\| = 0,$$

and, by the dominated convergence theorem,

$$\lim_{\substack{m\to\infty \\ n\to\infty}} V_2 = 0. \tag{2.3.16}$$

Formula (2.3.13) now follows from (2.3.11) and (2.3.14)–(2.3.16).

2.3.4. We now state the main properties of the stochastic integral.

Theorem 2.8 *If $B \in \mathbb{L}_2\big(\langle M\rangle \circ \mathbb{P}; \mathfrak{L}(\mathbb{H}, \mathbb{X})\big)$ and*

$$B \circ M^{ij}(t) := \int_{[0,t]} \big(B(s)h_i, e_j\big)_{\mathbb{X}} dM_i(s),$$

where $\{h_i\}$ is a CONS in $\mathbb{H}$ and $\{e_j\}$ is a CONS in $\mathbb{X}$, then there exists a random process $B \circ M \in \mathfrak{M}_2^c(\mathbb{R}_+, \mathbb{H})$ which is unique up to version and such that, for every $T \in \mathbb{R}_+$,

$$\lim_{\substack{m\to\infty \\ n\to\infty}} \mathbb{E} \sup_{t\leq T} \left\| B \circ M(t) - \sum_{i=1}^{n} \sum_{j=1}^{m} B \circ M^{ij}(t) e_j \right\|_{\mathbb{X}}^2 = 0,$$

and

$$\langle B \circ M\rangle_t = \int_{[0,t]} \| B(s) Q_M^{1/2}(s) \|^2 \, d\langle M\rangle_s. \tag{2.3.17}$$

The process $B \circ M$ *and the stochastic integral defined* $\mathbb{P}$- a.s. *by the equality*

$$\int_{[0,t]} B(s)dM(s) = B \circ M(t)$$

have properties (i)–(v) *listed below.*

(i) For every $B_1, B_2 \in \mathbb{L}_2\big(\langle M\rangle \circ \mathbb{P}, \mathfrak{L}(\mathbb{H}, \mathbb{X})\big)$, $\alpha_1\beta \in \mathbb{R}^1$, and $t \in \mathbb{R}_+$,

$$\int_{[0,t]} (\alpha\beta_1(s) + \beta B_2(s))dM(s)$$

$$= \alpha \int_{[0,t]} B_1(s)dM(s) + \beta \int_{[0,t]} B_2(s)dM(s), \ (\mathbb{P}\text{- a.s.}).$$

(ii) For every $s, u, t \in \mathbb{R}_+$ such that $0 \le s \le u \le t$ and $B \in \mathfrak{L}_2(\langle M\rangle \circ \mathbb{P}, \mathfrak{L}(\mathbb{H}, \mathbb{X}))$,

$$\int_{[0,t]} 1_{\{[s,u]\}}(r)B(r)dM(r) = \int_{[0,u]} B(r)dM(r) - \int_{[0,s]} B(r)dM(r)$$

$$:= \int_{[s,u]} B(r)dM(r).$$

(iii) If B and $B_n \in \mathbb{L}_2\big(\langle M\rangle \circ \mathbb{P}, \mathfrak{L}(\mathbb{H}, \mathbb{X})\big)$ and

$$\lim_{n\to\infty} \mathbb{E} \int_{[0,T]} \| Q_M^{1/2}(s)\big(B_n(s) - B(s)\big)\|^2 \, d\langle M\rangle_s = 0$$

for some $T \in \mathbb{R}_+$, then

$$\lim_{n\to\infty} \mathbb{E} \sup_{t\le T} \| B_n \circ M(t) - B \circ M(t)\|_{\mathbb{X}}^2 = 0.$$

(iv) If $B \in \mathbb{L}_2\big(\langle M\rangle \circ \mathbb{P}, \mathfrak{L}(\mathbb{H}, \mathbb{X})\big)$ and $A \in \mathfrak{L}(\mathbb{X}, \mathbb{Y})$, where $\mathbb{Y}$ is another Hilbert space, then

$$A \int_{[0,t]} B(s)dM(s) = \int_{[0,t]} AB(s)dM(s) \ (\mathbb{P}\text{- a.s.}).$$

(v) If $B \in \mathbb{L}_2\big(\langle M\rangle \circ \mathbb{P}, \mathfrak{L}(\mathbb{H},\mathbb{X})\big)$, $L \in \mathfrak{M}_2^c(\mathbb{R}_+,\mathbb{X})$, and $T \in \mathbb{R}_+$, then

$$\lim_{\substack{m\to\infty\\ n\to\infty}} \mathbb{E}\sup_{t\le T}\left|\langle B\circ M, L\rangle_t - \sum_{i=1}^{n}\sum_{j=1}^{m}\int_{[0,t]} \big(B(s)h_i, e_j\big)_{\mathbb{X}}\, d\langle M_i, L_j\rangle_s\right| = 0.$$

In particular, if $C \in \mathbb{L}_2\big(\langle M\rangle \circ \mathbb{P}, \mathfrak{L}(\mathbb{H},\mathbb{X})\big)$ then, for every $y, z \in \mathbb{X}$ and every $t \in \mathbb{R}_+$,

$$\big\langle (B\circ M, y)_{\mathbb{X}}, (C\circ M, z)_{\mathbb{X}}\big\rangle_t = \int_{[0,t]} \big(C(s)Q_M(s)B'(s)y, z\big)_{\mathbb{X}}\, d\langle M\rangle_s. \qquad \square$$

Proof Properties (i) and (ii) are obvious. Property (iii) follows from (2.3.10) and inequality (1.4.4).

Property (iv) follows form (2.3.13). Indeed, note that, for every $z \in \mathbb{Y}$, $B_{A'z} = (AB)_z$ (see the B_z notation on page 55). Then, in view of (2.3.13), we have, $\mathbb{P}$-a.s.,

$$\left(z, A\int_{[0,t]} B(s)dM(s)\right)_{\mathbb{Y}} = \left(A'z, \int_{[0,t]} B(s)dM(s)\right)_{\mathbb{X}}$$

$$= \int_{[0,t]} B_{A'z}(s)dM(s) = \int_{[0,t]} (AB(s))_z dM(s) = \left(z, \int_{[0,t]} AB(s)dM(s)\right)_{\mathbb{Y}}.$$

The first part of the statement of (v) follows from (2.3.9) and Proposition 2.3; equality

$$\langle B\circ M_{m,n}, L\rangle_t = \sum_{j=1}^{m}\left\langle \sum_{i=1}^{n} B\circ M^{ij}, L_j\right\rangle_t,$$

where $L_j(t) = \big(L(t), e_j\big)_{\mathbb{X}}$, follows from Lemma 2.2.

Hence, by property (v) of a martingale in $\mathbb{R}^1$ (see Sect. 1.4.16),

$$\langle B\circ M_{m,n}, L\rangle_t$$

$$= \sum_{i=1}^{n}\sum_{j=1}^{m}\int_{[0,t]} \big(B(s)h_i, e_j\big)_{\mathbb{X}}\, d\langle (M, h_i), (L, e_j)_{\mathbb{X}}\rangle_s \quad (\mathbb{P}\text{- a.s.}).$$

For the second part of the assertion, it is enough to consider $y = e_i$ and $z = e_j$. From the previous equality and Theorem 2.7 it follows that, for every $t \in \mathbb{R}_+$,

$$\begin{aligned}
&\big\langle (B \circ M_{m,n}, e_i)_{\mathbb{X}}, (C \circ M_{m,n}, e_j)_{\mathbb{X}} \big\rangle_t \\
&= \sum_{l,k=1}^{n} \int_{[0,t]} \big(B'(s)e_i, h_k\big)\big(C'(s)e_j, h_l\big)\, d\langle M_k, M_l\rangle_s \\
&= \int_{[0,t]} \Big(Q_M(s)\mathcal{P}_n\big(B'(s)e_i\big), \mathcal{P}_n\big(C'(s)e_j\big)\Big)\, d\langle M\rangle_s \quad (\mathbb{P}\text{- a.s.}).
\end{aligned}$$

Making use of Proposition 2.3, we pass to the limit as $m, n \to \infty$ and get ($\mathbb{P}$- a.s.)

$$\big\langle (B_y \circ M, e_i)_{\mathbb{X}}, (C \circ M, e_j)_{\mathbb{X}} \big\rangle_t = \int_{[0,t]} \big(Q_M(s)B'(s)e_i, C'(s)e_j\big)\, d\langle M\rangle_s. \qquad \square$$

2.3.5. Let $M \in \mathfrak{M}_2^c(\mathbb{R}_+, \mathbb{H})$ and $T \in \mathbb{R}_+$. Recall that $\mathscr{P}_T$ is the sigma-algebra generated by the sets of the form $(a, b] \times \Gamma$ with $a, b \in [0, T]$ and $\Gamma \in \mathscr{F}_a$; cf. Sect. 1.4.7. Similar to one-dimensional case in Sect. 1.4.15, we define the **Dolean measure** $\langle M\rangle \circ P_T$ on $\mathscr{P}_T$ and denote by $\overline{\mathscr{P}}_T$ the completion of $\mathscr{P}_T$ with respect to this measure.

Let

$$\mathbb{L}_2^T := \mathbb{L}_2\big([0, T] \times \Omega, \overline{\mathscr{P}}_T, \langle M\rangle \circ \mathbb{P}_T; \mathfrak{L}(\mathbb{H}, \mathbb{X})\big)$$

be the space (of equivalence classes) of $\overline{\mathscr{P}}_T$-measurable functions $B : [0, T] \times \Omega \to \mathfrak{L}(\mathbb{H}, \mathbb{X})$ such that

$$\mathbb{E} \int_{[0,T]} \|B(t)\|^2\, d\langle M\rangle_t < \infty.$$

Every element $B \in \mathbb{L}_2^T$ has a $\mathscr{P}_T$-measurable representative $\tilde{B}$ (cf. Sect. 1.3.8). The stochastic integral of $\tilde{B}$ with respect to M was defined in the previous section. By setting

$$B \circ M(t) := \tilde{B} \circ M(t),$$

the stochastic integral with respect to M becomes a strongly continuous linear operator from $\mathbb{L}_2^T$ to $\mathbb{L}_2(\Omega, \mathscr{F}_T, \mathbb{P}; \mathbb{X})$. In fact, the operator

$$\tilde{B} \mapsto \tilde{B} \circ M$$

has a version that is strongly continuous from the space $\mathbb{L}_2^T$ to $\mathbb{L}_2\big(\Omega, \mathscr{F}_T, \mathbb{P}; \mathbb{C}([0, T]; \mathbb{X})\big)$.

Because a strongly continuous (with respect to the operator norm) linear operator in Banach spaces is weakly continuous, we get the following result.

Theorem 2.9 *If $M \in \mathfrak{M}_2^c(\mathbb{R}_+, \mathbb{H})$, $B^n \in \mathbb{L}_2^T$, and the sequence $\{B^n,\ n \geq 1\}$ converges to $B \in \mathbb{L}_2^T$ weakly, then, as $n \to \infty$,*

$$\int\limits_{[0,T]} B^n(s)dM(s) \to \int\limits_{[0,T]} B(s)dM(s)$$

weakly in the space $\mathbb{L}_2(\Omega, \mathscr{F}_T, \mathbb{P}_T; \mathbb{X})$ (as $\mathbb{X}$-random variables), and $B^n \circ M \to B \circ M$ weakly in the space $\mathbb{L}_2\big(\Omega, \mathscr{F}_T, \mathbb{P}; \mathbb{C}([0, T]; \mathbb{X})\big)$ (as $\mathbb{X}$-processes). □

2.4 Stochastic Integral with Respect to a Local Martingale

2.4.1. The objective of this section is to extend the stochastic integral to more general integrands (operators B) and to more general integrators (local, as opposed to square-integrable, martingales M). We start by extending the class of integrands.

2.4.2. Let $\mathbb{H}$ and $\mathbb{X}$ be separable Hilbert spaces and let $M \in \mathfrak{M}_2^c(\mathbb{R}_+, \mathbb{H})$ be a continuous square-integrable martingale with correlation operator Q_M such that $Q_M(t) \in \mathfrak{L}_1(\mathbb{H})$. Denote by

$$\mathbb{L}_2\big(\langle M\rangle \circ \mathbb{P}, \mathfrak{L}_{Q_M}(\mathbb{H}, \mathbb{X})\big)$$

the collection of all mappings B from $\mathbb{R}_+ \times \Omega$ to the space of linear (but not necessarily bounded) operators from $\mathbb{H}$ to $\mathbb{X}$ such that $B Q_M^{1/2}$ is a predictable process with values in $\mathfrak{L}_2(\mathbb{H}, \mathbb{X})$ and, for all $t \in \mathbb{R}_+$,

$$\mathbb{E} \int\limits_{[0,t]} \|B(s)\|_{Q_M}\, d\langle M\rangle_s < \infty,$$

where

$$\|B(s)\|_{Q_M} = \|\!|B(s) Q_M^{1/2}(s)|\!\|.$$

The objective of this paragraph is to prove the following result.

Theorem 2.10 *For every $B \in \mathbb{L}_2\big(\langle M\rangle \circ \mathbb{P}, \mathfrak{L}_{Q_M}(\mathbb{H}, \mathbb{X})\big)$, there exists a sequence $B_n \in \mathbb{L}_2(\langle M\rangle \circ \mathbb{P}, \mathfrak{L}_2(\mathbb{H}, \mathbb{X}))$ and a stochastic process $B \circ M \in \mathfrak{M}_2^c(\mathbb{R}_+, \mathbb{X})$ such that, for every $T \in \mathbb{R}_+$,*

(a) $\lim_{n\to\infty} \mathbb{E} \sup_{t\le T} \|B\circ M(t) - B_n \circ M(t)\|^2_{\mathbb{X}} = 0$,

(b) $\langle B\circ M\rangle_T = \int_{[0,T]} \|B(s)\|^2_{Q_M}\, d\langle M\rangle_s$ ($\mathbb{P}$- a.s.),

(c) $\lim_{n\to\infty} \mathbb{E} \int_{[0,T]} \|B_n(s) - B(s)\|^2_{Q_M}\, d\langle M\rangle_s = 0$,

and the value of $B\circ M(T)$ *does not depend on the choice of the sequence* $B_n \in \mathbb{L}_2(\langle M\rangle \circ \mathbb{P}, \mathfrak{L}_2(\mathbb{H},\mathbb{X}))$ *as long as property* (a) *holds. The process* $B\circ M$ *and the stochastic integral defined by*

$$\int_{[0,t]} B(s)dM(s) := B\circ M(t),\ t\in\mathbb{R}_+,$$

have properties (i)–(v) *from Theorem 2.3.* □

To begin, we recall a well-known result from functional analysis (see e.g. [127, §106]).

Lemma 2.3 *Let* $Q\in\mathfrak{L}(\mathbb{H})$ *be a self-adjoint operator such that, for all* $h\in\mathbb{H}$,

$$m_0\|h\|^2 \le (Qh,h) \le m_1\|h\|^2.$$

If $f = f(t)$ *is a continuous non-negative real function on the interval* $[m_0, m_1]$ *and* $\{P_n(t) = a_0^n + a_1^n t + \ldots + a_k^n t^k\}$ *is a decreasing sequence of polynomials converging, as* $n\to\infty$, *to* $f(t)$, *for every* $t\in[m_0,\ m_1]$, *then the sequence of operators* $\{P_n(Q) = a_0^n + a_1^n Q + \ldots + a_k^n\, Q^k\}$ *converges strongly to the non-negative, bounded operator* $f(Q)$. *The operator* $f(Q)$ *does not depend on the choice of the approximating sequence* $\{P_n(\lambda)\}$. *Moreover, if* $Qh = \lambda h$, *then* $f(Q)h = f(\lambda)h$. □

Proof of Theorem 1 Let $B\in\mathbb{L}_2\big(\langle M\rangle\circ\mathbb{P};\mathfrak{L}_{Q_M}(\mathbb{H},\mathbb{X})\big)$. For $n\in\mathbb{N}$, define

$$B_n(s) := B(s)Q_M^{1/2}(s)\left(n^{-1}\mathbb{I} + Q_M^{1/2}(s)\right)^{-1}.$$

Note that, for every s and ω, $\left(n^{-1}\mathbb{I} + Q_M^{1/2}(s)\right)^{-1} \in \mathfrak{L}(\mathbb{H})$ and

$$\left\|\left(n^{-1}\mathbb{I} + Q_M^{1/2}(s)\right)^{-1}\right\| \le n,\quad \left\|Q_M^{1/2}(s)\left(n^{-1}\mathbb{I} + Q_M^{1/2}(s)\right)^{-1}\right\| \le 1.$$

Consequently

$$\|B_n(s)\| \le n\|B(s)\|_{Q_M},\quad \|B_n(s)\|_{Q_M} \le \|B(s)\|_{Q_M},$$

B_n is a predictable process with values in $\mathfrak{L}_2(\mathbb{H}, \mathbb{X})$, and the stochastic integral

$$\int_{[0,t]} B_n(s) dM(s)$$

is defined. Now we pass to the limit as $n \to \infty$.

For fixed s and ω, let $\{\varphi_i(s, \omega)\}$ be the CONS in $\mathbb{H}$ consisting of the eigenfunctions of $Q_M(s, \omega)$, and let $\{\lambda_i(s, \omega)\}$ be the corresponding eigenvalues. By Lemma 2.3,

$$\begin{aligned} \|B_n(s) - B(s)\|_{Q_M}^2 &= \sum_{i \geq 1} \|\big(B_n(s) - B(s)\big) Q_M^{1/2}(s) \varphi_i(s)\|_{\mathbb{X}}^2 \\ &= \sum_{i \geq 1} \|B(s) Q_M^{1/2}(s) \varphi_i(s)\|_{\mathbb{X}}^2 \frac{1}{(1 + n\sqrt{\lambda_i})^2}. \end{aligned}$$

If $\lambda_i = 0$, then $Q_M^{1/2}(s)\varphi_i(s) = 0$, and if $\lambda_i > 0$, then $(1 + n\sqrt{\lambda_i})^{-2} \to 0$, $n \to \infty$. Consequently, by the dominated convergence theorem, for every $(s, \omega) \in \mathbb{R}_+ \times \Omega$,

$$\lim_{n \to \infty} \|B_n(s) - B(s)\|_{Q_M}^2 = 0.$$

We also know that

$$\|B_n(s) - B(s)\|_{Q_M}^2 \leq 4 \|B(s)\|_{Q_M}^2.$$

Using (1.4.4), (2.3.17), and the dominated convergence theorem,

$$\lim_{n \to \infty} \mathbb{E} \int_{[0,t]} \|B_n(s) - B(s)\|_{Q_M}^2 \, d\langle M \rangle_s = 0, \ t \in \mathbb{R}_+,$$

and

$$\begin{aligned} &\lim_{\substack{m \to \infty \\ n \to \infty}} \mathbb{E} \sup_{t \leq T} \|B_n \circ M(t) - B_m \circ M(t)\|_{\mathbb{X}}^2 \\ &\quad \leq \lim_{\substack{m \to \infty \\ n \to \infty}} \mathbb{E} \int_{[0,T]} \|B_n(t) - B_m(t)\|_{Q_M}^2 \, d\langle M \rangle_t = 0. \end{aligned}$$

Thus, for every $B \in \mathbb{L}_2\big(\langle M \rangle \circ \mathbb{P}, \mathfrak{L}_{Q_M}(\mathbb{H}, \mathbb{X})\big)$, we can define the process $B \circ M \in \mathfrak{M}_2^c(\mathbb{R}_+, \mathbb{X})$ such that

$$\lim_{n \to \infty} \mathbb{E} \sup_{t \leq T} \|B \circ M(t) - B_n \circ M(t)\|_{\mathbb{X}}^2 = 0 \tag{2.4.1}$$

for every $T \in \mathbb{R}_+$.

Next, because

$$\langle B_n \circ M(t)\rangle_t = \int\limits_{[0,t]} \|B_n(s)\|_{Q_M}^2 \, d\langle M\rangle_s,$$

Proposition 1.2 implies

$$\langle B \circ M\rangle_t = \int\limits_{[0,t]} \|B(s)\|_{Q_M}^2 \, d\langle M\rangle_s. \tag{2.4.2}$$

Thus, for $B \in \mathbb{L}_2\big(\langle M\rangle \circ \mathbb{P}, \mathfrak{L}_{Q_M}(\mathbb{H}, \mathbb{X})\big)$, we define

$$\int\limits_{[0,t]} B(s)dM(s) := B \circ M(t), \; t \in \mathbb{R}_+.$$

This completes the proof of Theorem 1.9. □

If $B \in \mathbb{L}_2\big(\langle M\rangle \circ \mathbb{P}, \mathfrak{L}(\mathbb{H}, \mathbb{X})\big)$, then (2.4.1) and property (iii) from Theorem 2.8 imply that stochastic integral constructed in Theorem 1.9 coincides with the one constructed in the previous section.

2.4.3. We consider a particular case of the integral defined above. Let $b = b(s)$ be a predictable $\mathbb{H}$-process such that, for all $t \in \mathbb{R}_+$,

$$\mathbb{E} \int\limits_{[0,t]} \|b(s)\|^2 \, d\langle M\rangle_s < \infty.$$

For fixed $s \geq 0$, define the linear functional $\hat{b}(s)$ on $\mathbb{H}$ by

$$\hat{b}(s)h := \big(b(s), h\big).$$

Then $\hat{b} \in \mathbb{L}_2\big(\langle M\rangle \circ \mathbb{P}, \mathfrak{L}_2(\mathbb{H}, \mathbb{R}^1)\big)$ and, by Theorem 1.3, for every $t \in \mathbb{R}_+$,

$$\int\limits_{[0,t]} \hat{b}(s)dM(s) = \lim_{n\to\infty} \sum_{i=1}^{n} \int\limits_{[0,t]} \big(b(s), h_i\big)dM_i(s) \;\; (\mathbb{P}\text{- a.s.}). \tag{2.4.3}$$

Warning 2.6 *We will denote* $\int\limits_{[0,t]} \hat{b}(s)dM(s)$ *by* $\int\limits_{[0,t]} \big(b(s), dM(s)\big)$ *and call it the stochastic integral of* b *with respect to* M. *Equality* (2.4.3) *shows that this notation is natural.* □

2.4.4. Now we define the stochastic integral $B \circ M$ when

$$M \in \mathfrak{M}^c_{loc}(\mathbb{R}_+, \mathbb{H}), \text{ and } B \in \mathbb{L}_2\big(\langle M \rangle, \mathfrak{L}_{Q_M}(\mathbb{H}, \mathbb{X})\big),$$

that is, B is a mapping from $\mathbb{R}_+ \times \Omega$ to the space of linear (but not necessarily bounded) operators from $\mathbb{H}$ to $\mathbb{X}$ such that the process $BQ_M^{1/2}$ is predictable with values in $\mathfrak{L}_2(\mathbb{H}, \mathbb{X})$ and

$$\mathbb{P}\left(\int_{[0,t]} \|B(s)\|^2_{Q_M} \, d\langle M \rangle_s < \infty, \; t \in \mathbb{R}_+ \right) = 1.$$

Theorem 2.11 *If* $B \in \mathbb{L}_2\big(\langle M \rangle \circ \mathbb{P}, \mathfrak{L}_{Q_M}(\mathbb{H}, \mathbb{X})\big)$*, then there exists a stochastic process* $B \circ M \in \mathfrak{M}^c_{loc}(\mathbb{R}_+, \mathbb{X})$ *such that*

$$\langle B \circ M \rangle_t = \int_{[0,t]} \|B(s)\|^2_{Q_M} \, d\langle M \rangle_s.$$

The stochastic integral defined by the equality

$$\int_{[0,t]} B(s) dM(s) := B \circ M(t)$$

possesses properties (i), (ii), (iv) *from Theorem 2.3 and the following modification of property* (iii)*:*

(iii′) *Let* $B, B_n \in \mathbb{L}_2\big(\langle M \rangle, \mathfrak{L}_{Q_M}(\mathbb{H}, \mathbb{X})\big)$*,* $n \in \mathbb{N}$*, and, for some* $T \in \mathbb{R}_+$*,*

$$\lim_{n \to \infty} \int_{[0,T]} \|B(s) - B_n(s)\|^2_{Q_M} \, d\langle M \rangle_s = 0$$

in probability. Then

$$\lim_{n \to \infty} \sup_{t \le T} \|B^n \circ M(t) - B \circ M(t)\|_{\mathbb{X}} \to 0$$

in probability. □

Proof Given $B \in \mathbb{L}_2\big(\langle M \rangle, \mathfrak{L}_{Q_M}(\mathbb{H}, \mathbb{X})\big)$ and $n \in \mathbb{N}$, define the set

$$\Gamma_n(t, \omega) = \left\{ (t, \omega) : \int_{[0,t]} \|B(s)\|^2_{Q_M} \, d\langle M \rangle_s > n \right\},$$

the stopping time

$$\tau_n = \begin{cases} \inf_{t\in\mathbb{R}_+} \Gamma_n(t), & \text{if } \Gamma_n(t) \neq \emptyset, \\ \infty, & \text{if } \Gamma_n(t) = \emptyset, \end{cases}$$

and write

$$B^n(t) := B(t)1_{\{[0,\tau_n]\}}(t),\ n \in \mathbb{N}.$$

Note that τ_n is indeed a stopping time for every n (see Sect. 1.4.8) and, by assumptions on B, $\tau_n \uparrow \infty$ ($\mathbb{P}$-a.s.). Next, the process $t \mapsto 1_{\{[0,\tau_n]\}}(t)$, $t \in \mathbb{R}_+$, is predictable because it is left-continuous and $\mathscr{F}_t$-adapted. As a result, $B^n \in \mathbb{L}_2\big(\langle M\rangle \circ \mathbb{P}, \mathfrak{L}_{Q_M}(\mathbb{H},\mathbb{X})\big)$ for every n, and the stochastic integral $\int_{[0,t]} B^n(s)dM(s)$ is defined.

Let $\tau_0 = 0$ and define

$$B\circ M(0) = 0,\ B\circ M(t) = \sum_{i=1}^{\infty} B^i \circ M(t)1_{\{(\tau_{i-1},\ \tau_i]\}}(t),\ t > 0.$$

By construction, $B\circ M \in \mathfrak{M}^2_{loc}(\mathbb{R}_+,\mathbb{H})$, because $B\circ M$ is continuous, $\mathscr{F}_t$-adapted, and $B\circ M(t\wedge\tau_n) = B^n \circ M(t)$, $n\in\mathbb{N}$, ($\mathbb{P}$-a.s.). Also,

$$\langle B\circ M\rangle_t = \int_{[0,t]} \|B(s)\|^2_{Q_M}\, d\langle M\rangle_s \ \ (\mathbb{P}\text{- a.s.}).$$

Define the stochastic integral by

$$\int_{[0,t]} B(s)dM(s) := B\circ M(t) \ \ (\mathbb{P}\text{- a.s.}).$$

Properties (i), (ii), and (iv) from Theorem 2.8 continue to hold, but, by Corollary 2.2, property (iii) becomes (iii'). Note that property (v) does not make sense unless the operator B is bounded. □

Example 2.1 Let G be a domain in $\mathbb{R}^{\mathrm{d}}$ and let $b = b(t)$ be a predictable process with values in $\mathbb{L}_2(G)$ and such that

$$\mathbb{P}\left(\int_{[0,t]} \|b(s)\|^2_{\mathbb{L}_2(G)}ds < \infty,\ t\in\mathbb{R}_+\right) = 1.$$

Let $w = w(t)$ be a one-dimensional standard Wiener process. For each $s \in \mathbb{R}_+$, define the mapping $B(s)$ from $\mathbb{R}^1$ to $\mathbb{L}_2(G)$ by $B(s)r = b(s)r$, $r \in \mathbb{R}^1$. Then, by the previous theorem, the stochastic integral

$$\int_{[0,t]} b(s)\,dw(s) = \int_{[0,t]} B(s)\,dw(s)$$

exists and is a continuous local martingale with values in $\mathbb{L}_2(G)$. □

Remark 2.4 By [26, Chapter II, Theorem 11.17], there exists a $\mathscr{P} \otimes \mathscr{B}(G)$-measurable function $\hat{b} = \hat{b}(t, x, \omega)$ such that, for $\mathfrak{l} \times \mathbb{P}$-a.a. (t, ω), $\hat{b} = b$ as elements of $\mathbb{L}_2(G)$ and

$$\int_G \int_{[0,t]} \hat{b}^2(s, x)\,ds\,dx < \infty, \ t \in \mathbb{R}_+.$$

The stochastic integral $\int_{[0,t]} \tilde{b}(s, x)\,dw(s)$ is defined for $\mathfrak{l}_d$-a.a. $x \in G$ (cf. *Sect. 1.5.2*) and is a version of the stochastic integral defined in the example. □

2.4.5.

Remark 2.5 Suppose $b = b(t)$ is a predictable $\mathbb{H}$-process such that

$$\mathbb{P}\left(\int_{[0,t]} \|b(s)\|^2\,d\langle M\rangle_s < \infty, \ t \in \mathbb{R}_+\right) = 1.$$

Then, arguing as in Sect. 2.2, we define the stochastic integral of b with respect to M by the equality

$$\int_{[0,t]} \big(b(s), dM(s)\big) := \int_{[0,t]} \hat{b}(s) dM(s).$$ □

Warning 2.7 *In this section, we distinguish between a stochastic integral and its continuous version, and, in particular, use different notations for them. From now on, we only consider the continuous version of stochastic integrals, and use any of the available notations, such as* $B \circ M(t)$, $\int_{[0,t]} B(s)dM(s)$, *or* $\int_{[0,t]} \big(B(s), dM(s)\big)$.

□

2.4.6.

Remark 2.6 With obvious changes, all the results of this section extend to processes on an interval $[T_0, T]$, $T_0, T \in \mathbb{R}_+$. □

2.5 An Energy Equality in a Rigged Hilbert Space

2.5.1. Embedding theorems establish connections between various function spaces and are widely used in the study of differential equations. Here is a simple example of an embedding theorem.

Let $f = f(t)$, $t \in (0, T)$, be a real-valued function and let f' be the generalized derivative of f:

$$\int_0^T f'(t)g(t)\,dt = -\int_0^T f(t)g'(t)\,dt$$

for every $g \in \mathbb{C}_0^\infty\big((0,T);\mathbb{R}^1\big)$. Denote by $\mathbb{W}_2^1\big((0,T);\mathbb{R}^1\big)$ the collection of all functions $f \in \mathbb{L}_2\big((0,T);\mathbb{R}^1\big)$ such that $f' \in \mathbb{L}_2\big((0,T);\mathbb{R}^1\big)$.

Proposition 2.4 *If $f \in \mathbb{W}_2^1\big((0,T);\mathbb{R}^1\big)$, then f can be modified on a subset of $(0, T)$ with zero Lebesgue measure to become continuous on $(0, T)$.* □

In other words, the proposition means that every equivalence class of functions in $\mathbb{W}_2^1\big((0,T);\mathbb{R}^1\big)$ contains a continuous function, or, in a sense, the space $\mathbb{W}_2^1\big((0,T);\mathbb{R}^1\big)$ is embedded (included) into the space $\mathbb{C}\big((0,T);\mathbb{R}^1\big)$ of continuous functions on $(0, T)$.

The intuition behind this result is simple: by the Cauchy–Schwarz inequality,

$$|f(t) - f(s)| = \left| \int_{[s,t]} f'(r)dr \right| \le |t-s|^{1/2}\, \|f'\|_{\mathbb{L}_2((0,T);\mathbb{R}^1)}.$$

Definition 2.7 We say that a Hilbert space $\mathbb{X}$ is **normally embedded** into a Hilbert space $\mathbb{Y}$ if the embedding is dense and continuous, that is, $\mathbb{X}$ is a dense subset of $\mathbb{Y}$ (in the topology generated by the norm $\|\cdot\|_{\mathbb{Y}}$), and there exists a number $N > 0$ such that, for all $x \in \mathbb{X}$,

$$\|x\|_{\mathbb{Y}} \le N\|x\|_{\mathbb{X}}.$$

In other words, normal embedding of $\mathbb{X}$ in $\mathbb{Y}$ means that $\mathbb{X}$ is a smaller space with a bigger norm. By switching to an equivalent norm in $\mathbb{X}$, one can take $N = 1$. Also, as all norms are equivalent in finite dimensions, all spaces are assumed to be infinite-dimensional.

As usual, we will have a specially designated separable Hilbert space $\mathbb{H}$ with inner product $(\cdot, \cdot)$ and norm $\|\cdot\|$. For all other spaces, the corresponding notation will come with a subscript, as in $\|\cdot\|_{\mathbb{X}}$.

The next step is an analog of Proposition 2.4 for functions with values in Hilbert spaces.

Let $\mathbb{X}$ and $\mathbb{H}$ be Hilbert spaces such that $\mathbb{X}$ is normally embedded into $\mathbb{H}$ and $\mathbb{H}$ is identified with its topological dual $\mathbb{H}^*$. Let $\mathbb{X}^*$ be the topological dual of $\mathbb{X}$, relative to the inner product in $\mathbb{H}$, that is, $\mathbb{H} \subset \mathbb{X}^*$ and $hx = (x, h)$ for every $x \in \mathbb{X},\ h \in \mathbb{H}$. The Hilbert space $\mathbb{H}$ equipped with such a pair $\mathbb{X}, \mathbb{X}^*$ is called a **rigged Hilbert space**.

The following result is a Hilbert space version of Proposition 2.4; cf. [91, Section 1.2.2].

Theorem 2.12 *With the Hilbert spaces $\mathbb{X} \subset \mathbb{H} \subset \mathbb{X}^*$ as above, let $u = u(t),\ t \in [0, T]$, be an $\mathbb{X}$-valued function such that $u \in \mathbb{L}_2\big((0, T), \overline{\mathscr{B}((0, T))}, \mathfrak{l}; \mathbb{X}\big)$, that is,*

$$\int_{[0,T]} \|u(t)\|_{\mathbb{X}}^2 dt < \infty,$$

and there exist $\varphi \in \mathbb{H}$ and $f \in \mathbb{L}_2\big((0, T), \overline{\mathscr{B}((0, T))}, \mathfrak{l}; \mathbb{X}^*)$ such that

$$u(t) = \varphi + \int_0^t f(s)\, ds$$

*in $\mathbb{L}_2\big((0, T), \overline{\mathscr{B}((0, T))}, \mathfrak{l}; \mathbb{X}^*\big)$. Then the values of $u(t)$ can be changed on a subset of $[0, T]$ with zero Lebesgue measure so that the resulting function $\tilde{u}$ belongs to $\mathbb{C}([0, T]; \mathbb{H})$ and satisfies the **energy equality***

$$\|\tilde{u}(t)\|^2 = \|\tilde{u}(0)\|^2 + 2 \int_{[0,t]} f(s)u(s)\, ds,\ t \in [0, T]. \qquad \square$$

Our objective in this section is to extend Theorem 2.12 to $\mathbb{X}$-valued semimartingales. The resulting generalization of the energy equality can be considered the Itô formula for the square of the norm and is a key tool in the study of stochastic evolution systems.

2.5.2. First, we elaborate a bit more on the topic of rigged Hilbert spaces.

Definition 2.8 The triple $(\mathbb{X}, \mathbb{H}, \mathbb{X}')$ of separable Hilbert spaces is called **normal** if $\mathbb{X} \subset \mathbb{H} \subset \mathbb{X}'$, with all the embeddings normal, and there exists a number N such that, for all $x \in \mathbb{X}$ and $y \in \mathbb{H}$,

$$\big|(x, y)\big| \le N\|x\|_{\mathbb{X}} \cdot \|y\|_{\mathbb{X}'}. \qquad \square$$

If $(\mathbb{X}, \mathbb{H}, \mathbb{X}')$ is a normal triple and $y \in \mathbb{X}'$, then there exists a sequence $\{y_n\}$ from $\mathbb{X}$ such that $\|y - y_n\|_{\mathbb{X}'} \to 0$ as $n \to \infty$. Write

$$[x, y] := \lim_{n\to\infty} (x, y_n),\ x \in \mathbb{X}.$$

The definition of the normal triple implies that the limit on the right-hand side of the equality exists and does not depend on the choice of the sequence $\{y_n\}$ approximating y. The resulting mapping

$$[\cdot, \cdot] : \mathbb{X} \times \mathbb{X}' \to \mathbb{R}$$

is a bilinear form and is called the **canonical bilinear functional** (CBF) of the normal triple $(\mathbb{X}, \mathbb{H}, \mathbb{X}')$.

The following properties of the CBF are immediate consequences of the definition:

(i) $\big|[x, y]\big| \le N\ \|x\|_{\mathbb{X}} \cdot \|y\|_{\mathbb{X}'}$ for every $x \in \mathbb{X}$, $y \in \mathbb{X}'$;
(ii) $[x, h] = (x, h)$ for every $h \in \mathbb{H}$;
(iii) if $f \in \mathbb{L}_1\big((0, T), \mathscr{B}((0, T)), \mathfrak{l}; \mathbb{X}'\big)$, then, for every $x \in \mathbb{X}$,

$$\left[x, \int_{[0,T]} f(s)\, ds\right] = \int_{[0,T]} [x, f(s)] ds.$$

Remark 2.7 It can be shown [64, Section 4.5.10] that if a Hilbert space $\mathbb{X}$ is normally embedded in a Hilbert space $\mathbb{X}'$, then there exists a space $\mathbb{H}$ such that $(\mathbb{X}, \mathbb{H}, \mathbb{X}')$ is a normal triple and the CBF of this triple gives an isometric isomorphism between $\mathbb{X}'$ and $\mathbb{X}^*$:

$$\mathbb{X}' \ni x' \leftrightarrow [\cdot, x'] \in \mathbb{X}^*.$$

So, considering the CBF as a duality between $\mathbb{X}$ and $\mathbb{X}'$ relative to the inner product in $\mathbb{H}$, we will identify $\mathbb{X}'$ and $\mathbb{X}^*$. □

2.5.3. Throughout the rest of this section, we fix the following objects:

- A normal triple $(\mathbb{X}, \mathbb{H}, \mathbb{X}')$;
- A stochastic basis $\mathbb{F} = (\Omega, \mathscr{F}, \{\mathscr{F}_t\}_{t\in[0,T]}, \mathbb{P})$ with the usual assumptions;
- A continuous local martingale $M = M(t)$, relative to $\{\mathscr{F}_t\}$, with values in $\mathbb{H}$, and $M(0) = 0$;
- Predictable functions $x : \mathbb{R}_+ \times \Omega \to \mathbb{X}$ and $x' : \mathbb{R}_+ \times \Omega \to \mathbb{X}'$ such that

$$\mathbb{P}\left(\int_{[0,T]} \left(\|x(t)\|^2_{\mathbb{X}} + \|x'(t)\|^2_{\mathbb{X}'}\right) dt < \infty,\ T > 0 \right) = 1;$$

- An $\mathscr{F}_0$-measurable random variable $x(0)$ with values in $\mathbb{H}$;
- A stopping time τ.

The notation x' is, in a sense, suggestive, not only because x' takes values in $\mathbb{X}'$, but also because x' will be related to the time derivative of x.

The objective is to prove the following result.

Theorem 2.13 *Assume that, for every $\eta \in \mathbb{X}$, the equality*

$$\big(\eta, x(t)\big) = \big(\eta, x(0)\big) + \int_{[0,t]} [\eta, x'(s)]ds + \big(\eta, M(t)\big) \tag{2.5.1}$$

holds for $\ell \times \mathbb{P}$-a.a. (t, ω) satisfying $t < \tau(\omega)$. Then there exists a continuous adapted version of x with values in $\mathbb{H}$.

More precisely, there exist a set $\tilde{\Omega} \subset \Omega$ with $\mathbb{P}(\tilde{\Omega}) = 1$ and a function $\tilde{x} = \tilde{x}(t, \omega)$ with values in $\mathbb{H}$ such that

(i) *The function $\tilde{x}$ is continuous in t on $[0, \tau)$ for every $\omega \in \tilde{\Omega}$, $\tilde{x}(t, \omega) = x(t, \omega)$ for $\ell \times \mathbb{P}$-a.a. (t, ω) satisfying $t < \tau(\omega)$, and, for every Borel set $A \subset \mathbb{H}$ and every $t > 0$,*

$$\{\omega : \tilde{x}(t, \omega) \in A,\ t < \tau(\omega)\} \in \mathscr{F}_t;$$

(ii) *For every $\omega \in \tilde{\Omega}$, $t < \tau(\omega)$, and $\eta \in \mathbb{X}$,*

$$\big(\eta, \tilde{x}(t)\big) = \big(\eta, x(0)\big) + \int_{[0,t]} [\eta, x'(s)]ds + \big(\eta, M(t)\big); \tag{2.5.2}$$

(iii) *For every $\omega \in \tilde{\Omega}$ and $t < \tau(\omega)$,*

$$\|\tilde{x}(t)\|^2 = \|x(0)\|^2 + 2\int_{[0,t]} [x(s), x'(s)]ds + 2\int_{[0,t]} \big(\tilde{x}(s), dM(s)\big) + \langle M\rangle_t; \tag{2.5.3}$$

(iv) *If* (2.5.1) *holds* $\mathbb{P}$-a.s. *on the set $\{\omega : t_0 < \tau(\omega)\}$ for some $t_0 > 0$ and every $\eta \in \mathbb{X}$, then $x(t_0) = \tilde{x}(t_0)$,* $\mathbb{P}$-a.s. *on the set $\{\omega : t_0 < \tau(\omega)\}$.* □

Remark 2.8 As far as the stochastic integral $\int_{[0,t]} (\tilde{x}(s),\ dM(s))$, see Remarks 2.5 and 2.6. The stochastic process $\tilde{x}(t)$ is predictable because it is continuous in $\mathbb{H}$ and $\mathscr{F}_t$-adapted.

To make the relation between this theorem and Theorem 2.12 clearer, we note that if x coincides ($\ell \times \mathbb{P}$-a.s.) with the $\mathbb{X}'$-valued semimartingale

$$y(t) = x(0) + \int_{[0,t]} x'(s)ds + M(t),$$

then (2.5.1) holds. If $\mathbb{X} = \mathbb{H} = \mathbb{X}'$, then (2.5.3) is the Itô formula for the square of the norm of $y(t)$. □

The proof of Theorem 2.13 is long and will take the rest of this section. To begin, we show that, with no loss of generality, we can assume that τ and $\int_{[0,t]} \left(\|x(s)\|_{\mathbb{X}}^2 + \|x'(s)\|_{\mathbb{X}'}^2 \right) ds$ are uniformly bounded, and M is a continuous square-integrable (as opposed to local) martingale.

Indeed, define

$$R(t) = \|x(0)\|^2 + \int_{[0,t]} \|x(s)\|_{\mathbb{X}}^2 \, ds + \int_{[0,t]} \|x'(s)\|_{\mathbb{X}}^2 \, ds + \langle M \rangle_t.$$

Since the process R is predictable, $\tau(n) := \inf\{t \geq 0 : R(t) \geq n\} \wedge \tau \wedge n$ is a stopping time for every n and $\tau(n) \uparrow \tau$ as $n \to \infty$.

If we prove the theorem when τ is replaced by $\tau(n)$ with arbitrary n, then we get the corresponding collection of sets $\tilde{\Omega}_n$ with $\mathbb{P}(\tilde{\Omega}_n) = 1$ and take

$$\tilde{\Omega} := \bigcap_n \tilde{\Omega}_n.$$

Thus, without loss of generality we assume that there exists a $T > 0$ such that $\tau(\omega) \leq T$ for all ω. Then Proposition 2.1 allows us to assume that $M \in \mathfrak{M}_2^c([0, T], \mathbb{H})$.

2.5.4.

Lemma 2.4 *There exists a sequence of nested partitions* $\{0 = t_0^n < t_1^n < \ldots < t_{k(n)+1}^n = T\}$ *of the interval* $[0, T]$, *with* $\lim_n \max_i |t_{i+1}^n - t_i^n| = 0$, *and a set* $\Omega' \subset \Omega$ *having the following properties:*

1) $\mathbb{P}(\Omega') = 1$ *and, for all* $\omega \in \Omega'$, $t < \tau(\omega)$, $t \in \mathcal{I} := \{t_i^n, \; i = 1, 2, \ldots, k(n), \; n \in \mathbb{N}\}$, *and* $\eta \in \mathbb{X}$, *equality* (2.5.1) *holds.*
2) *For each* n, *define two processes,* $x_n^1 = x_n^1(t)$ *and* $x_n^2 = x_n^2(t)$ *as follows:* $x_n^1(t) := x(t_i^n)$ *for* $t \in [t_i^n, t_{i+1}^n)$ $i = 1, 2, \ldots, k(n)$, $x_n^1(t) := 0$ *for* $t \in [0, t_1^n)$; $x_n^2(t) := x(t_{i+1}^n)$ *for* $t \in [t_i^n, t_{i+1}^n)$, $i = 0, 1, \ldots, (k(n) - 1)$, $x_n^2(t) := 0$ *for* $t \in [t_{k(n)}^n, T)$.

Then

$$\mathbb{E}\sup_{t\le T}\|x_n^j(t)\|_{\mathbb{X}}^2 < \infty,\ j = 1, 2, \tag{2.5.4}$$

and

$$\lim_{n\to\infty}\mathbb{E}\int\limits_{[0,T]}\|x(t) - x_n^i(t)\|_{\mathbb{X}}^2\,dt = 0,\ j = 1, 2. \tag{2.5.5}$$

□

Proof We start with the second statement and prove it using a well-known method developed by J. Doob (cf. [25, Ch. IX, §5]).

Let $f : \mathbb{R}^1 \to \mathbb{X}$ be a measurable function with compact support and such that $\int\limits_{\mathbb{R}^1}\|f(s)\|_{\mathbb{X}}^2\,ds < \infty$. Then

$$\lim_{\delta\to 0}\int\limits_{\mathbb{R}^1}\|f(s+\delta) - f(s)\|_{\mathbb{X}}^2\,ds = 0. \tag{2.5.6}$$

Indeed, given $\varepsilon > 0$, let $f_\varepsilon : \mathbb{R}^1 \to \mathbb{X}$ be a continuous function with compact support and such that

$$\int\limits_{\mathbb{R}^1}\|f(s) - f_\varepsilon(s)\|_{\mathbb{X}}^2\,ds \le \varepsilon^2.$$

Then, by the triangle inequality,

$$\limsup_{\delta\to 0}\left(\int\limits_{\mathbb{R}^1}\|f(s+\delta) - f(s)\|_{\mathbb{X}}^2\,ds\right)^{1/2}$$

$$\le \limsup_{\delta\to 0}\left(\int\limits_{\mathbb{R}^1}\|f_\varepsilon(s+\delta) - f_\varepsilon(s)\|_{\mathbb{X}}^2\,ds\right)^{1/2} + 2\varepsilon = 2\varepsilon,$$

which implies (2.5.6). In fact, one more approximation shows that f does not have to be compactly supported.

Now define

$$\hat{x}(t,\omega) = \begin{cases} x(t,\omega), & t \in [0,T]; \\ 0, & t \notin [0,T]. \end{cases}$$

From (2.5.6) it follows that, with probability one,

$$\lim_{\delta \to 0} \int_{\mathbb{R}^1} \|\hat{x}(t+\delta) - \hat{x}(t)\|_{\mathbb{X}}^2 \, dt = 0.$$

Denote by $[a]$ the integer part of the number a, and define $\chi^1(n,t) := 2^{-n}[2^n t]$, $\chi^2(n,t) := \chi^1(n,t) + 2^{-n}$. Then, for every $t \geq 0$,

$$\lim_{n \to \infty} \int_{\mathbb{R}^1} \left\| \hat{x}\big(\chi^j(n,t) + s\big) - \hat{x}(t+s) \right\|_{\mathbb{X}}^2 ds = 0, \ j = 1, 2 \ (\mathbb{P}\text{- a.s.}).$$

Therefore, in view of the boundedness of the process $R = R(t)$, the dominated convergence theorem implies

$$\lim_{n \to \infty} \mathbb{E} \int_{\mathbb{R}^1 \times \mathbb{R}^1} \left\| \hat{x}\big(\chi^j(n,t) + s\big) - \hat{x}(t+s) \right\|_{\mathbb{X}}^2 dsdt = 0, \quad j = 1, 2.$$

From this it follows that there exists a sequence of integers r_n such that, for $\mathfrak{l}$-a.a. s,

$$\lim_{n \to \infty} \mathbb{E} \int_{\mathbb{R}^1} \left\| \hat{x}\big(\chi^j(r_n,t) + s\big) - \hat{x}(t+s) \right\|_{\mathbb{X}}^2 dt = 0, \quad j = 1, 2. \tag{2.5.7}$$

Separability of $\mathbb{X}$ and the Fubini theorem imply the existence of a set $S \subset [0, T]$ such that $[0, T] \backslash S$ is at most countable and, for all $t \in S$ and $\eta \in \mathbb{X}$, equality (2.5.1) holds $\mathbb{P}$-a.s. on the set $\{\omega : t < \tau(\omega)\}$.

By construction, for $\mathfrak{l}$-a.a. $s \in [0, T]$, the intersection of the range of the function $t \mapsto \chi^j(r_n, t - s) + s$ with $[0, T]$ is a subset of S. Thus we can choose s for which (2.5.7) holds and such that all the values of the function $\chi^j(r_n, t-s)+s$, for $j = 1, 2, \ \ n \geq 1$, and $t \in [0, T]$, inside $[0, T]$ belong to S. For fixed n define $\{t_i^n\}$ to be the set of the values $\chi^1(r_n, \ t - s) + s$ inside $[0, T]$ for this s and $t \in [0, T]$, together with 0 and T; the set $\mathcal{I}$ is the collection of all such t_i^n, $n \geq 1$. Then Ω' is the subset of Ω where equality (2.5.1) holds for all $\eta \in \mathbb{X}$ and $t := t_i^n < \tau(\omega)$ for $i = 1, 2, \ldots, k(n)$, $n \in \mathbb{N}$.

By construction, Ω' and $\mathcal{I}$ ensure that Part 1) of the lemma and equality (2.5.5) hold. Finally, (2.5.4) is equivalent to

$$\mathbb{E} \int_{[0,T]} \|x_n^i(t)\|_{\mathbb{X}}^2 \, dt < \infty,$$

which, for large n, follows from (2.5.5). Because partitions are nested, it also holds for small n. □

2.5.5.

Lemma 2.5 *Let $\widetilde{M}(t) := x(0)+M(t)$ and let Ω' and $\mathcal{I}$ be the sets from the previous lemma. Then, for $\omega \in \Omega'$ and $t, s \in \mathcal{I}$ such that $s \le t < \tau(\omega)$,*

$$\|x(t)\|^2 = 2 \int_{[0,t]} [x(t), x'(u)]du + \|\widetilde{M}(t)\|^2 - \|x(t) - \widetilde{M}(t)\|^2, \tag{2.5.8}$$

and

$$\|x(t)\|^2 - \|x(s)\|^2 = 2 \int_{[s,t]} [x(t), x'(u)]du + 2\big(x(s), M(t) - M(s)\big)$$
$$+ \|M(t) - M(s)\|^2 - \big\|x(t) - x(s) - (M(t) - M(s))\big\|^2. \tag{2.5.9}$$

□

Proof It is enough to establish (2.5.8); then (2.5.9) follows by simple algebra.

From Lemma 2.1 it follows that $x(t, \omega) \in \mathbb{X}$ for all $t \in \mathcal{I}$ and $\omega \in \Omega'$ such that $t < \tau(\omega)$. Then we use $\eta = x(t)$ in (2.5.1) to get

$$\|x(t)\|^2 = \int_{[0,t]} [x(t), x'(u)]du + \big(x(t), \widetilde{M}(t)\big).$$

From this it follows that, for the same ω and t,

$$\|\widetilde{M}(t)\|^2 - \|x(t) - \widetilde{M}(t)\|^2 = 2\big(x(t), \widetilde{M}(t)\big) - \|x(t)\|^2$$
$$= \|x(t)\|^2 - 2 \int_{[0,t]} [x(t), x'(u)]du,$$

which proves (2.5.8). □

2.5.6. Define the set

$$\Omega'' := \Omega' \bigcap \Big\{\omega : \sup_{t\in\mathcal{I},\, t<\tau} \|x(t)\|^2 < \infty\Big\}.$$

Lemma 2.4 implies $\mathbb{P}(\Omega'') = 1$.

The next auxiliary result is

Lemma 2.6 *There exists a function $\tilde{x} = \tilde{x}(t, \omega)$ defined for $\omega \in \Omega''$ and $t < \tau(\omega)$ and taking values in $\mathbb{H}$, such that $\tilde{x}$ is weakly continuous in t and satisfies* (2.5.2) *for all $\eta \in \mathbb{X}$, $\omega \in \Omega''$ and $t < \tau(\omega)$. Moreover, $\tilde{x}(t)$ is $\mathscr{F}_t$-measurable on the set $\{\omega : t < \tau(\omega)\}$, and the equality $\tilde{x}(t) = x(t)$ holds for all $\omega \in \Omega''$, $t \in \mathcal{I}$ satisfying $t < \tau(\omega)$, as well as for $\mathfrak{l} \times \mathbb{P}$- a.a. (t, ω) from the set $\{(t, \omega) : t < \tau(\omega)\}$.* □

Proof Since, for $\eta \in \mathbb{X}$, $\omega \in \Omega''$, $t \in \mathcal{I}$, $t < \tau$, the function $t \mapsto= (\eta, x(t))$ coincides with the right-hand side of (2.5.1), this function is continuous in t. Thus, for every $s \leq \tau$, we can choose a sequence of points $t_i \in \mathcal{I}$ not exceeding τ, converging to s, and such that $\lim_{i\to\infty} (\eta, x(t_i))$ exists. Since the function $\|x(t)\|$ is bounded on the set $\mathcal{I} \cap \{t < \tau\}$, it follows that the sequence $\{x(t_i)\}$ has a weak limit in $\mathbb{H}$, which we denote by $\tilde{x}(s)$. By construction, $\tilde{x}$ satisfies (2.5.2) for all $\omega \in \Omega''$, $t < \tau(\omega)$, and $\eta \in \mathbb{X}$, completing the proof of the lemma. □

2.5.7. So far, we have $\tilde{x}$ defined only for $\omega \in \Omega''$ and $t < \tau(\omega)$. Let $\tilde{x}(t) = \tilde{x}(\tau)$ for $t \geq \tau$ and $\tilde{x}(t) = 0$ for $\omega \notin \Omega''$. If $t < \tau$ and $t \notin \mathcal{I}$, the value of $\tilde{x}(t)$ is defined as a weak limit of $x(t_i)$ over some sequence of $t_i \in \mathcal{I}$.

$$\sup_{t\leq T} \|\tilde{x}(t)\| = \sup_{t\in\mathcal{I},\, t<\tau} \|x(t)\| < \infty. \tag{2.5.10}$$

The function $\tilde{x}$ is predictable, being weakly continuous and $\mathscr{F}_t$-adapted.

By (2.5.10) the stochastic integral

$$\int_{[0,t]} (\tilde{x}(s), dM(s)), \ t \leq T,$$

is defined (see Sect. 2.5).

To investigate this stochastic integral, consider the sequence of processes $\tilde{x}_n$ with $\tilde{x}_n(t) := \tilde{x}(t_i^n)$ for $t \in [t_i^n, t_{i+1}^n)$, $i = 0, 1, \ldots, k(n)$.

Lemma 2.7 *We have*

$$\lim_{n\to\infty} \sup_{t\leq T} \left| \int_{[0,t]} (x_n(s) - x(s), dM(s)) \right| = 0 \tag{2.5.11}$$

in probability. □

Proof Let $\{h_i,\ i \in \mathbb{N}\}$ be a CONS in $\mathbb{H}$ and denote by $\mathcal{P}_k$ the orthogonal projection on the linear span of $h_1, \ldots, h_k$. By Lemma 2.6, $t \mapsto \tilde{x}(t)$ is a weakly continuous function in $\mathbb{H}$ for every $\omega \in \Omega''$. Therefore, $t \mapsto \mathcal{P}_k(\tilde{x}(t))$ is strongly continuous in $\mathbb{H}$ for every $\omega \in \Omega''$. As a result, for every $k \in \mathbb{N}$,

$$\mathbb{P}\left(\lim_{n\to\infty} \int_{[0,t]} \|\mathcal{P}_k(\tilde{x}_n(s)) - \mathcal{P}_k(\tilde{x}_n(s))\|^2 \, d\langle M\rangle_s = 0 \right) = 1,$$

and, to prove the lemma, it remains to show that, for every $\varepsilon > 0$,

$$\lim_{k\to\infty} \sup_n \mathbb{P}\left(\sup_{t\le T} \left| \int_{[0,t]} \Big((\mathbb{I} - \mathcal{P}_k)\big(\tilde{x}_n(s)\big), dM(s) \Big) \right| \ge \varepsilon \right) = 0$$

and

$$\lim_{k\to\infty} \mathbb{P}\left(\sup_{t\le T} \left| \int_{[0,t]} \Big((\mathbb{I} - \mathcal{P}_k)\big(\tilde{x}(s)\big), dM(s) \Big) \right| \ge \varepsilon \right) = 0.$$

The second equality follows from (2.5.10) and property (iii′) in Theorem 3.3. To prove the first, define

$$M_k^n(t) := \int_{[0,t]} \Big((\mathbb{I} - \mathcal{P}_k)\, \big(\tilde{x}_n(s)\big), dM(s) \Big) = \int_{[0,t]} \Big(\tilde{x}_n(s), d(\mathbb{I} - \mathcal{P}_k)\big(M(s)\big) \Big).$$

By Corollary 2.2, for every $C, \delta \in \mathbb{R}_+$,

$$\begin{aligned}
&\mathbb{P}\left(\sup_{t\le T} |M_k^n(t)| \ge \varepsilon \right) \\
&\le \frac{3\delta}{\varepsilon} + \mathbb{P}\left(\left(\int_{[0,T]} \|\tilde{x}_n(s)\|^2 \, d\big\langle (\mathbb{I} - \mathcal{P}_k)\big(M(s)\big) \big\rangle_s \right)^{1/2} \ge \delta \right) \\
&\le \frac{3\delta}{\varepsilon} + \mathbb{P}\big(\sup_{t\le T} \|\tilde{x}_n(s)\| \ge C \big) + \frac{C}{\delta} \left(\mathbb{E} \left\| (\mathbb{I} - \mathcal{P}_k)\big(M(T)\big) \right\|^2 \right)^{1/2}.
\end{aligned}$$

From (2.5.10) it follows that

$$\sup_n \mathbb{P}\big(\sup_{t\le T} \|\tilde{x}_n(s)\| \ge C \big) = \mathbb{P}\big(\sup_{t\le T} \|\tilde{x}(s)\| \ge C \big);$$

the latter probability can be made arbitrarily small with a suitable choice of C. To complete the proof, recall that M is a square-integrable martingale with values in $\mathbb{H}$, so that

$$\lim_{k\to\infty} \mathbb{E} \left\| (\mathbb{I} - \mathcal{P}_k)\big(M(T)\big) \right\|^2 = 0. \qquad \square$$

2.5.8. Now we will construct the set $\tilde{\Omega}$ using Theorem 2.13. First, choose a subsequence $\{x_{n'}\}$ such that relation (2.5.11) holds with probability one, and denote by Ω_1 the corresponding set of full probability.

Next, denote by $\mathbb{H}_k$, $k = 1, 2, \ldots$, the linear span of $h_1, \ldots, h_k$. By the results of Sect. 2.2 we have that, for every k, t

$$\lim_{n\to\infty} \sum_{t^n_{j+1}\le t} \left\| (\mathbb{I} - \mathcal{P}_k)\big(M(t^n_{j+1}) - M(t^n_j)\big) \right\|^2 = \big\langle (\mathbb{I} - \mathcal{P}_k)(M) \big\rangle_t \tag{2.5.12}$$

in probability. Once again, we extract a subsequence along which the convergence in (2.5.12) is with probability one, denote by $\Omega_{k,t}$ the corresponding set of full probability, and define

$$\Omega_2 := \bigcap_{k=1}^{\infty} \bigcap_{t\in\mathcal{I}} \Omega_{k,t}.$$

By construction, $\mathbb{P}(\Omega_2) = 1$.

Finally, two more similar constructions: we select a subsequence along which convergence in (2.5.5) is with probability one, for both $j = 1$ and $j = 2$, and denote by Ω_3 the corresponding set of full probability; then, using

$$\lim_{k\to\infty} \mathbb{E}\big\langle (\mathbb{I} - \mathcal{P}_k)(M) \big\rangle_T = \lim_{k\to\infty} \mathbb{E}\left\| (\mathbb{I} - \mathcal{P}_k)\big(M(T)\big) \right\|^2 = 0,$$

we pass to a suitable subsequence to get the above convergence with probability one, and denote by Ω_4 the corresponding set of full probability.

Now define

$$\tilde{\Omega} = \Omega'' \bigcap \left(\bigcap_{i=1}^{4} \Omega_i \right),$$

where Ω'' is defined in Sect. 2.5. By construction, $\mathbb{P}(\tilde{\Omega}) = 1$, and the next lemma confirms that $\tilde{\Omega}$ is indeed the right set for the conclusion of Theorem 2.13.

Lemma 2.8 *If $\omega \in \tilde{\Omega}$, $t, s \in \mathcal{I}$, and $s < t < \tau(\omega)$, then*

$$\begin{aligned} \|\tilde{x}(t)\|^2 = \|x(0)\|^2 &+ 2\int_{[0,t]} [x(u), x'(u)]du \\ &+ 2\int_{[0,t]} \big(\tilde{x}(u), dM(u)\big) + \langle M\rangle_t \end{aligned} \tag{2.5.13}$$

and

$$\|\tilde{x}(t)-\tilde{x}(s)\|^2 = 2\int_{[s,t]} [x(u)-x(s), x'(u)]du + 2\int_{[s,t]} \big(\tilde{x}(u)-\tilde{x}(s), dM(u)\big) + \langle M\rangle_t - \langle M\rangle_s. \tag{2.5.14}$$

□

Proof For $\omega \in \tilde{\Omega}$ and $t = t_i^n < \tau(\omega)$ we show that

$$\|\tilde{x}(t)\|^2 = \|x(0)\|^2 + 2\int_{[0,t]} [x_n^2(u),\ x'(u)]du + 2\int_{[0,t]} \big(\tilde{x}_n(u), dM(u)\big) + \sum_{t_{j+1}^n \le t} \|M(t_{j+1}^n) - M(t_j^n)\|^2 - \sum_{t_{j+1}^n \le t} \big\|\tilde{x}(t_{j+1}^n) - \tilde{x}(t_j^n)) - \big(M(t_{j+1}^n) - M(t_j^n)\big)\big\|^2. \tag{2.5.15}$$

To prove this, we note that, for $j = 2, \ldots, k(n)$,

$$\|\tilde{x}(t_j^n)\|^2 - \|\tilde{x}(t_{j-1}^n)\|^2 = 2\int_{[t_{j-1}^n, t_j^n]} [x(t_j^n),\ x'(u)]du + 2\big(\tilde{x}(t_{j-1}^n), M(t_j^n) - M(t_{j-1}^n)\big) + \|M(t_j^n) - M(t_{j-1}^n)\|^2 - \big\|\big(\tilde{x}(t_j^n) - \tilde{x}(t_{j-1}^n)\big) - \big(M(t_j^n) - M(t_{j-1}^n)\big)\big\|^2. \tag{2.5.16}$$

Next, using (2.5.8) and $\tilde{M}(0,\omega) = x(0,\omega) = \tilde{x}(0,\omega)$ for all $\omega \in \Omega$, we get

$$\|\tilde{x}(t_1^n)\|^2 = 2\int_{[0,t_1^n]} [x(t_1^n), x'(u)]du + \|\tilde{M}(0)\|^2 + 2\big(\tilde{x}(0), \tilde{M}(t_1^n) - \tilde{M}(0)\big) + \|\tilde{M}(t_1^n) - \tilde{M}(0)\|^2 - \|(\tilde{x}(t_1^n) - \tilde{x}(0)) - (\tilde{M}(t_1^n) - \tilde{M}(0))\|^2. \tag{2.5.17}$$

Summing equality (2.5.16) over all j and adding the resulting equality to (2.5.17) we get (2.5.15).

Next, we show that the last term in the right-hand part of (2.5.15) tends to zero as $n \to \infty$. Define

$$I_n := \sum_{t^n_{j+1} \le t} \big\| \big(\tilde{x}(t^n_{j+1}) - \tilde{x}(t^n_j)\big) - \big(M(t^n_{j+1}) - M(t^n_j)\big) \big\|^2,$$

and then re-write I_n as follows:

$$I_n = \sum_{t^n_{j+1} \le t} \Big(\tilde{x}(t^n_{j+1}) - \tilde{x}(t^n_j) - M(t^n_{j+1}) + M(t^n_j),$$

$$\tilde{x}(t^n_{j+1}) - \tilde{x}(t^n_j) - \mathcal{P}_k\big(M(t^n_{j+1}) - M(t^n_j)\big)\Big)$$

$$- \sum_{t^n_{j+1} \le t} \Big(\tilde{x}(t^n_{j+1}) - \tilde{x}(t^n_j) - M(t^n_{j+1}) + M(t^n_j),$$

$$\big(\mathbb{I} - \mathcal{P}_k\big)\big(M(t^n_{j+1}) - M(t^n_j)\big)\Big) := I^{(1)}_n + I^{(2)}_n.$$

Since $\mathbb{X}$ is dense in $\mathbb{H}$ we can assume that each $h_i \in \mathbb{X}$ and then $\mathcal{P}_k\big(M(t^n_{j+1}) - M(t^n_j)\big) \in \mathbb{X}$ for all $k,\ j$, and n.

For brevity, write $\mathcal{P}_k\big(M(t)\big) := v_k(t)$. Starting from v_k, we construct the functions $v^j_{k,n}$, $j = 1, 2$, exactly the same way as x^j_n were constructed for x in Lemma 2.4. From (2.5.1) with

$$\eta = v(t^n_{j+1}) - v(t^n_j) - \mathcal{P}_k\big(M(t^n_{j+1}) - M(t^n_j)\big),$$

we get

$$I^{(1)}_n = \int_{[0,t]} [x^2_n(u) - x^1_n(u), x'(u)]du - \int_{[0,t]} [v^2_{k,n}(u) - v^1_{k,n}(u), x'(u)]du. \tag{2.5.18}$$

The first term in this equality tends to zero as $n \to \infty$, because, in view of the properties of a normal triple and the Cauchy–Schwarz inequality, we have

$$\int_{[0,t]} \Big| [x^2_n(u) - x^1_n(u), x'(u)] \Big| du$$

$$\le N \int_{[0,t]} \|x^2_n(u) - x^1_n(u)\|_{\mathbb{X}} \cdot \|x'(u)\|_{\mathbb{X}'} \, du$$

$$\le N \left(\int_{[0,T]} \|x^2_n(u) - x^1_n(u)\|^2_{\mathbb{X}} \, du \right)^{1/2} \left(\int_{[0,T]} \|x'(u)\|^2_{\mathbb{X}'} \, du \right)^{1/2},$$

where the last term tends to zero as $n \to \infty$ since $\omega \in \Omega_3$.

The function $\mathcal{P}_k(M)$, with values in $\mathbb{X}$, is strongly continuous in t. Hence $\|v_{k,n}^2(u) - v_{k,n}^1(u)\|_{\mathbb{X}} \to 0$, as $n \to \infty$, uniformly in $u \in [0, T]$. Therefore, making use of estimates similar to the ones given above, it is easy to show that the second term on the right-hand side of (2.5.18) also tends to zero. Hence

$$\lim_{n\to\infty} I_n^{(1)} = 0.$$

On the other hand,

$$I_n^{(2)} \le \left(\sum_{t_{n+1}^n \le t} \left\| \left(\tilde{x}(t_{j+1}^n) - \tilde{x}(t_j^n)\right) - \left(M(t_{j+1}^n) - M(t_j^n)\right) \right\|^2 \right)^{1/2}$$

$$\times \left(\sum_{t_{j+1}^n \le t} \|(\mathbb{I} - \mathcal{P}_k)\, (M(t_{j+1}^n) - M(t_j^n)\|^2 \right)^{1/2}.$$

Coming back to the definition of I_n, we conclude that

$$I_n - I_n^{(1)} \le \sqrt{I_n} \left(\sum_{t_{j+1}^n \le t} \|(\mathbb{I} - \mathcal{P}_k)\, (M(t_{j+1}^n) - M(t_j^n)\|^2 \right)^{1/2}.$$

Taking $\limsup_{n\to\infty}$ on both side of the above inequality, and writing $\bar{I} = \limsup_n I_n$, we conclude that

$$\sqrt{\bar{I}} \le \sqrt{\langle (\mathbb{I} - \mathcal{P}_k)(M) \rangle_t}.$$

Taking into account that $\langle (\mathbb{I} - \mathcal{P}_k)M \rangle_t \to 0$ as $k \to \infty$, we get from the last inequality that $\lim_{n\to\infty} I_n = 0$ for all $\omega \in \Omega$.

Finally, since $\omega \in \tilde{\Omega} \subset \Omega_3 \cap \Omega_1$, we have

$$\lim_{n\to\infty} \int_{[0,t]} [x_n^2(u), x'(u)]du = \int_{[0,t]} [x(u), x'(u)]du$$

and

$$\lim_{n\to\infty} \int_{[0,t]} \left(\tilde{x}_n(u), dM(u)\right) = \int_{[0,t]} \left(\tilde{x}(u), dM(u)\right).$$

Thus, passing to the limit in inequality (2.5.15) we get (2.5.13). Equality (2.5.14) can be derived from (2.5.13) by the application of the relations

$$(a-b)^2 = a^2 + b^2 - 2ab = a^2 - b^2 - 2b(a-b)$$

and the identity

$$-\Big(\tilde{x}(s),\ \big(\tilde{x}(t) - \tilde{x}(s)\big)\Big) = -2\int\limits_{[s,t]} [x(s), x'(u)]du - 2\int\limits_{[s,t]} \big(\tilde{x}(s), dM(u)\big).$$

This completes the proof of Lemma 2.8. □

2.5.9. To prove items (i)–(iii) of the theorem, it remains to verify that, for every $\omega \in \tilde{\Omega}$, the function $\tilde{x} = \tilde{x}(t,\omega)$ is strongly continuous in $\mathbb{H}$ with respect to t. In a Hilbert space, a weakly continuous function with a continuous norm is strongly continuous, and so it is sufficient to prove the validity of (2.5.13) for all $t < \tau(\omega)$, where $\omega \in \widetilde{\Omega}$. When $t = 0$, equality (2.5.13) is obvious. Let us fix $t > 0$ such that $t < \tau(\omega)$ for $\omega \in \widetilde{\Omega}$.

For every sufficiently large n we can find $j := j(n)$ such that $0 < t_j^n \le t < t_{j+1}^n$. Write $t(n) = t_{j(n)}^n$ and note that $t(n) \uparrow t$, $n \to \infty$.

Next,

$$\lim_{n\to\infty} \int\limits_{[t(n),t]} \Big|[x(u) - x(t(n)), x'(u)]\Big|\, du$$

$$\le \lim_{n\to\infty} \int\limits_{[0,T]} \|x(u) - x_n^1(u)\|_{\mathbb{X}} \cdot \|x'(u)\|_{\mathbb{X}'}\, du$$

$$\le \lim_{n\to\infty} \left(\int\limits_{[0,T]} \|x(u) - x_n^1(u)\|_{\mathbb{X}}^2\, du\right)^{1/2} \left(\int\limits_{[0,T]} \|x'(u)\|_{\mathbb{X}'}^2\, du\right)^{1/2} = 0,$$

$$\lim_{n\to\infty} \Big(\langle M\rangle_t - \langle M\rangle_{t(n)}\Big) = 0,$$

and

$$\lim_{n\to\infty} \left| \int\limits_{[t(n),s]} \Big(\tilde{x}(u) - \tilde{x}(t(n)), dM(u)\Big) \right|$$

$$= \lim_{n\to\infty} \sup_{s\le t} \left| \int\limits_{[0,s]} \Big(\tilde{x}(u) - \tilde{x}_n(u), dM(u)\Big)\right.$$

$$- \int_{[0,t(n)]} \Big(\tilde{x}(u) - \tilde{x}_n(u),\ dM(u)\Big)\Bigg|$$

$$\leq 2 \lim_{n\to\infty} \sup_{s\leq t} \Bigg| \int_{[0,s]} \Big(\tilde{x}(u) - \tilde{x}_n(u),\ dM(u)\Big) \Bigg| = 0.$$

Thus there exists a subsequence $n(k)$ such that, for $s(k) = t(n(k))$, we have

$$\sum_{k=1}^{\infty} \Bigg\{ \Bigg(\int_{[s(k),s(k+1)]} \Big| [x(u) - x(s(k)), x'(u)] \Big| du \Bigg)^{1/2} + \Big(\langle M \rangle_{s(k+1)} - \langle M \rangle_{s(k)} \Big)^{1/2} \Bigg\}$$

$$+ \Bigg(\int_{[s(k),s(k+1)]} \Big(\tilde{x}(u) - \tilde{x}(s(k)), dM(u) \Big) \Bigg)^{1/2} < \infty.$$

Making use of (2.5.14), we obtain

$$\sum_{k=1}^{\infty} \|\tilde{x}(s(k+1)) - \tilde{x}(s(k))\| < \infty.$$

Therefore $\tilde{x}(s(k))$ has a strong limit as $k \to \infty$. Since $s(k) \to t$, the sequence $\tilde{x}(s(k))$ converges in $\mathbb{H}$ weakly, hence strongly, to $\tilde{x}(t)$. Now, in (2.5.13), we replace t with the corresponding $s(k)$ and then pass as $k \to \infty$. The result is (2.5.13) for this t.

To prove (iv), note that, by (2.5.1) and (2.5.2), we obtain that $(\eta, \tilde{x}(t)) = (\eta, \tilde{x}(t))$, ($\mathbb{P}$-a.s.), on the set $\{\omega : t < \tau(\omega)\}$ for every $\eta \in \mathbb{X}$. From this, in view of the separability of $\mathbb{X}$, it follows that $\big(\eta, \tilde{x}(t)\big) = \big(\eta, x(t)\big)$ for all η simultaneously ($\mathbb{P}$-a.s. on the set $\{\omega : t < \tau(\omega)\}$). Since $\mathbb{X}$ is dense in $\mathbb{H}$, the statement of item (iv) of the theorem follows. □

Chapter 3
Linear Stochastic Evolution Systems in Hilbert Spaces

3.1 Introduction

3.1.1. Fix $T \in \mathbb{R}_+$ and consider a stochastic basis $\mathbb{F} = (\Omega, \mathscr{F}, \{\mathscr{F}_t\}_{t\in[0,T]}, \mathbb{P})$ with the usual assumptions. Let $(\mathbb{X}, \mathbb{H}, \mathbb{X}')$ be a normal triple of separable Hilbert spaces with canonical bi-linear functional $[\cdot,\cdot]$, and let $\mathbb{Y}$ be another separable Hilbert space. As before, $\|\cdot\|$ is the norm in $\mathbb{H}$ and $(\cdot,\cdot)$ is the inner product in $\mathbb{H}$; the subscripts, such as $\|\cdot\|_{\mathbb{X}}$, are used for all other Hilbert space.

Let $W = W(t)$, $t \in [0,T]$, be a Wiener process with values in $\mathbb{Y}$ and with covariance operator $Q \in \mathfrak{L}_1(\mathbb{Y})$; cf. Sect. 2.2.14. Denote by $\mathfrak{L}_Q(\mathbb{Y},\mathbb{H})$ the collection of linear, possibly unbounded, operators B from $\mathbb{Y}$ to $\mathbb{H}$ such that $BQ^{1/2} \in \mathfrak{L}_2(\mathbb{Y},\mathbb{H})$ and write

$$\|B\|_Q = \|\!|BQ^{1/2}|\!\|;$$

cf. Sect. 2.4.1.

The objective of this chapter is to study **linear stochastic evolution systems** (LSESs) of the form

$$u(t) = \varphi + \int_{[0,t]} \Big(Au(s) + f(s)\Big)\, ds + \int_{[0,t]} Bu(s)\, dW(s) + M(t), \qquad (3.1.1)$$

where

- $\varphi = \varphi(\omega)$ is an $\mathscr{F}_0$-measurable random variable with values in $\mathbb{H}$;
- $A = A(t,\omega)$ is a predictable process with values in $\mathfrak{L}(\mathbb{X},\mathbb{X}')$;

B. L. Rozovsky, S. V. Lototsky, *Stochastic Evolution Systems*, Probability Theory and Stochastic Modelling 89, https://doi.org/10.1007/978-3-319-94893-5_3

- $f = f(t, \omega)$ is a predictable process with values $\mathbb{X}'$ and such that

$$\mathbb{P}\left(\int_{[0,T]} \|f(t)\|_{\mathbb{X}'} dt < \infty\right) = 1;$$

- $B = B(t, \omega)$ is a predictable process with values in $\mathfrak{L}_Q(\mathbb{Y}, \mathbb{H})$;
- $M \in \mathfrak{M}_2^c([0, T], \mathbb{H})$.

The integrals in the equality (3.1.1) are understood in the sense of Sects. 1.3.12 and 2.4.2.

System (3.1.1) is called **coercive** if it satisfies the following condition:

(A) *There exists a real number K and a positive number $\delta > 0$ such that, for all $(t, \omega) \in [0, T] \times \Omega$ and all $x \in \mathbb{X}$,*

$$2[x, A(t, \omega)x] + \|B(t, \omega)x\|_Q^2 + \delta\|x\|_{\mathbb{X}}^2 \leq K\|x\|^2.$$

If $\delta = 0$, then the system is called **dissipative**; the corresponding condition becomes

(A′) *There exists a real number $K \in \mathbb{R}$ such that, for all $(t, \omega) \in [0, T] \times \Omega$ and all $x \in \mathbb{X}$,*

$$2[x, A(t, \omega)x] + \|B(t, \omega)x\|_Q^2 \leq K\|x\|^2.$$

The coercive and dissipative systems are the subject of this chapter. In general, an operator $U : \mathbb{X} \to \mathbb{X}^*$ is called coercive if $xUx \geq \delta\|x\|_{\mathbb{X}}^2$, $x \in \mathbb{X}$, and dissipative if $xUx \geq 0$. Thus, if $\mathbb{Y} = \mathbb{R}^1$, $B = 0$, $S := K\mathbb{I} - 2A$, $\mathbb{X}^* := \mathbb{X}'$, and $[\cdot, \cdot]$ is the duality relation between $\mathbb{X}$ and $\mathbb{X}^*$, then condition (A) is equivalent to coercivity of the operator S and condition (A') is equivalent to dissipativity of S.

3.1.2. The structure of the chapter is as follows:

In Sect. 3.2 we consider coercive LSESs. For such systems we prove the existence and uniqueness of a solution, as well as additional results about regularity of the solution.

In Sect. 3.3 we extend the construction of a normal triple to a Hilbert scale and use the result to study dissipative LSESs.

In Sect. 3.4 we consider weaker conditions for the uniqueness of a solution of LSES (3.1.1), establish a Markov property of the solution, and investigate methods of approximating the solution of this system.

In Sect. 3.5 we apply the results of the previous section to the Itô stochastic partial differential equation of arbitrary finite order. For this equation we consider the first boundary problem. In the same section we discuss briefly a variety of scales of Sobolev spaces which are important examples of Hilbert scales

3.2 Coercive Systems

3.2.1. In this section we study LSES (3.1.1) when the operators A and B satisfy the coercivity condition (A). We prove the existence and uniqueness theorem, and also discuss conditions which ensure that the solution of (3.1.1) has better regularity than what is guaranteed by the existence theorem.

3.2.2. In what follows, we assume that the family of the operators $A = A(t, \omega)$ is uniformly bounded in (t, ω), that is,

(B) *There exists a positive real number K such that for all $(t, \omega) \in [0, T] \times \Omega$ and $x \in \mathbb{X}$,*

$$\|A(t, \omega)x\|_{\mathbb{X}'} \leq K \|x\|_{\mathbb{X}}.$$

Note that conditions (A) and (B), together with property (i) of the CBF from Sect. 2.5.2, imply that family of operator $B = B(t, \omega)$ is also uniformly continuous:

$$\|B(t, \omega)x\|_{Q} \leq K \|x\|_{\mathbb{X}} \tag{3.2.1}$$

for all $(t, \omega) \in [0, T] \times \Omega$ and $x \in \mathbb{X}$.

3.2.3. Given a Hilbert space $\mathbb{V}$,

- $\mathbb{L}_2([0, T], \mathscr{P}; \mathbb{V})$ is the set of predictable representatives of the classes of functions from $\mathbb{L}_2([0, T] \times \Omega, \mathscr{B}([0, T]) \otimes \mathscr{F}, \mathfrak{l} \times \mathbb{P}; \mathbb{V})$; cf. Sect. 1.3.14;
- $\mathbb{L}_2^{\omega}([0, T], \mathscr{P}; \mathbb{V})$ is set of predictable processes $x = x(t, \omega)$ with values in $\mathbb{V}$ and such that

$$\mathbb{P}\left\{\omega : \int_{[0,T]} \|x(t, \omega)\|_{\mathbb{V}}^2 \, dt < \infty\right\} = 1;$$

- $\mathbb{C}([0, T], \mathscr{P}; \mathbb{V})$ is the set of predictable strongly continuous $\mathbb{V}$-processes.

Definition 3.1 A **solution of LSES** (3.1.1) is a stochastic process $u = u(t)$ such that $u \in \mathbb{L}_2^{\omega}([0, T]), \mathscr{P}; \mathbb{X})$ and, for all y from a dense (in the strong topology) subset of $\mathbb{X}$, the equality

$$\begin{aligned}\big(y, u(t)\big) = (y, \varphi) + \int_{[0,t]} \big[y, Au(s) + f(s)\big] \, ds \\ + \Big(y, \int_{[0,t]} Bu(s) \, dW(s) + M(t)\Big)\end{aligned} \tag{3.2.2}$$

holds for $\mathfrak{l} \times \mathbb{P}$- a.a. $(t, \omega) \in [0, T] \times \Omega$. □

If conditions (A) and (B) are fulfilled, then (3.2.1) holds and all the integrals in (3.2.2) are well-defined.

Remark 3.1 If $\mathbb{X}' = \mathbb{X}^*$ and $\mathbb{H}$ is a Hilbert space rigged by the pair $\mathbb{X}$, $\mathbb{X}^*$ (see Sect. 2.5.1), then, according to the definition, the solution of the LSES (3.1.1) is a function $u \in \mathbb{L}_2^{\omega}([0,T]); \mathscr{P}, \mathbb{X})$ such that (3.1.1) holds in $\mathbb{X}^*$ for $\mathfrak{l} \times \mathbb{P}$- a.a. $(t,\omega) \in [0,T]$. □

3.2.4.

Definition 3.2 We say that a solution u of LSES (3.1.1) has a continuous version in $\mathbb{H}$ if there exists a function $\tilde{u} \in \mathbb{C}\big([0,T]; \mathscr{P}; \mathbb{H}\big)$ such that

(i) $\tilde{u}(t,\omega) = u(t,\omega)$, $\mathfrak{l} \times \mathbb{P}-$a.s.;
(ii) there exists a set $\Omega' \subset \Omega$ such that $\mathbb{P}(\Omega') = 1$ and, for all $(t,\omega) \in [0,T] \times \Omega'$, $y \in \mathbb{X}$,

$$\begin{aligned}\big(y, \tilde{u}(t,\omega)\big) &= \big(y, \varphi(\omega)\big) + \int\limits_{[0,t]} \big[y, A\tilde{u}(s,\omega) + f(s)\big]\, ds \\ &\quad + \Bigg(y, \int\limits_{[0,t]} B\tilde{u}(s,\omega)\, dW(s) + M(t,\omega)\Bigg).\end{aligned} \tag{3.2.3}$$

□

Remark 3.2 From Proposition 1.2 it follows that $1_{\{\tilde{u}\in\mathbb{X}\}}$ is a predictable (real) stochastic process. Thus, the integrals of $A\tilde{u}$ and $B\tilde{u}$ in (3.2.3), considered as integrals of the functions $1_{\{\tilde{u}\in\mathbb{X}\}}\, A\tilde{u}$ and $1_{\{\tilde{u}\in\mathbb{X}\}}\, B\tilde{u}$, are well-defined. □

An immediate consequence of Theorem 2.13 is

Proposition 3.1 *If it exists, a solution of LSES (3.1.1) has a continuous version in* $\mathbb{H}$. □

Warning 3.1 *In what follows, a solution of an LSES will be identified with the corresponding continuous version. Also, as long as there is no danger of confusion, there will be no explicit reference to the underlying normal triple of Hilbert spaces.* □

3.2.5. The main result of this section is the following theorem.

Theorem 3.1 *Suppose that conditions* (A) *and* (B) *are fulfilled, and also*

(C) $\quad \mathbb{E}\|\varphi\|^2 < \infty, \qquad \mathbb{E}\int\limits_{[0,t]} \|f(t)\|_{\mathbb{X}'}^2\, dt < \infty.$

Then LSES (3.1.1) has a unique solution $u \in \mathbb{L}_2([0, T], \mathscr{P}; \mathbb{X})$, and

$$\mathbb{E} \sup_{t \le T} \|u(t)\|^2 + \mathbb{E} \int_{[0,T]} \|u(t)\|_{\mathbb{X}}^2 \, dt \\ \le C\, \mathbb{E}\Big(\|\phi\|^2 + \int_{[0,T]} \|f(t)\|_{\mathbb{X}'}^2 \, dt + \langle M \rangle_T \Big), \tag{3.2.4}$$

where the positive number C depends only on K, T, and δ. □

Proof We start by proving uniqueness. Let u_1 and u_2 be two solutions of (3.1.1). Define $U = u_1 - u_2$. By linearity, U is a solution of (3.1.1) with $\varphi \equiv f \equiv M \equiv 0$.

Define the set $\Gamma_n(\omega) := \{t : t \le T, \ \|U(t, \omega)\| > n\}$, $n \in \mathbb{N}$, and the stopping times

$$\tau_n(\omega) := \begin{cases} \inf_t \Gamma_n(\omega), & \text{if } \Gamma_n(\omega) \ne \emptyset; \\ T, & \text{if } \Gamma_n(\omega) = \emptyset. \end{cases}$$

By Theorem 2.13, for all $t \in [0, T]$, we have

$$\|U(t \wedge \tau_n)\|^2 = \int_{[0,t\wedge\tau_n]} \Big(2[U(s), AU(s)] + \|BU(s)\|_Q^2 \Big) \, ds \\ + 2 \int_{[0,t\wedge\tau_n]} \Big(U(s), d\big((BU) \circ W(s)\big) \Big) \tag{3.2.5}$$

with probability one.

The stochastic integral on the right-hand part of equality (3.2.5) belongs to $\mathfrak{M}_{loc}^c([0, T]; \mathbb{R}^1)$, and its quadratic variation is

$$\int_{[0,t\wedge\tau_n]} \|U(s)\|^2 \cdot \|BU(s)\|_Q^2 \, ds.$$

In view of (3.2.1),

$$\int_{[0,t\wedge\tau_n]} \|U(s)\|^2 \cdot \|BU(s)\|_Q^2 \, ds \le K^2 \int_{[0,t\wedge\tau_n]} \|U(s)\|^4 \, ds \le K^2 n^4 T.$$

By Proposition 2.1(iii), the stochastic integral considered above is a martingale and its expectation is equal to zero. Hence, taking expectations of both parts of equality (3.2.5) and using condition (A), we find

$$\mathbb{E}\|U(t \wedge \tau_n)\|^2 \le K \int_{[0,t]} \mathbb{E}\|U(s \wedge \tau_n)\|^2 \, ds.$$

The Gronwall–Bellman lemma then implies $\mathbb{E}\|U(t \wedge \tau_n)\|^2 = 0$ for all t and n. Since $U \in \mathbb{C}([0, T], \mathscr{P}; \mathbb{H})$, we have $\tau_n \uparrow T$ ($\mathbb{P}$- a.s.), so that, by the Fatou lemma,

$$\mathbb{P}\Big(\sup_{t \le T} \|u_1(t) - u_2(t)\| > 0 \Big) = 0,$$

proving uniqueness of solution.

Next, we prove existence of solution by Galerkin's method.

Let us fix a CONS $\{h_i\}$ in $\mathbb{H}$ and another CONS $\{y_i\}$ in $\mathbb{Y}$. Since $\mathbb{X}$ is dense in $\mathbb{H}$ we will assume that $h_i \in \mathbb{X}$.

For every $n \in \mathbb{N}$, consider the following system of the Itô equations in $\mathbb{R}^n$:

$$\begin{aligned} u_n^i(t) &= (h_i, \varphi) + \int_{[0,t]} \left[h_i,\ A(s) \sum_{j=1}^n h_j u_n^j(s) + f(s) \right] ds \\ &+ \sum_{k=1}^n \int_{[0,t]} \Big(h_i, \Big(B(s) \sum_{j=1}^n h_j u_n^j(s)\Big) y_k\Big) dw^k(s) + M_i(t),\ i = 1, \ldots, n, \end{aligned} \tag{3.2.6}$$

where $w^k(s) := (w(s), y_k)_{\mathbb{Y}},\ M_i(t) := (M(t), h_i)$.

By condition (B), for all $i \le n$, $\mathbb{E}(h_i, \varphi)^2 < \infty$, $\mathbb{E} \int_{[0,t]} \Big| [h_i, f(s)] \Big|^2 ds < \infty$.

Thus system (3.2.6) has a unique solution that is continuous in t, and $\mathbb{E} \sup_{t \le T} \sum_{i=1}^n |u_n^i(t)|^2 < \infty$; see Sect. 1.5.4. □

3.2.6. We now derive more precise estimates for the solution of (3.2.6). Define

$$u_n(t) := \sum_{i=1}^n u_n^i(t) h^i.$$

Lemma 3.1 *There exists a constant N, independent of n, such that*

$$\begin{aligned} &\sup_{t \le T} \mathbb{E}\|u_n(t)\|^2 + \mathbb{E} \int_{[0,T]} \|u_n(t)\|_{\mathbb{X}}^2 \, dt \\ &\le N\, \mathbb{E} \left(\|\varphi\|^2 + \int_{[0,T]} \|f(t)\|_{\mathbb{X}'}^2 \, dt + \langle M \rangle_T \right). \end{aligned} \tag{3.2.7}$$

□

Proof Denote by $\mathcal{P}_n$ the orthogonal projection on the linear span of $\{h_1, \ldots, h_n\}$, and denote by $\tilde{\mathcal{P}}_n$ the orthogonal projection on the linear span of $\{y_1, \ldots, y_n\}$.

Next, define

$$L_n(t) := \int_{[0,t]} \mathcal{P}_n(Bu_n(s))\, d\tilde{\mathcal{P}}_n(W(s))$$

and

$$\breve{M}_n(t) := L_n(t) + \mathcal{P}_n(M(t)).$$

By (2.5.3), for all $t \in [0, T]$ and ω from the same set of full probability, we have

$$\|u_n(t)\|^2 = \sum_{i=1}^{n} |u_n^i(t)|^2 = \|\mathcal{P}_n(\varphi)\|^2 + 2\int_{[0,t]} [u_n(s), Au_n(s) + f(s)]\, ds + 2\int_{[0,t]} \big(u_n(s), d\breve{M}_n(s)\big) + \langle \breve{M}_n \rangle_t. \tag{3.2.8}$$

Also, by definition,

$$\langle L_n \rangle_t = \int_{[0,t]} \|\mathcal{P}_n\big(Bu_n(s)\big)\tilde{\mathcal{P}}_n\|_Q^2\, ds.$$

Theorem 1.8 and inequality (3.2.1), together with $|ab| \le a^2 + b^2$, imply

$$\begin{aligned}
&\mathbb{E}\left(\int_{[0,T]} \|u_n(s)\|^2\, d\langle \breve{M}_n \rangle_s\right)^{1/2} \\
&\le N\, \mathbb{E} \sup_{t \le T} \|u_n(s)\| \left(\int_{[0,T]} \|\mathcal{P}_n Bu_n(s)\tilde{\mathcal{P}}_n\|_Q^2\, ds + \langle \mathcal{P}_n M \rangle_T\right)^{1/2} \\
&\le N\, \mathbb{E}\left(\sup_{t \le T} \|u_n(t)\|^2 + \int_{[0,T]} \|Bu_n(s)\|_Q^2\, ds + \langle M \rangle_T\right) < \infty.
\end{aligned}$$

Thus the stochastic integral in (3.2.8) is a martingale (see Sect. 2.3.4). Making use of this fact, we obtain from (3.2.8) that

$$\mathbb{E}\|u_n(t)\|^2 = \mathbb{E}\|\mathcal{P}_n(\varphi)\|^2 + \mathbb{E}\int\limits_{[0,t]} \Big(2[u_n(s), Au_n(s) + f(s)] \\ + \|\mathcal{P}_n Bu_n(s)\tilde{\mathcal{P}}_n\|_Q^2\Big)\, ds + 2\mathbb{E}\langle L_n, \mathcal{P}_n(M)\rangle_t + \mathbb{E}\langle \mathcal{P}_n(M)\rangle_t. \tag{3.2.9}$$

Using property (i) of a CBF (see Sect. 2.5.2) and the elementary inequality

$$2ab \leq \varepsilon a^2 + \varepsilon^{-1}\, b^2, \quad \varepsilon > 0, \tag{3.2.10}$$

we get

$$2\mathbb{E}[u_n(s), f(s)] \leq \varepsilon\mathbb{E}\|u_n(s)\|_{\mathbb{X}}^2 + \varepsilon^{-1}\mathbb{E}\|f(s)\|_{\mathbb{X}'}^2, \tag{3.2.11}$$

and

$$2\mathbb{E}\langle L_n, \mathcal{P}_n(M)\rangle_t = 2\mathbb{E}\Big(L_n(t), \mathcal{P}_n(M(t))\Big) \\ \leq \varepsilon \int\limits_{[0,t]} \|\mathcal{P}_n Bu_n(s)\tilde{\mathcal{P}}_n\|_Q^2\, ds + \varepsilon^{-1}\mathbb{E}\|\mathcal{P}_n(M(t))\|^2. \tag{3.2.12}$$

Because the norm of a projection operator does not exceed 1, and in view of the coercivity property (A), we find from (3.2.9), (3.2.11), and (3.2.12) that

$$\mathbb{E}\|u_n(t)\|^2 + \delta\mathbb{E}\int\limits_{[0,t]} \|u_n(s)\|_{\mathbb{X}}^2\, ds \leq \mathbb{E}\|\varphi\|^2 + \varepsilon\mathbb{E}\int\limits_{[0,t]} \|u_n(s)\|_{\mathbb{X}}^2\, ds \\ + \varepsilon\mathbb{E}\int\limits_{[0,t]} \|Bu_n(s)\|_Q^2\, ds + \varepsilon^{-1}\mathbb{E}\int\limits_{[0,t]} \|f(s)\|_{\mathbb{X}'}^2\, ds \\ + (1+\varepsilon^{-1})\mathbb{E}\|M(t)\|^2. \tag{3.2.13}$$

From (3.2.1) we obtain

$$\varepsilon\mathbb{E}\int\limits_{[0,t]} \|B(s)u_n(s)\|_Q^2\, ds \leq \varepsilon\, K\, \mathbb{E}\int\limits_{[0,t]} \|u_n(s)\|^2\, ds. \tag{3.2.14}$$

Combining (3.2.13), (3.2.14) and taking ε sufficiently small, we find that, for some $\delta_1 > 0$ and all $t \in [0, T]$,

$$\begin{aligned} &\mathbb{E}\|u_n(t)\|^2 + \delta_1 \mathbb{E} \int_{[0,t]} \|u_n(s)\|_{\mathbb{X}}^2 \, ds \\ &\leq N\, \mathbb{E}\Big(\|\varphi\|^2 + \int_{[0,t]} \|f(s)\|_{\mathbb{X}'}^2 \, ds + \|M(t)\|^2 + \int_{[0,t]} \|u_n(s)\|^2 \, ds \Big). \end{aligned} \tag{3.2.15}$$

From this, by the Gronwall–Bellman lemma, we see that

$$\sup_{t \leq T} \mathbb{E}\|u_n(t)\|^2 \leq N\mathbb{E}\Big(\|\varphi\|^2 + \int_{[0,T]} \|f(t)\|_{\mathbb{X}'}^2 \, dt + \langle M \rangle_t \Big), \tag{3.2.16}$$

which along with (3.2.15) completes the proof of the lemma. □

3.2.7. By construction, for every $n \in \mathbb{N}$, the function $u_n = u_n(t)$ is predictable with values in $\mathbb{X}$. Moreover, from Lemma 3.1, it follows that some subsequence of u_n converges weakly in the space $\mathbb{L}_2\big([0, T] \times \Omega, \overline{\mathscr{P}}_T, \ell \times \mathbb{P}; \mathbb{X}\big)$ (see the notation of Sect. 1.4.7) to some function u. To simplify the presentation, we identify this subsequence with the original sequence. From Proposition 1.1 it follows that the limit u has a $\mathscr{P}$-measurable version, which will also be denoted by u. Hence $u \in \mathbb{L}_2([0, T], \mathscr{P}; \mathbb{X})$.

Let η be an arbitrary bounded random variable on $(\Omega, \mathscr{F})$ and ψ be an arbitrary bounded Lebesgue-measurable function on $[0, T]$.

By (1.6), for $n \in \mathbb{N}$ and $h_i \in \{h_j\}$ with $i \leq n$, we have

$$\begin{aligned} \mathbb{E} \int_{[0,T]} \eta\psi(t) \big(h_i, u_n(t)\big) \, dt &= \mathbb{E} \int_{[0,T]} \eta\psi(t) \Big\{ (h_i, \varphi) + \int_{[0,t]} [h_i, Au_n(s) + f(s)] \, ds \\ &\quad + \int_{[0,t]} \Big(h_i, Bu_n(s)\tilde{\mathcal{P}}_n dW(s)\Big) + M_i(t) \Big\} dt. \end{aligned}$$

As $n \to \infty$,

$$\mathbb{E} \int_{[0,T]} \eta\psi(t)\big(h_i, u_n(t)\big) \, dt \to \mathbb{E} \int_{[0,T]} \eta\psi(t)\big(h_i, u(t)\big) \, dt. \tag{3.2.17}$$

Next, in view of condition (B) and Lemma 3.1,

$$\mathbb{E}\left|\eta \int\limits_{[0,t]} [h_i, Au_n(s)]\, ds\right| < N < \infty, \tag{3.2.18}$$

where N does not depend on n.

It is also clear that, for every $t \in [0, T]$,

$$\lim_{n\to\infty} \mathbb{E} \int\limits_{[0,t]} \eta[h_i, Au_n(s)]\, ds = \mathbb{E} \int\limits_{[0,t]} \eta[h_i, Au(s)]\, ds. \tag{3.2.19}$$

Using (3.2.18) and (3.2.19) together with Fubini's theorem and the dominated convergence theorem,

$$\begin{aligned}\mathbb{E} \int\limits_{[0,T]} \eta\psi(t) \int\limits_{[0,t]} [h_i, Au_n(s)]\, ds\, dt &= \int\limits_{[0,T]} \psi(t)\mathbb{E} \int\limits_{[0,t]} \eta[h_i, Au_n(s)]\, ds\, dt \\ &\longrightarrow \int\limits_{[0,T]} \psi(t)\mathbb{E} \int\limits_{[0,t]} \eta[h_i, Au(s)]\, ds\, dt,\ n \to \infty.\end{aligned} \tag{3.2.20}$$

From (3.2.1) and Lemma 3.1 it follows that

$$\mathbb{E}\left|\eta \int\limits_{[0,t]} \left(h_i, \mathcal{P}_n Bu_n(s)\tilde{\mathcal{P}}_n dW(s)\right)\right| < N < \infty, \tag{3.2.21}$$

where N does not depend on n.

By Theorem 2.9 we find that for every $t \in [0, T]$

$$\mathbb{E}\eta \int\limits_{[0,t]} (h_i, \mathcal{P}_n Bu_n(s)\tilde{\mathcal{P}}_n dW(s)) \to \mathbb{E}_n \int\limits_{[0,t]} \big(h_i, Bu(s) dW(s)\big). \tag{3.2.22}$$

Therefore,

$$\begin{aligned}&\lim_{n\to\infty} \mathbb{E} \int\limits_{[0,T]} \eta\psi(t) \int\limits_{[0,t]} \left(h_i, Bu_n(s)\tilde{\mathcal{P}}_n dW(s)\right) dt \\ &= \mathbb{E} \int\limits_{[0,T]} \eta\psi(t) \int\limits_{[0,t]} \big(h_i, Bu(s) dW(s)\big)\, dt.\end{aligned} \tag{3.2.23}$$

It is also clear that

$$\lim_{n\to\infty} \mathbb{E} \int\limits_{[0,T]} \eta\psi(t)\big(h_i, \mathcal{P}_n M(t)\big)\, dt = \mathbb{E} \int\limits_{[0,T]} \eta\psi(t)\big(h_i, M(t)\big)\, dt. \tag{3.2.24}$$

Combining (3.2.17), (3.2.20), (3.2.23), and (3.2.24) we obtain that, $\mathfrak{l} \times \mathbb{P}$-a.s.,

$$\begin{aligned}\big(h_i, u(t)\big) = (h_i, \varphi) + \int\limits_{[0,t]} [h_i, Au(s) + f(s)]\, ds \\ + \int\limits_{[0,t]} \big(h_i, Bu(s) dW(s)\big) + \big(h_i, M(t)\big).\end{aligned}$$

This proves the existence of the solution for system (3.1.1). In view of Proposition 3.1, we will consider this solution as a continuous function of t with values in $\mathbb{H}$.

To complete the proof of the theorem it only remains to establish (3.2.4).

Define

$$L(t) := 2 \int\limits_{[0,t]} Bu(s) dW(s), \quad \breve{M}(t) := L(t) + M(t).$$

By Theorem 2.13, there exists a set $\Omega' \subset \Omega$ with $\mathbb{P}(\Omega') = 1$ such that, for all $(t, \omega) \in [0, T] \times \Omega'$, we have

$$\begin{aligned}\|u(t)\|^2 = \|\varphi\|^2 + 2 \int\limits_{[0,t]} [u(s), Au(s) + f(s)]\, ds \\ + 2 \int\limits_{[0,t]} \big(u(s), \breve{M}(s)\big) + \langle \breve{M} \rangle_t.\end{aligned} \tag{3.2.25}$$

For $n \in \mathbb{N}$, define the set

$$\Gamma_n(\omega) := \{t : t \le T, \ \|u(t, \omega)\|^2 \ge n\}$$

and the stopping time

$$\tau_n(\omega) := \begin{cases} \inf\limits_t \Gamma_n(w), & \text{if } \Gamma_n(\omega) \ne \emptyset, \\ T, & \text{if } \Gamma_n(\omega) = \emptyset. \end{cases}$$

Then

$$\mathbb{E}\int_{[0,t\wedge\tau_n]}\|u(s)\|^2\,d\langle\breve{M}\rangle_s \le N\,\mathbb{E}\int_{[0,t\wedge\tau_n]}\|u(s)\|^2\cdot\|Bu(s)\|_Q^2\,ds$$

$$+N\,\mathbb{E}\int_{[0,t\wedge\tau_n]}\|u(s)\|^2\,d\langle M\rangle_s<\infty,$$

and, by Proposition 2.1,

$$\mathbb{E}\int_{[0,t\wedge\tau_n]}\big(u(s),d\breve{M}(s)\big)=0.$$

It now follows from (3.2.25) that

$$\mathbb{E}\|u(t\wedge\tau_n)\|^2=\mathbb{E}\|\varphi\|^2+2\mathbb{E}\int_{[0,t\wedge\tau_n]}[u(s),Au(s)+f(s)]ds$$

$$+\,\mathbb{E}\langle\breve{M}\rangle_{t\wedge\tau_n}.$$

From the last equality, exactly in the same way as in Lemma 3.1, we derive that

$$\begin{aligned}\sup_{t\le T}\mathbb{E}\|u(t\wedge\tau_n)\|^2+\mathbb{E}\int_{[0,T\wedge\tau_n]}\|u(s)\|_{\mathbb{X}}^2\,ds\\ \le N\,\mathbb{E}\Big(\|\varphi\|^2+\int_{[0,T]}\|f(s)\|_{\mathbb{X}'}^2\,ds+\langle M\rangle_T\Big).\end{aligned}\tag{3.2.26}$$

Since $\|u\|$ is a continuous function of t we have $\tau_n\uparrow T$ as $n\to\infty$. Passing to the limit in (3.2.26), with Fatou's lemma and the monotone convergence theorem in mind, we get

$$\begin{aligned}\sup_{t\le T}\mathbb{E}\|u(t)\|^2+\mathbb{E}\int_{[0,T]}\|u(s)\|_{\mathbb{X}}^2\,ds\\ \le N\,\mathbb{E}\Big(\|\varphi\|^2+\int_{[0,T]}\|f(s)\|_{\mathbb{X}'}^2\,ds+\langle M\rangle_T\Big).\end{aligned}\tag{3.2.27}$$

3.2.8. It remains to establish a version of (3.2.27) with $\mathbb{E}\sup_{t\le T}\|u(t)\|^2$ instead of $\sup_{t\le T}\mathbb{E}\|u(t)\|^2$.

By definition of $\breve{M}$,

$$\langle \breve{M} \rangle_t = \langle M \rangle_t + \int\limits_{[0,t]} \|Bu(s)\|_Q^2 \, ds + 2\Big\langle M, \int\limits_{[0,\cdot]} Bu(s)dW(s)\Big\rangle_t. \tag{3.2.28}$$

From (3.2.25), in view of condition (A), it follows that

$$\begin{aligned}\mathbb{E} \sup_{t \le T} \|u(t \wedge \tau_n)\|^2 &\le 2\mathbb{E} \int\limits_{[0,T]} \Big| [u(s), f(s)] + K\|u(s)\|^2 \Big| \, ds \\ &+ 2\mathbb{E} \sup_{t \le T} \left| \int\limits_{[0,t\wedge\tau_n]} \big(u(s), d\breve{M}(s)\big) \right| \\ &+ 2\mathbb{E}\left(M(T), \int\limits_{[0,T]} Bu(s)dW(s) \right) + \mathbb{E}\langle M \rangle_T.\end{aligned} \tag{3.2.29}$$

By the Burkholder–Davis inequality (Sect. 2.2.8) and (3.2.10),

$$\begin{aligned}\mathbb{E} \sup_{t \le T} \left| \int\limits_{[0,t\wedge\tau_n]} \big(u(s), d\breve{M}(s)\big) \right| &\le 3\mathbb{E}\Big(\int\limits_{[0,t\wedge\tau_n]} \|u(s)\|^2 d\langle \breve{M} \rangle_s \Big)^{1/2} \\ &\le 3\mathbb{E}\Big(\sup_{t \le T} \|u(t \wedge \tau_n)\| \langle \breve{M} \rangle_T^{1/2} \Big) \\ &\le 3\varepsilon \, \mathbb{E} \sup_{t \le T} \|u(t \wedge \tau_n)\|^2 + \frac{3}{\varepsilon} \mathbb{E}\langle \breve{M} \rangle_T, \quad \varepsilon > 0.\end{aligned} \tag{3.2.30}$$

From (3.2.1) it follows that

$$\mathbb{E}\Big(M(t), \int\limits_{[0,T]} Bu(s)dW(s) \Big) \le \frac{1}{2}\Big(\mathbb{E}\langle M \rangle_T + K\mathbb{E} \int\limits_{[0,T]} \|u(s)\|_{\mathbb{X}}^2 \, ds \Big). \tag{3.2.31}$$

By the Kunita–Watanabe inequality and (3.2.1),

$$\mathbb{E}\langle \breve{M} \rangle_T \le \Big(\mathbb{E}\langle M \rangle_T + \mathbb{E} \int\limits_{[0,T]} \|u(s)\|_{\mathbb{X}}^2 \, ds \Big). \tag{3.2.32}$$

Next, combine (3.2.29)–(3.2.32), and then take ε small enough to get

$$\mathbb{E}\sup_{t\le T}\|u(t\wedge\tau_n)\|^2 \le N\,\mathbb{E}\Big(\langle M\rangle_T + \int_{[0,T]}\|u(s)\|^2_{\mathbb{X}}\,ds\Big).$$

Passing to the limit $n\to\infty$ in the above inequality and combining the result with (3.2.27) yields (3.2.4) and completes the proof of the theorem. □

Analysis of the proof shows that the following result is true.

Proposition 3.2 *If conditions* (A′), (B), *and* (C) *hold and if (3.1.1) has a solution that is an element of* $\mathbb{L}_2([0,T]\times\Omega;\mathscr{P};\mathbb{X})$, *then there exists a number* N, *which depends only on* K *and* T, *such that*

$$\mathbb{E}\sup_{t\le T}\|u(t)\|^2 \le N\,\mathbb{E}\Big(\|\varphi\|^2 + \int_{[0,T]}\|f(t)\|^2_{\mathbb{X}'}\,dt + \langle M\rangle_T\Big).$$ □

3.2.9. Under some additional conditions, the solution of (3.1.1) has better regularity than what is guaranteed by the existence theorem, and the objective of this paragraph is to investigate those conditions. More specifically, we want the solution to take values in a space that is normally embedded into $\mathbb{X}$.

Let $(\mathbb{V},\mathbb{U},\mathbb{V}')$ be another normal triple with CBF $[\cdot,\cdot]_{\mathbb{U}}$, and such that $\mathbb{V}\subset\mathbb{X}$, $\mathbb{U}\subset\mathbb{H}$, and $\mathbb{V}'\subset\mathbb{X}'$, with all the embeddings normal; cf. Sect. 2.5.1.

Recall the following classical result from functional analysis (see e.g. [64, Chapter IV, §1.10]).

Proposition 3.3 *Then there exists a unique positive, self-adjoint operator* Λ *on* $\mathbb{H}$ *with domain* $\mathbb{U}$ *such that,*

- *For every* $h\in\mathbb{U}$, $\|h\|_{\mathbb{U}}=\|\Lambda h\|_{\mathbb{H}}$;
- *For every* $\alpha>0$, *the domain of the operator* $\Lambda^{1+\alpha}$ *is dense in* $\mathbb{U}$ *(in the topology generated by* $\|\cdot\|_{\mathbb{U}}$). □

With the operator Λ as above, let $\mathbb{Z}$ be the collection of all $x\in\mathbb{V}$ such that x is in the domain of Λ^2 and $\Lambda^2 x\in\mathbb{X}$:

$$\mathbb{Z}=\big\{x\in\mathbb{V}: x\in \mathrm{Dom}(\Lambda),\ \Lambda^2 x\in\mathbb{X}\big\},$$

and assume that $\mathbb{Z}\neq\emptyset$.

Lemma 3.2 *If* $x\in\mathbb{Z}$, *and* $y\in\mathbb{V}'$, *then* $[x,y]_{\mathbb{U}}=[\Lambda^2 x,y]$. □

Proof Let $\{y_n\}$, $n\in\mathbb{N}$, be a sequence of elements of $\mathbb{V}$ such that $\lim_{n\to\infty}\|y_n-y\|_{\mathbb{V}'}=0$. Then, by above the proposition,

$$[x,y]_{\mathbb{U}}=\lim_{n\to\infty}(x,y_n)_{\mathbb{U}}=\lim_{n\to\infty}(\Lambda^2 x,y_n)=[\Lambda^2 x,y].$$ □

In addition to the assumptions made in the introduction and in Sect. 3.2, we also assume that $M \in \mathfrak{M}_2^c([0,T],\mathbb{U})$ and, for all $(t,\omega) \in [0,T] \times \Omega$,

$$A(t,\omega) \in \mathfrak{L}(\mathbb{V},\mathbb{V}'),\ B(t,\omega) \in \mathfrak{L}\big(\mathbb{V},\mathfrak{L}_Q(\mathbb{Y},\mathbb{U})\big),\ \varphi(\omega) \in \mathbb{U}.$$

The norm in $\mathfrak{L}_Q(\mathbb{Y},\mathbb{U})$ will be denoted by $\|\cdot\|_{Q,\mathbb{U}}$.

Theorem 3.2 *Suppose that $\Lambda^2(\mathbb{Z})$ is dense in $\mathbb{X}$ (with respect to the topology generated by $\|\cdot\|_{\mathbb{X}}$), and, there exist a real numbers K' and a positive number δ' such that, for all $(t,\omega) \in [0,T] \times \Omega$ and $x \in \mathbb{V}$,*

(A$_1$) $2[x, A(t,\omega)x]_{\mathbb{U}} + \|B(t,\omega)\|^2_{Q,\mathbb{U}} + \delta'\|x\|^2_{\mathbb{V}} \le K'\|x\|^2_{\mathbb{U}}$;

(B$_1$) $\|A(t,\omega)x\|_{\mathbb{V}'} \le K'\|x\|_{\mathbb{V}}$.

If

$$\mathbb{E}\|\varphi\|^2_{\mathbb{U}} < \infty \text{ and } \mathbb{E}\int_{[0,T]} \|f(t)\|^2_{\mathbb{V}'}\,dt < \infty,$$

then the solution u of (3.1.1) belongs to $\mathbb{L}_2([0,T];\mathscr{P};\mathbb{V}) \cap \mathbb{C}([0,T];\mathscr{P};\mathbb{U})$, and there exists a positive number C, depending only on K', δ', and T, such that

$$\mathbb{E}\sup_{t\le T}\|u(t)\|^2_{\mathbb{U}} + \mathbb{E}\int_{[0,T]}\|u(t)\|^2_{\mathbb{V}}\,dt \le C\,\mathbb{E}\Big(\|\varphi\|^2_{\mathbb{U}} + \int_{[0,T]}\|f(t)\|^2_{\mathbb{V}'}\,dt + \|M(T)\|^2_{\mathbb{U}}\Big). \tag{3.2.33}$$

□

Proof Consider Eq. (3.1.1) in $(\mathbb{V},\mathbb{U},\mathbb{V}')$. Observe that, for every $x \in \mathbb{V}$, the processes $Ax = A(t)x$ and $Bx = B(t)x$ are predictable with values in $\mathbb{V}'$ and $\mathfrak{L}_Q(\mathbb{Y},\mathbb{U})$, respectively, f is predictable with values in $\mathbb{V}'$, and φ is $\mathscr{F}_0$-measurable with values in $\mathbb{U}$. Indeed, consider, for example, the process Bx. It suffices to show that, for every y and h from dense subsets of $\mathbb{Y}$ and $\mathbb{H}$, respectively, the real-valued process $(BxQ^{1/2}y, h)_{\mathbb{U}}$ is predictable (see Sects. 1.3.9 and 1.3.7). We take as these sets $\mathbb{Y}$ itself and the domain of Λ^2 in $\mathbb{U}$. It follows from the proposition that

$$\big((B(t)x)y', h\big)_{\mathbb{U}} = \big((B(t)x)y', \Lambda^2 h\big),\quad y' \in Q^{1/2}(\mathbb{Y}).$$

Since Bx is a predictable $\mathfrak{L}_Q(\mathbb{Y},\mathbb{H})$-process, the right-hand side of the last equality is a predictable real-valued process. Thus Theorem 3.1 is applicable to (3.1.1) in the normal triple $(\mathbb{V},\mathbb{U},\mathbb{V}')$. Therefore, (3.1.1) has a solution u in this normal triple, and u satisfies (3.2.33).

It remains to prove that this solution is also a solution in the original normal triple $(\mathbb{X}, \mathbb{H}, \mathbb{X}')$.

By Proposition 1.2(i) and using that $\mathbb{V}$ is normally embedded in $\mathbb{H}$, we conclude that $u \in \mathbb{L}_2([0, T] \times \Omega, \mathscr{P}; \mathbb{X})$. Furthermore, by Theorem 3.1, for all $y \in \mathbb{V}$ and $\ell \times \mathbb{P}$- a.a. t, ω, the following equality holds:

$$(y, u(t))_{\mathbb{U}} = (y, \varphi)_{\mathbb{U}} + \int_{[0,t]} [y, Au(s) + f(s)]_{\mathbb{U}}\, ds + (y, \tilde{M}(t))_{\mathbb{U}},$$

where $\tilde{M}(t) := M(t) + \int_{[0,t]} Bu(s) dW(s)$. From this, making use of the proposition and the lemma from this paragraph, we find that, for all $y \in \mathbb{V}$,

$$(\Lambda^2 y, u(t)) = (\Lambda^2 y, \varphi) + \int_{[0,t]} [\Lambda^2 y, Au(s) + f(s)]\, ds + (\Lambda^2 y, \tilde{M}(t)),$$

$\ell \times \mathbb{P}$- a.s. Since $\Lambda^2(\mathbb{Z})$ is dense in $\mathbb{X}$, the last equality completes the proof of the theorem. □

3.3 Dissipative Systems

3.3.1. In this section we consider the LSES (3.1.1) in a scale of Hilbert spaces. A family of Hilbert spaces $\{\mathbb{H}^\alpha, \ \alpha \in \mathbb{R}^1\}$ is called a **Hilbert scale** if

(i) For every $\beta > \alpha$ the space $\mathbb{H}^\beta$ is normally embedded in $\mathbb{H}^\alpha$;
(ii) For every $\alpha < \beta < \gamma$ and $x \in \mathbb{H}^\gamma$,

$$\|x\|_\beta \le \|x\|_\alpha^{(\gamma-\beta)/(\gamma-\alpha)} \cdot \|x\|_\gamma^{(\beta-\alpha)/(\gamma-\alpha)};$$

here and below,

$$\|\cdot\|_\alpha := \|\cdot\|_{\mathbb{H}^\alpha}.$$

Instead of assuming coercivity of (3.1.1) in the original normal triple, we will assume dissipativity in a different normal triple, coming from a suitable Hilbert scale, and establish the corresponding existence and uniqueness result.

3.3.2. Let $\mathbb{X}$ and $\mathbb{H}$ be Hilbert spaces such that $\mathbb{X}$ is normally embedded in $\mathbb{H}$. Recall that $\|\cdot\|$ and $(\cdot, \cdot)$ are the norm and inner product in $\mathbb{H}$.

There exists a special Hilbert scale containing spaces $\mathbb{X}$ and $\mathbb{H}$; see e.g. [64, IV.1.10] or [43, IV.9.1]. Below is a outline of the construction.

We start by applying Proposition 3.3 to the spaces $\mathbb{X}$ and $\mathbb{H}$, and then, using the spectral decomposition of the corresponding operator Λ, we define Λ^α, $\alpha \in \mathbb{R}^1$, by

$$\Lambda^\alpha x = \int_{[1,\infty]} \lambda^\alpha dE_\lambda x, \tag{3.3.1}$$

where E_λ is the spectral measure of Λ.

For $\alpha > 0$, denote by $\mathbb{H}^\alpha$ the domain of Λ^α in $\mathbb{H}$. Then $\mathbb{H}^\alpha$ is a Hilbert space with inner product

$$(x, y)_\alpha := (\Lambda^\alpha x, \Lambda^\alpha y). \tag{3.3.2}$$

For $\alpha < 0$, the space $\mathbb{H}^\alpha$ is the completion of $\mathbb{H}$ with respect to the norm

$$\|\cdot\|_\alpha = \|\Lambda^\alpha \cdot\|.$$

Representation (3.3.1) implies that the family of Hilbert spaces $\{\mathbb{H}^\alpha\}$ constructed in this way is indeed a Hilbert scale; we call this scale the Hilbert scale connecting $\mathbb{X}$ and $\mathbb{H}$. Note that

(a) it is the unique scale satisfying $\mathbb{H}^1 = \mathbb{X}$ and $\mathbb{H}^0 = \mathbb{H}$,
(b) the space

$$\mathbb{H}^\infty := \bigcap_{\beta\in\mathbb{R}^1} \mathbb{H}^\beta$$

is dense in $\mathbb{H}^\alpha$ for every $\alpha \in \mathbb{R}^1$.

It follows that $x \in \mathbb{H}^\alpha$ if and only if $x = \Lambda^{-\alpha} y$ for some $y \in \mathbb{H} = \mathbb{H}^0$, and then, for every $\beta \in \mathbb{R}^1$, $\Lambda^\beta x = \Lambda^{\beta-\alpha} y$ defines a linear homeomorphism from $\mathbb{H}^\alpha$ to $\mathbb{H}^{\alpha-\beta}$.

Given $\alpha > \beta \in \mathbb{R}^1$, let $\gamma := 2\beta - \alpha$ (so that $\alpha - \beta = \beta - \gamma$) and consider the triple of spaces

$$(\mathbb{H}^\alpha, \mathbb{H}^\beta, \mathbb{H}^\gamma).$$

Then, for every $x \in \mathbb{H}^\alpha$ and $y \in \mathbb{H}^\beta$,

$$|(x, y)_\beta| = |(\Lambda^\beta x, \Lambda^\beta y)_0| = |(\Lambda^{\beta-\alpha}\Lambda^\alpha x), \Lambda^\beta y)_0| = \|(\Lambda^\alpha x, \Lambda^\gamma y)_0\|$$
$$\leq \|x\|_\alpha \; \|y\|_\gamma.$$

Thus the triple $(\mathbb{H}^\alpha, \mathbb{H}^\beta, \mathbb{H}^{2\beta-\alpha})$ is normal; its CBF will be denoted by $[\cdot, \cdot]_{\alpha,\beta}$. By property (i) of CBF (Sect. 2.5.2), the mapping $\mathcal{J} : y \to [\cdot, y]_{\alpha,\beta}$ sends $\mathbb{H}^\gamma$ to $(\mathbb{H}^\alpha)^*$.

In fact, a stronger result holds [42, IV.1.10].

Proposition 3.4 *The mapping $\mathcal{J}$ is an isometric isomorphism between $\mathbb{H}^\gamma$ and $(\mathbb{H}^\alpha)^*$.* □

Remark 3.3 To summarize, CBF $[\cdot,\cdot]_{\alpha,\beta}$ establishes a duality between $\mathbb{H}^\alpha$ and $\mathbb{H}^{2\beta-\alpha}$. In particular, for every $\alpha > 0$, $[\cdot,\cdot]_{\alpha,0}$ established a duality between $\mathbb{H}^\alpha$ and $\mathbb{H}^{-\alpha}$.

It was mentioned in Remark 2.7 that if $\mathbb{X}$ is normally embedded into $\mathbb{H}$, then there exists a normal triple $(\mathbb{X}, \mathbb{H}, \mathbb{X}')$ such that its CBF establishes a duality between $\mathbb{X}$ and $\mathbb{X}'$. We now have an explicit construction: set $\mathbb{X}' = \mathbb{H}^{-1}$, where $\{\mathbb{H}^\alpha\}$ is the Hilbert scale connecting $\mathbb{X}$ and $\mathbb{H}$, and $[\cdot,\cdot] = [\cdot,\cdot]_{1,0}$.

3.3.3. Let $(\mathbb{X}, \mathbb{H}, \mathbb{X}')$ be the original normal triple and let $\{\mathbb{H}^\alpha\}$ be the Hilbert scale connecting $\mathbb{X}$ and $\mathbb{H}$ so that $\mathbb{H}^1 = \mathbb{X}$, $\mathbb{H}^0 = \mathbb{H}$, and $\mathbb{H}^{-1} = \mathbb{X}'$. In addition to the assumptions from the beginning of this chapter, let us fix a number $\lambda \geq 1$ and suppose that $M \in \mathfrak{M}_2^c([0,T], \mathbb{H}^{\lambda+1})$ and, for all t, ω, $A(t,\omega) \in \mathfrak{L}(\mathbb{H}^{\lambda+1}, \mathbb{H}^{\lambda-1})$, $B(t,\omega) \in \mathfrak{L}(\mathbb{H}^{\lambda+1}, \mathfrak{L}_Q(\mathbb{Y}, \mathbb{H}^\lambda))$, $f(t,\omega) \in \mathbb{H}^\gamma$, $\varphi(\omega) \in \mathbb{H}^\gamma$.

Denote by $\|\cdot\|_{Q,\alpha}$ the norm in $\mathfrak{L}_Q(\mathbb{Y}, \mathbb{H}^\alpha)$.

Theorem 3.3 *Suppose that there exists a number $K \in \mathbb{R}_+$ such that, for all $x \in \mathbb{H}^{\lambda+1}$ and $(t,\omega) \in [0,T] \times \Omega$,*

(A_2) $\quad 2[x, A(t,\omega)x]_\lambda + \|B(t,\omega)x\|^2_{\lambda,Q} \leq K\|x\|^2_\lambda;$

(B_2) $\quad \|A(t,\omega)x\|_{\alpha-1} \leq K\|x\|_{\alpha+1}, \quad \alpha = \lambda, 0;$

$\qquad\quad \|B(t,\omega)x\|_{Q,\alpha-1} \leq K\|x\|_\alpha, \quad \alpha = \lambda, 1.$

If

$$\mathbb{E}\|\varphi\|^2_\lambda < \infty \quad \textit{and} \quad \mathbb{E}\int_{[0,T]} \|f(t)\|^2_\lambda \, dt < \infty,$$

then Eq. (3.1.1) has a unique solution u in $(\mathbb{H}^1, \mathbb{H}^0, \mathbb{H}^{-1})$. This solution belongs to $\mathbb{L}_2([0,T]\mathscr{P}; \mathbb{H}^\lambda)$ and, for some $N \in \mathbb{R}_+$, depending only on K and T,

$$\mathbb{E}\int_{[0,T]} \|u(t)\|^2_\lambda \, dt \leq N\,\mathbb{E}\left(\|\varphi\|^2_\lambda + \int_{[0,T]} \|f(t)\|^2_\lambda dt + \|M(T)\|^2_{\lambda+1}\right). \tag{3.3.3}$$

□

Proof First of all note that

$$\|x\|^2_{\lambda+1} = [x, \Lambda^2 x]_{\lambda+1,\lambda}, \quad x \in \mathbb{H}^{\lambda+1}. \tag{3.3.4}$$

Indeed, if $x \in \mathbb{H}^{\lambda+2}$, then $\Lambda^2 x \in \mathbb{H}^\lambda$ and, by property (ii) of CBF (Sect. 2.5.2),

$$[x, \Lambda^2 x]_{\lambda+1,\lambda} = (x, \Lambda^2 x)_\lambda = \|x\|^2_{\lambda+1}.$$

If $x \in \mathbb{H}^{\lambda+1}$, then choose a sequence $\{x_n\}$ from $\mathbb{H}^{\lambda+2}$ such that $\lim_{n\to\infty} \|x - x_n\|_{\lambda+1} = 0$. As a result,

$$\lim_{n\to\infty} \|\Lambda^2 x_n - \Lambda^2 x\|_{\lambda-1} = \lim_{n\to\infty} \|\Lambda^{\lambda+1}(x_n - x)\|_0 = \lim_{n\to\infty} \|x_n - x\|_{\lambda+1} = 0,$$

which implies (3.3.4) for all $x \in \mathbb{H}^{\lambda+1}$.

Fix $\varepsilon > 0$ and define

$$A_\varepsilon(t,\omega) := A(t,\omega) + \varepsilon\Lambda^2.$$

From (3.3.4) and assumption (A_2), it follows that, for all $(t,\omega) \in [0,T] \times \Omega$ and $x \in \mathbb{H}^{\lambda+1}$,

$$2[x, A_\varepsilon(t,\omega)x]_{\lambda+1,\lambda} + \|B(t,\omega)x\|^2_{Q,\lambda} + \varepsilon\|x\|^2_{\lambda+1} \le K\|x\|^2_\lambda \tag{3.3.5}$$

and

$$\|A_\varepsilon(t,\omega)x\|_{\lambda-1} \le (K+\varepsilon)\,\|x\|_{\lambda+1}.$$

Similar to Sect. 3.2, we can show that the processes $A_\varepsilon x$, f, and Bx, are predictable with values in $\mathbb{H}^{\lambda-1}$, $\mathbb{H}^{\lambda-1}$, and $\mathfrak{L}_Q(\mathbb{Y}, \mathbb{H}^\lambda)$, respectively, and that φ is an $\mathscr{F}_0$-measurable random variable with values in $\mathbb{H}^\lambda$.

Applying Theorem 3.1 to the equation

$$u(t) = \varphi + \int_{[0,t]} \Big(A_\varepsilon u(s) + f(s)\Big)\, ds + \int_{[0,t]} Bu(s)dW(s) + M(t) \tag{3.3.6}$$

in the normal triple $(\mathbb{H}^{\lambda+1}, \mathbb{H}^\lambda, \mathbb{H}^{\lambda-1})$, we conclude that (3.3.6) has a unique solution $u^\varepsilon \in \mathbb{L}_2([0,T], \mathscr{P}; \mathbb{H}^{\lambda+1}) \cap \mathbb{C}([0,T], \mathscr{P}; \mathbb{H}^\lambda)$, and there exists a number $N \in \mathbb{R}_+$ such that

$$\begin{aligned}\mathbb{E}\sup_{t\le T} \|u^\varepsilon(t)\|^2_\lambda + \mathbb{E}\int_{[0,t]} \|u^\varepsilon(t)\|^2_{\lambda+1} dt \\ \le N\,\mathbb{E}\Bigg(\|\varphi\|^2_\lambda + \int_{[0,T]} \|f(s)\|^2_{\lambda-1}\, ds + \|M(T)\|^2_\lambda\Bigg).\end{aligned} \tag{3.3.7}$$

In general, the number N in this inequality depends on ε. Accordingly, our next step is to derive an alternative estimate that is uniform in ε.

3.3.4.

Lemma 3.3 *The solution* u^ε *of the LSES* (3.3.6) *satisfies*

$$\sup_{t\le T}\mathbb{E}\|u^\varepsilon(t)\|_\lambda^2 \le N\,\mathbb{E}\Big(\|\varphi\|_\lambda^2 + \int_{[0,T]}\|f(s)\|_\lambda^2\,ds + \|M(T)\|_\lambda^2\Big), \tag{3.3.8}$$

where the number N *does not depend on* ε. □

Proof From Theorem 2.13 it follows that, for all $t\in[0,T]$ and ω from some set of probability one,

$$\|u^\varepsilon(t)\|_\lambda^2 = \|\varphi\|_\lambda^2 + 2\int_{[0,t]}[u^\varepsilon(s), A_\varepsilon u^\varepsilon(s) + f(s)]_{\lambda+1,\lambda}\,ds + 2\int_{[0,t]}\Big(u^\varepsilon(s), d\big((Bu^\varepsilon)\circ W(s) + M(s)\Big)_\lambda + \langle (Bu^\varepsilon)\circ W + M\rangle_t. \tag{3.3.9}$$

By estimate (3.3.7) and Proposition 2.1, the stochastic integrals in (3.3.9) are continuous martingales. Thus from (3.3.9) we obtain

$$\mathbb{E}\|u^\varepsilon(t)\|_\lambda^2 = \mathbb{E}\|\varphi\|_\lambda^2 + 2\mathbb{E}\int_{[0,t]}[u^\varepsilon(s), A_\varepsilon u^\varepsilon(s) + f(s)]_{\lambda+1,\lambda}\,ds + 2\mathbb{E}\|(Bu^\varepsilon)\circ W(t) + M(t)\|. \tag{3.3.10}$$

Since $f(s,\omega)\in\mathbb{H}^\lambda$ for all (s,ω), we have $[u^\varepsilon(s), f(s)]_{\lambda+1,\lambda} = \big(u^\varepsilon(s), f(s)\big)_\lambda$. This equality implies

$$\mathbb{E}\int_{[0,t]}\Big|[u^\varepsilon(s), f(s)]_{\lambda+1,\lambda}\Big|\,ds \le \frac{1}{2}\Big(\mathbb{E}\int_{[0,t]}\|u^\varepsilon(s)\|_\lambda^2\,ds + \mathbb{E}\int_{[0,t]}\|f(s)\|_\lambda^2\,ds\Big). \tag{3.3.11}$$

Next, by direct computation,

$$\|(Bu^\varepsilon)\circ W(t) + M(t)\|_\lambda^2 = \int_{[0,t]}\|Bu^\varepsilon(s)\|_{Q,\lambda}^2\,ds + 2\Big((Bu^\varepsilon)\circ W(t), M(t)\Big)_\lambda + \|M(t)\|_\lambda^2. \tag{3.3.12}$$

From (3.3.2) it follows that

$$\left(\int\limits_{[0,t]} Bu^{\varepsilon}(s)dW(s), M(t)\right)_{\lambda} = \left(\int\limits_{[0,t]} \Lambda^{\lambda-1}Bu^{\varepsilon}(s)dW(s),\ \Lambda^{\lambda+1}M(t)\right)_{0}$$

$$\leq \frac{1}{2}\left(\int\limits_{[0,t]} \|\Lambda^{\lambda-1}Bu^{\varepsilon}(s)dW(s)\|_{0}^{2} + \|M(t)\|_{\lambda+1}^{2}\right). \tag{3.3.13}$$

By the assumption (B_2),

$$\mathbb{E}\int\limits_{[0,t]} \|\Lambda^{\lambda-1}Bu^{\varepsilon}(s)\|_{Q,0}^{2}\,ds = \mathbb{E}\int\limits_{[0,t]} \|Bu^{\varepsilon}(s)\|_{Q,\lambda-1}^{2}\,ds \tag{3.3.14}$$

$$\leq K\,\mathbb{E}\int\limits_{[0,t]} \|u^{\varepsilon}(s)\|_{\lambda}^{2}\,ds.$$

Combining (3.3.11)–(3.3.14) and making use of (3.3.5), we obtain from (3.3.10) that, for every $t \in [0, T]$,

$$\mathbb{E}\|u^{\varepsilon}(t)\|_{\lambda}^{2} \leq N\,\mathbb{E}\left(\|\varphi\|_{\lambda}^{2} + \int\limits_{[0,t]} \|f(s)\|_{\lambda}^{2}\,ds + \|M(t)\|_{\lambda+1}^{2} + \int\limits_{[0,t]} \|u^{\varepsilon}(s)\|_{\lambda}^{2}\,ds\right),$$

where the constant N does not depend on ε. From this, by the Gronwall–Bellman lemma we obtain (3.3.8). □

3.3.5. Because u^{ε} is a solution of (3.3.6) in $(\mathbb{H}^{\lambda+1}, \mathbb{H}^{\lambda}, \mathbb{H}^{\lambda-1})$, the following equality holds $\mathfrak{l} \times \mathbb{P}$- a.s. on $[0, T] \times \Omega$:

$$\left(y, u^{\varepsilon}(t)\right)_{\lambda} = (y, \varphi)_{\lambda} + \int\limits_{[0,t]} [y, A_{\varepsilon}u^{\varepsilon}(s) + f(s)]_{\lambda+1,\lambda}\,ds$$

$$+\left(y, \int\limits_{[0,t]} Bu^{\varepsilon}(s)dW(s) + M(t)\right)_{\lambda},\quad y \in \mathbb{H}^{\infty}. \tag{3.3.15}$$

In Lemma 3.2, take $\mathbb{V} = \mathbb{H}^{\lambda+1}$, $\mathbb{U} = \mathbb{H}^{\lambda}$, $\mathbb{V}' = \mathbb{H}^{\lambda-1}$ and replace Λ with Λ^{λ}. Since $\mathbb{H}^{\infty}$ is dense in every $\mathbb{H}^{\lambda}$, it follows that, for all s, ω,

$$[A_{\varepsilon}u^{\varepsilon}(s) + f(s), y]_{\lambda+1,\lambda} = [A_{\varepsilon}u^{\varepsilon}(s) + f(s), \Lambda^{2\lambda}y]_{1,0}.$$

From this and (3.3.2), making use of the self-adjointness of Λ^{λ}, we obtain that equality (3.3.15) is equivalent to

$$\begin{aligned}\big(\Lambda^{2\lambda}y, u^{\varepsilon}(t)\big)_0 &= (\Lambda^{2\lambda}y, \varphi)_0 + \int\limits_{[0,t]} [\Lambda^{2\lambda}y, A_{\varepsilon}u^{\varepsilon}(s) + f(s)]_{1,0}\, ds \\ &\quad + \left(\Lambda^{2\lambda}y, \int\limits_{[0,t]} Bu^{\varepsilon}(s)dW(s) + M(t)\right)_0 .\end{aligned} \tag{3.3.16}$$

From Proposition 1.2(a) it follows that $u^{\varepsilon} \in \mathbb{L}_2([0,T], \mathscr{P}; \mathbb{H}^1)$, and therefore u^{ε} is indeed a solution of the LSES (3.1.1) in $(\mathbb{H}^1, \mathbb{H}^0, \mathbb{H}^{-1})$; recall that $\{\Lambda^2 y,\ y \in \mathbb{H}^{\infty}\}$ is a dense subset of $\mathbb{H}^1$.

Lemma 3.3 implies that there exist a $u \in \mathbb{L}^2([0,T], \mathscr{P}; \mathbb{H}^1)$ and a subsequence $\{\varepsilon_n\}$ converging to zero as $n \to \infty$ such that $u^{\varepsilon_n} \to u$ weakly in

$$\mathbb{L}_2\big([0,T] \times \Omega, \overline{\mathscr{B}([0,T]) \otimes \mathscr{F}}, \ell \times \mathbb{P}; \mathbb{H}^1\big).$$

Arguing in the same way as in Theorem 3.1, we pass to the limit in (3.3.16) (over the sequence u^{ε_n}) and prove that u is the solution of the LSES (3.1.1) in $(\mathbb{H}^1, \mathbb{H}^0, \mathbb{H}^{-1})$.

Since the norm of a weak limit is bounded by the lower limit of the norms, inequality (3.3.8) implies (3.3.3), completing the proof of Theorem 3.3. □

Analysis of the proof shows that the solution is weakly continuous in $\mathbb{H}^{\lambda}$. In the next section, we will take a closer look at the results of this type.

3.4 Uniqueness and the Markov Property

3.4.1. The assumptions of Theorem 3.1 ensure the existence and uniqueness of a solution of (3.1.1) in the normal triple $(\mathbb{X}, \mathbb{H}, \mathbb{X}')$. However, if we only consider the problem of uniqueness, these assumptions are too restrictive. In this section we prove the uniqueness theorem under more general conditions, in particular, those satisfied by dissipative systems.

In addition to the uniqueness theorem, we also discuss in this section approximation of the solution of (3.1.1) by solutions of other systems of the same type. We also prove that if the coefficients, initial conditions and free terms of (3.1.1) do not depend on ω, then the solution of this system has a Markov property.

3.4.2. We start with a uniqueness result.

Theorem 3.4 *Let $K = K(t,\omega)$ and $N = N(t,\omega)$ be $\mathscr{F}_t$-adapted stochastic processes with values in $\mathbb{R}_+$ such that*

$$\mathbb{P}\left(\sup_{t\leq T} K(t,\omega) < \infty \quad \int\limits_{[0,T]} N(t,\omega)\,dt < \infty\right) = 1.$$

Suppose that, for all $(t,\omega) \in [0,T] \times \Omega$ and $x \in \mathbb{X}$,

(A_ω) $\quad 2\,[x, A(t,\omega)x] + \|B(t,\omega)x\|_Q^2 \leq N(t,\omega)\,\|x\|^2,$

(B_ω) $\quad \|A(t,\omega)x\|_{\mathbb{X}'} \leq K(t,\omega)\|x\|_{\mathbb{X}}.$

Then Eq. (3.1.1) has at most one solution. □

Proof We first observe that, under the above assumptions, the integrals in (3.1.1) are well-defined. Indeed, from property (i) of a CBF (see Sect. 2.5.2) and conditions (A_ω), (B_ω) it follows that, for all $(t,\omega) \in [0,T] \times \Omega$ and $x \in \mathbb{X}$,

$$\|B(t,\omega)x\|_Q^2 \leq K(t,\omega)\|x\|_{\mathbb{X}}^2 + N(t,\omega)\,\|x\|^2. \tag{3.4.1}$$

In particular, for every $(t,\omega) \in [0,T] \times \Omega$, we have $A(t,\omega) \in \mathfrak{L}(\mathbb{X},\mathbb{X}')$ and $B(t,\omega) \in \mathfrak{L}(\mathbb{X}, \mathfrak{L}_Q(\mathbb{Y},\mathbb{H}))$, so that the operators A and B map predictable processes to predictable processes. Moreover, from the assumptions of the theorem and (3.4.1), it follows that, for every $x \in \mathbb{L}_2^\omega([0,T],\mathscr{P};\mathbb{X})$,

$$\int\limits_{[0,T]} \left(\|Ax(t,\omega)\|_{\mathbb{X}'} + \|Bx(t,\omega)\|_Q^2\right) dt < \infty \ (\mathbb{P}\text{-a.s.}),$$

and then the results of Sects. 1.3.12 and 2.4.4 confirm that the integrals in (3.1.1) are well-defined.

Now assume that u_1 and u_2 are two solutions of (3.1.1) in $(\mathbb{X},\mathbb{H},\mathbb{X}')$ and define $\bar{u} := u_1 - u_2$. By linearity, $\bar{u}$ is a solution of (3.1.1) with $\varphi = 0$, $f = 0$, $M = 0$.

Next, define

$$r(t,\omega) := \|\bar{u}(t,\omega)\|^2 + \int\limits_{[0,t]} \left(N(s,\omega) + K(s,\omega)\,\|\bar{u}(s,\omega)\|_{\mathbb{X}}^2\right) ds,$$

$$\Gamma_n(\omega) := \{t : t \leq T,\ r(t,\omega) \geq n\},\ n \in \mathbb{N},$$

$$\tau_n(\omega) := \begin{cases} \inf\limits_t \Gamma_n(\omega), & \text{if } \Gamma_n(\omega) \neq \emptyset; \\ T, & \text{if } \Gamma_n(\omega) = \emptyset. \end{cases}$$

By construction, τ_n is a stopping time and $\tau_n \uparrow T$ as $n \to \infty$. Moreover, the process

$$\int_{[0,t\wedge\tau_n]} B\bar{u}(s)dW(s), \ t \in [0, T],$$

is a continuous square-integrable martingale with values in $\mathbb{H}$, because

$$\int_{[0,t\wedge\tau_n]} \|B\bar{u}(s)\|_Q^2 ds \leq N_n < \infty, \ n \in \mathbb{N}, \ \ (\mathbb{P}\text{- a.s.}), \tag{3.4.2}$$

where, in view of (3.4.1), the number N_n does not depend on ω.

Thus $(B\bar{u}) \circ W$ is a continuous local martingale with localizing sequence $\{\tau_n\}$.

From Theorem 2.13 it follows that there exists a set Ω' with $\mathbb{P}(\Omega') = 1$ such that, for all $t \leq T$ and $\omega \in \Omega'$,

$$\|\bar{u}(t\wedge\tau_n)\|^2 \exp\Big\{-\int_{[0,t\wedge\tau_n]} N(s)\|\bar{u}(s)\|^2 ds\Big\} = \int_{[0,t\wedge\tau_n]} \exp\Big\{-\int_{[0,s]} N(r)\|\bar{u}(r)\|^2 dr\Big\}$$
$$\times \Big(2[\bar{u}(s), A\bar{u}(s)] + \|B\bar{u}(s)\|_Q^2 - N(s)\|\bar{u}(s)\|^2\Big)\, ds \tag{3.4.3}$$
$$+2\int_{[0,t\wedge\tau_n]} \exp\left(-\int_{[0,s]} N(r)\|\bar{u}(r)\|^2 dr\right) \Big(\bar{u}(s), d\big((B\bar{u})\circ W(s)\big)\Big).$$

Next, in view of (3.4.2),

$$\mathbb{E}\Bigg(\int_{[0,t\wedge\tau_n]} \exp\left(-2\int_{[0,s]} N(r)\|\bar{u}(r)\|^2 dr\right) \|\bar{u}(s)\|^2 \|B(s)\bar{u}(s)\|_Q^2\, ds\Bigg)^{1/2}$$
$$\leq \mathbb{E} \sup_{s\leq t\wedge\tau_n} \|\bar{u}(s)\| \Big(\int_{[0,t\wedge\tau_n]} \|B\bar{u}(s)\|^2 ds\Big)^{1/2} \leq \sqrt{n\, N_n}.$$

Hence the stochastic integral in (3.4.3) is a martingale (see Proposition 2.1) and so its expectation is equal to zero. In view of this and condition (A_ω), we obtain by taking expectations on both sides of (3.4.3) that

$$\mathbb{E}\|\bar{u}(t\wedge\tau_n)\|^2 \exp\Big\{-\int_{[0,t\wedge\tau_n]} N(s)\|u(s)\|^2 ds\Big\} \leq 0,$$

and consequently

$$\mathbb{P}\Big(\|\bar{u}(t \wedge \tau_n)\| = 0\Big) = 1.$$

After passing to the limit in the last equality as $\tau_n \uparrow T$ and using continuity of $\|\bar{u}\|$ in t,

$$\mathbb{P}\big(\sup_{t \le T} \|u_1(t) - u_2(t)\| = 0\big) = 1,$$

which completes the proof of the theorem. □

3.4.3. Suppose that the collection $\{A(t, \omega),\ B(t, \omega),\ \varphi(\omega),\ f(t, \omega),\ M(t, \omega)\}$ satisfies the assumptions from Sects. 3.1.1 and 3.2.2. Consider a sequence

$$\{A_n(t, \omega),\ B_n(t, \omega),\ \varphi(\omega),\ f_n(t, \omega),\ M_n(t, \omega);\ n \in \mathbb{N}\}$$

satisfying the same assumptions, but possibly with different bounds.

Next, we assume that (3.1.1) and each of the following equations

$$u^n(t) = \varphi_n + \int\limits_{[0,t]} (A_n u^n(s) + f_n(s))\, ds + \int\limits_{[0,t]} B_n u^n(s) dW(s) + M_n(t), \tag{3.4.4}$$

is uniquely solvable in the same normal triple $(\mathbb{X}, \mathbb{H}, \mathbb{X}')$, with solutions belonging to $\mathbb{L}_2([0, T], \mathscr{P}; \mathbb{X})$.

Theorem 3.5 *Suppose that Eqs. (3.4.4) satisfy the assumptions of Proposition 3.2 uniformly in n and*

a) $\lim\limits_{n\to\infty} \mathbb{E}\Big(\|\varphi_n - \varphi\|^2 + \int\limits_{[0,T]} \|f_n(t) - f(t)\|^2_{\mathbb{X}'}\, dt + \langle M_n - M\rangle_T\Big) = 0;$

b) *for every* $x \in \mathbb{L}_2([0, T], \mathscr{P}; \mathbb{X})$,

$$\lim_{n\to\infty} \mathbb{E} \int\limits_{[0,T]} \Big(\|(A_n - A)x(t)\|^2_{\mathbb{X}'} + \|(B_n - B)x(t)\|^2_Q\Big)\, dt = 0.$$

Then

$$\lim_{n\to\infty} \mathbb{E} \sup_{t \le T} \|u^n(t) - u(t)\|^2 = 0.$$

If, in addition, the coercivity condition (A) *holds both for (3.1.1) and for each of Eqs. (3.4.4) uniformly in n, then*

$$\lim_{n\to\infty} \mathbb{E} \int\limits_{[0,T]} \|u^n(t) - u(t)\|^2_{\mathbb{X}}\, dt = 0.$$

□

Proof We only prove the first statement; the proof of the second statement is an easy exercise for the interested reader.

Define $v^n := u^n - u$ so that v^n is a solution, in $(\mathbb{X}, \mathbb{H}, \mathbb{X}')$, of

$$v^n(t) = \tilde{\varphi}_n + \int\limits_{[0,t]} \Big(A_n v^n(s) + \tilde{f}_n(s)\Big)\, ds + \int\limits_{[0,t]} B_n v^n(s) dW(s) + \tilde{M}_n(t), \tag{3.4.5}$$

where

$$\tilde{\varphi}_n := \varphi_n - \varphi,\ \tilde{f}_n(s) := (A_n - A)u(s) + \tilde{f}_n(s) - f(s),$$

$$\tilde{M}_n(t) := M_n(t) - M(t) + \int\limits_{[0,t]} (B_n - B)u(s) dW(s).$$

Then (3.4.5) also satisfies the assumptions of Proposition 3.2, uniformly in n, and consequently there exists a positive number N independent of n and such that

$$\mathbb{E} \sup_{t \le T} \|v^n\|^2 \le N\, \mathbb{E}\Big(\|\tilde{\varphi}_n\|^2 + \int\limits_{[0,T]} \|\tilde{f}_n(t)\|^2_{\mathbb{X}'}\, dt + \|\tilde{M}_n(T)\|^2\Big).$$

Passing to the limit $n \to \infty$ leads to the first part of the statement of the theorem. □

3.4.4. In this paragraph we assume that

(a) The operators A, B and the function f do not depend on ω.
(b) The martingale M has the form

$$M(t) = \int\limits_{[0,t]} g(s) dW(s),$$

where $g \in \mathbb{L}_2\big([0, T], \overline{\mathscr{B}([0, T])}, \ell; \mathfrak{L}_Q(\mathbb{Y}, \mathbb{H})\big)$.

We also suppose that Eq. (3.1.1) is well-posed in the normal triple $(\mathbb{X}, \mathbb{H}, \mathbb{X}')$, that is, (3.1.1) has a unique solution and the solution depends continuously on the input φ, f, and M; cf. Theorem 3.1 or Theorem 3.3 and Proposition 3.2.

We will see that, under these assumptions, the solution of (3.1.1) has a Markov property: for every Borel set $\Gamma \subset \mathbb{H}$ and all $s,\ t \in [0, T]$ such that $s \le t$, the following equality holds $\mathbb{P}$-a.s.:

$$\mathbb{P}\Big(u(t) \in \Gamma \Big| \mathscr{F}_s(u)\Big) = \mathbb{P}\Big(u(t) \in \Gamma \Big| u(s)\Big). \tag{3.4.6}$$

Here and below, $\mathscr{F}_s(u)$ is the σ-algebra generated by the solution $u(r, \omega)$ of (3.1.1) for $r \in [0, s]$.

To state the corresponding theorem, we need to discuss equations of the type (3.1.1) on time intervals other than $[0, T]$.

Let $y(s) \in \mathbb{L}_2(\Omega, \mathscr{F}_s, \mathbb{P}; \mathbb{H})$. Consider the LSES

$$
\begin{aligned}
u(t, y(s), s) = y(s) + \int_{[s,t]} \Big(A(r)u\big(t, y(s), s\big) + f(r)\Big)dr \\
+ \int_{[s,t]} \Big(B(r)u\big(r, y(s), r\big) + g(r)\Big)dW(r)
\end{aligned}
\tag{3.4.7}
$$

for $t \in [s, T]$. As in Definition 3.1, we say that a predictable process $u = u\big(t, y(s), s\big)$ belonging, for $\mathbb{P}$-a.a. ω, to $\mathbb{L}_2([s,t], \mathscr{B}([s,t]), \mathfrak{l}; \mathbb{X})$, is a solution of the LSES (3.4.7) in $(\mathbb{X}, \mathbb{H}, \mathbb{X}')$, if, for $\mathfrak{l} \times \mathbb{P}$-a.a. $(t, \omega) \in [s, T] \times \Omega$ and all x from a dense (in the strong topology) subset of $\mathbb{X}$, the following equality holds:

$$
\begin{aligned}
\Big(x, u\big(t, y(s), s\big)\Big) = (x, y(s)) + \int_{[s,t]} \Big[x, A(r)u\big(r, y(s), s\big) + f(r)\Big]dr \\
+ \left(x, \int_{[s,t]} \Big(B(r)u\big(r, y(s), s\big) + g(r)\Big)dW(r)\right).
\end{aligned}
$$

Theorem 3.6 *Let $u = u(t)$ be the solution of* (3.4.7) *on* $[0, T]$ *and let $F : \mathbb{H} \to \mathbb{R}^1$ be a bounded Borel function. Define $\mathscr{F}_t(u) := \sigma\big(u(r), r \le t\big)$ and*

$$
\Phi_{t,s}(x) := \mathbb{E}\big[F\big(u(t)\big)\big|u(s) = x\big],\ x \in \mathbb{H},\ 0 \le s < t \le T.
$$

Then, for $\mathbb{P}$- a.a. ω,

$$
\mathbb{E}\Big[F\big(u(t)\big)|\mathscr{F}_s(u)\Big] = \Phi_{t,s}\big(u(s)\big). \tag{3.4.8}
$$

□

Proof With no loss of generality, assume that F is continuous.

Fix $0 \le s < t \le T$ and let $u = u(t, x, s)$ be the solution of (3.4.7) with $y(s) = x$. To begin, note that, because x is non-random, the random variable $u(t, x, s)$ and the σ-algebra $\mathscr{F}_s(u)$ are independent. Indeed, by uniqueness, analysis of the proofs of Theorems 3.1 and 3.3 shows that $u(t, x, s)$ is a limit of solutions of finite-dimensional systems, for which this independence is well-known; cf. [70, Chapter II, §9]. It remains to pass to the corresponding limit.

Next, let $u(s)$ be the solution of (3.4.7) at time s and let $\{u_n(s, \omega),\ n \in \mathbb{N}\}$ be a sequence of $\mathscr{F}_s$-measurable functions with finitely many values in $\mathbb{H}$ such that $u_n(s, \omega)$ converge to $u(s, \omega)$ in the strong topology of $\mathbb{H}$ for all (s, ω). Denote by

Γ_n the set of values of $u_n(s,\omega)$. By linearity,

$$v(s) = \sum_{x \in \Gamma_n} 1_{\{u_n(s)=x\}} u(t,x,s)$$

is the solution of (3.4.7) with $y(s) = u_n(s)$, and, by uniqueness, $v(s,x)$ must coincide with $u\big(t, u_n(s), s\big)$. Since $u(t,x,s)$ does not depend on $\mathscr{F}_s(u)$, we get

$$\begin{aligned}\mathbb{E}\Big[F\Big(u\big(t,u_n(s),s\big)\Big)\Big|\mathscr{F}_s(u)\Big] &= \mathbb{E}\left[\sum_{x\in\Gamma_n} 1\{u_n(s)=x\}\, F\big(u(t,x,s)\big)\Big|\mathscr{F}_s(u)\right] \\ &= \sum_{x\in\Gamma_n} 1\{u_n(s)=x\}\, \mathbb{E}F\big(u(t,x,s)\big) = \Phi_{t,s}\big(u_n(s)\big),\end{aligned}$$

that is,

$$\mathbb{E}\Big[F\Big(u\big(t,u_n(s),s\big)\Big)\Big|\mathscr{F}_s(u)\Big] = \Phi_{t,s}\big(u_n(s)\big).$$

To derive (3.4.8), it remains to pass to the limit $n \to \infty$ in the above equality using continuity of F and $\Phi_{t,s}$. □

3.5 The First Boundary Value Problem for Itô Partial Differential Equations

3.5.1. Recall that

- $x = (x^1, \ldots, x^{\mathrm{d}})$, a point in $\mathbb{R}^{\mathrm{d}}$, $dx := dx^1 \cdots dx^{\mathrm{d}}$;
- α, β, γ: d-dimensional **multi-indices**, that is, d-dimensional vectors with coordinates from $\mathbb{N} \cup \{0\}$;
- $|\alpha| = \alpha^1 + \cdots + \alpha^{\mathrm{d}}$;
- $f_\alpha := \dfrac{\partial^{\alpha^1+\ldots+\alpha^{\mathrm{d}}} f}{\partial(x^1)^{\alpha^1}\ldots\partial(x^{\mathrm{d}})^{\alpha^{\mathrm{d}}}}$, $\alpha := (\alpha^1, \ldots, \alpha^{\mathrm{d}})$, with the convention $f_\alpha = f$ if $\alpha = (0, \ldots, 0)$ is the zero multi-index.

Introduce the following objects:

- G, a domain in $\mathbb{R}^{\mathrm{d}}$;
- $M \in \mathfrak{M}_2^c\big([0,T], \mathbb{L}_2(G)\big)$;
- $w = \big(w^1(t), \ldots, w^{\mathrm{d}_1}(t)\big)$, a standard Wiener process in $\mathbb{R}^{\mathrm{d}_1}$.

In this section we consider the following equation:

$$u(t,x,\omega) = \varphi(x,\omega) - \int\limits_{[0,t]} \sum_{|\alpha|\le m} (-1)^{|\alpha|}\Bigg(\sum_{|\beta|\le m} a^{\alpha\beta}(s,x,\omega)u_\beta(s,x,\omega) + f^\alpha(s,x,\omega)\Bigg)_\alpha ds + \int\limits_{[0,t]} \sum_{l=1}^{\mathrm{d}_1} \sum_{|\beta|\le m} b^\beta(s,x,\omega)u_\beta(s,x,\omega)\,dw^l(s,\omega) + M(t,x,\omega), \tag{3.5.1}$$

in $[0,T]\times G$ with boundary conditions

$$u_\gamma(t,x,\omega)\Big|_{(t,x)\in S} = 0 \tag{3.5.2}$$

for all γ such that $|\gamma| \le \mathrm{d}-1$, where S is the lateral surface of the cylinder $[0,T]\times G$.

Making use of Theorem 3.1, we show that, under general assumptions concerning the coefficients, the initial condition, and free terms, this problem has a unique solution in a suitable Sobolev space.

3.5.2 Sobolev spaces are of great importance in the theory of partial differential equations. So we begin with a short review of some key results concerning these spaces. Full proofs of the results given below in Sects. 3.5.2–3.5.8 can be found e.g. in [117, 147].

Fix numbers m and p such that $m \in \mathbb{N}$ and $1 \le p < \infty$.

Definition 3.3 The space (of equivalence classes) of real functions on G belonging, together with their generalized derivatives up to and including the order of m, to $\mathbb{L}_p(G)$ is called the **Sobolev space** $\mathbb{W}^m_p(G)$. □

Theorem 3.7 *The Sobolev space $\mathbb{W}^m_p(G)$ endowed with the norm*

$$\|u\|_{G,m,p} = \Big(\sum_{|\alpha|\le m}\int\limits_G |u_\alpha(x)|^p\,dx\Big)^{1/p}$$

is a separable and, for $p>1$, reflexive Banach space. For $p=2$, it is a Hilbert space with inner product

$$(f,g)_{G,m,2} = \sum_{|\alpha|\le m}\int\limits_G f_\alpha(x)g_\alpha(x)\,dx.$$

□

Warning 3.2 *If $G = \mathbb{R}^{\mathrm{d}}$, then we omit G in the corresponding notation. For example, $\|\cdot\|_{4,5} = \|\cdot\|_{\mathbb{R}^{\mathrm{d}},4,5}$ and $\mathbb{W}^1_2 = \mathbb{W}^1_2(\mathbb{R}^{\mathrm{d}})$.* □

3.5.3. When $p = 2$ and $G = \mathbb{R}^{\mathrm{d}}$, there exists another very useful definition of the Sobolev spaces, which we now present.

Let

$$\boldsymbol{\Delta} = \sum_{i=1}^{\mathrm{d}} \frac{\partial^2}{\partial(x^i)^2}$$

be the Laplace operator and define $\Lambda^2 := \mathbb{I} - \boldsymbol{\Delta}$. Then Λ^2 is a positive self-adjoint unbounded operator on $\mathbb{H}^0 := \mathbb{L}_2(\mathbb{R}^{\mathrm{d}})$. Using the operator Λ^2 on $\mathbb{H}^0$, we now construct the Hilbert scale $\{\mathbb{H}^s,\ s \in \mathbb{R}^1\}$ similar to Sect. 3.4.2. Namely, for $s > 0$, the space $\mathbb{H}^s$ is the domain of Λ^s in $\mathbb{H}^0$; for $s \le 0$, the space $\mathbb{H}^s$ is the completion of $\mathbb{H}^0$ in the norm $\|\cdot\|_s := \|\Lambda^s \cdot\|_{\mathbb{L}_2(\mathbb{R}^{\mathrm{d}})}$. As was mentioned in Sect. 3.4.2, each $\mathbb{H}^s$ is a Hilbert space with inner product

$$(f, g)_s = (\Lambda^s f, \Lambda^s g)_{\mathbb{L}_2(\mathbb{R}^{\mathrm{d}})}.$$

Theorem 3.8 *If $m \in \mathbb{N}$, then the spaces $\mathbb{W}^m_2 = \mathbb{W}^m_2(\mathbb{R}^{\mathrm{d}})$ and $\mathbb{H}^m$ are equivalent, that is, $\mathbb{H}^m = \mathbb{W}^m_2$ as sets, and the norms $\|\cdot\|_m$ and $\|\cdot\|_{m,2}$ are equivalent, i.e. there exist positive numbers N_1, N_2 such that, for every $u \in \mathbb{H}^m$,*

$$N_1\|u\|_m \le \|u\|_{m,2} \le N_2\|u\|_m.$$

In particular,

$$\|u\|_1^2 = \|u\|_0^2 + \sum_{i=1}^{\mathrm{d}} \|u_i\|_0^2.$$ □

Warning 3.3 *In view of the last theorem, we will, in the future, identify the spaces $\mathbb{W}^m_2$ and $\mathbb{H}^m$ for $m \in \mathbb{N}$, and use the notation $\mathbb{H}^m$, $\|\cdot\|_m$, $(\cdot,\cdot)_m$ for these spaces.* □

From the properties of a Hilbert scale we obtain the following result.

Corollary 3.1 *For every $k, m \in \mathbb{N}$, the spaces $\mathbb{H}^{m+k}$, $\mathbb{H}^m$, and $\mathbb{H}^{m-k}$ form a normal triple, and the mapping*

$$\mathbb{H}^{m-k} \ni x \longleftrightarrow [\cdot, x]_{m+k,m} \in (\mathbb{H}^{m+k})^*,$$

where $[\cdot,\cdot]_{m+k,m}$ is the CBF of this triple, is an isometric isomorphism between $(\mathbb{H}^{m+k})^*$ and $\mathbb{H}^{m-k}$. □

As was discussed in Sect. 3.4.2, the operator Λ^r, $r \in \mathbb{R}^1$, extends to every H^s, and the following result holds.

Proposition 3.5 (i) *For every $r, s \geq 0$ and $x \in \mathbb{H}^{s+2r}$, $y \in \mathbb{H}^{s+r}$,*

$$\Lambda^r\left(\mathbb{H}^s\right) = \mathbb{H}^{s-r} \quad \textit{and} \quad (\Lambda^{2r}x, y)_s = (x, y)_{s+r}.$$

(ii) *The space $\mathbb{C}_0^\infty(\mathbb{R}^{\mathrm{d}})$ is dense in every $\mathbb{H}^s$.* □

3.5.4. Our next step is to define Sobolev spaces in domains.

Definition 3.4 The boundary Γ of the domain $G \subset \mathbb{R}^{\mathrm{d}}$ is said to be regular if there exist a finite open covering $\{\Gamma_i\}$ of the boundary, a finite collection of open, bounded cones $\{K_j\}$, and a number $\varepsilon > 0$ such that

(a) For every point from Γ, the ball with center at that point and radius ε lies entirely in some set Γ_i from the covering.
(b) For every point from $\Gamma_i \cap G$, the cone with vertex at that point, obtained by the parallel transfer of some K_j, lies entirely in G. □

In what follows, we always assume that G is a bounded domain with regular boundary.

Lemma 3.4 *Let G be a bounded domain in $\mathbb{R}^{\mathrm{d}}$ and let $\boldsymbol{R}(G)$ be the set of functions which are restrictions of $\mathbb{C}_0^\infty(\mathbb{R}^{\mathrm{d}})$ functions to G. Then $\boldsymbol{R}(G)$ is dense in $\mathbb{W}_p^m(G)$ (in the strong topology of $\mathbb{W}_p^m(G)$).* □

3.5.5. The theorem presented below is among the most important results in the theory of Sobolev spaces. Recall that $\mathbb{C}^n(G \cup \Gamma)$ is a Banach space with norm

$$\|f\|_{\mathbb{C}^n(G\cup\Gamma)} = \sum_{|\alpha|\leq n} \sup_{x\in G\cup\Gamma} |f_\alpha(x)|.$$

Theorem 3.9 *Let G be a bounded domain with regular boundary.*
If

$$m > \frac{\mathrm{d}}{p} - \frac{\mathrm{d}}{2},$$

then $\mathbb{W}_p^m(G)$ is normally embedded into $\mathbb{L}_2(G)$.
If $m \in \mathbb{N}$, $n \in \mathbb{N} \cup \{0\}$, and

$$m > n + \frac{\mathrm{d}}{p},$$

then $\mathbb{W}_p^m(G)$ is normally embedded into $\mathbb{C}^n(G \cup \Gamma)$. □

Remark 3.4 The above theorem presents two examples of the Sobolev embedding theorems. The first part contains the familiar fact that $\mathbb{L}_p(G) \subset \mathbb{L}_2(G)$ for $p > 2$. The second part shows that, as the power of integrability p grows, the number of generalized derivatives approaches the number of classical derivatives. □

3.5.6

Definition 3.5 The closure of $\mathbb{C}_0^\infty(G)$ with respect to the norm $\|\cdot\|_{G,1,2}$ will be called the Sobolev space $\mathring{\mathbb{H}}^1(G)$. □

Theorem 3.10 *The space $\mathring{\mathbb{H}}^1(G)$ is a separable Hilbert space and is normally embedded in $\mathbb{L}_2(G)$.* □

The Hilbert scale connecting $\mathbb{X} = \mathring{\mathbb{H}}^1(G)$ and $\mathbb{H} = \mathbb{L}_2(G)$ will be denoted by $\{\mathring{\mathbb{H}}^s(G),\ s \in \mathbb{R}^1\}$. The existence of this scale follows from the above theorem.

Proposition 3.6 *For every $s \geq 0$, the space $\mathring{\mathbb{H}}^s(G)$ is the closure of $\mathbb{C}_0^\infty(G)$ with respect to the norm $\|\cdot\|_s$. The norms $\|\cdot\|_{\mathring{\mathbb{H}}^m}$ and $\|\cdot\|_{G,2,m}$ are equivalent for every positive integer m.* □

Warning 3.4 *The norm, the scalar product, and the CBF for spaces from the scale $\{\mathring{\mathbb{H}}^s(G)\}$ will be denoted in the same way as corresponding objects for the scale $\{\mathbb{H}^s\}$. Since one scale is related to the bounded domain G and the other to the whole space $\mathbb{R}^{\mathrm{d}}$, there should be no confusion.* □

From the properties of the Hilbert scale we obtain the following result.

Corollary 3.2 *For all integers $k, m \in \mathbb{N}$ the spaces $\mathring{\mathbb{H}}^{m+k}$, $\mathring{\mathbb{H}}^m$, and $\mathring{\mathbb{H}}^{m-k}$ form a normal triple, and the mapping*

$$\mathring{\mathbb{H}}^{m-k}(G) \ni x \longleftrightarrow [\cdot, x]_{m+k,m} \in \left(\mathring{\mathbb{H}}^{m+k}(G)\right)^*$$

is an isometric isomorphism between $\mathring{\mathbb{H}}^{m-k}$ and $(\mathring{\mathbb{H}}^{m+k})^*$. □

Lemma 3.5 (Friedrichs) *Let m be a positive integer and let G be a bounded domain with regular boundary.*

There exists a positive number N, depending on G and m, such that, for every $u \in \mathring{\mathbb{H}}^m(G)$,

$$\|u\|_m \leq N \sum_{|\alpha|=m} \|u_\alpha\|_{\mathbb{L}_2(G)}.$$ □

3.5.7. Analysis of the above constructions shows that $\mathring{\mathbb{H}}^s(\mathbb{R}^{\mathrm{d}}) = \mathbb{H}^s$, $s \geq 0$. It is therefore not surprising that many results that hold for $u \in \mathbb{H}^s$ also hold for $u \in \mathring{\mathbb{H}}^s(G)$.

Recall the notation

$$f_i = \frac{\partial f}{\partial x^i}.$$

Proposition 3.7 *Let m be a non-negative integer and let G be a bounded domain with a regular boundary. Denote by X^m either $\mathbb{H}^m$ or $\mathring{\mathbb{H}}^m(G)$.*

(i) *There exists a positive number N such that, for all* $u \in X^m$,

$$\|u_i\|_{m-1} \leq \mathbb{N}\|u\|_m.$$

(ii) *A function u belongs to* X^{m-1} *if and only if this function can be represented as*

$$u = \sum_{|\alpha|\leq 1} (v^\alpha)_\alpha$$

with $v^\alpha \in X^m$.

(iii) *If* $v \in \mathbb{C}_b^1(\mathbb{R}^d)$ *or* $v \in \mathbb{H}^1$, *and* $u \in \mathbb{H}^1$, *then*

$$\int_{\mathbb{R}^d} v(x)u_i(x)dx = -\int_{\mathbb{R}^d} v_i(x)u(x)dx.$$

(iv) *If* $v \in \mathbb{C}_b^1(G)$ *or* $v \in \mathbb{W}_2^1(G)$, *and* $u \in \mathring{\mathbb{H}}^m(G)$, *then*

$$\int_G v(x)u_i(x)dx = -\int_G v_i(x)u(x)dx. \qquad \square$$

3.5.8 When it is necessary to consider bounded or slowly growing at infinity functions, the spaces $\mathbb{W}_p^m$ are of no use and must be replaced with weighted Sobolev spaces.

There are several different constructions of weighted Sobolev spaces. Below is the construction used in this book.

Denote by $\mathbb{L}_p(r)$, $r \in \mathbb{R}^1$, $p \geq 1$, the space consisting of the equivalence classes of $\mathscr{B}(\mathbb{R}^d)$-measurable functions on $\mathbb{R}^d$ such that

$$\int_{\mathbb{R}^d} (1+|x|^2)^{pr/2}\,|f(x)|^p dx < \infty.$$

Next, let $\mathbb{W}_p^m(r)$, $m \in \mathbb{N} \cup \{0\}$ be the space of real-valued functions on $\mathbb{R}^d$ belonging, together with their generalized derivatives up to and including mth order, to $\mathbb{L}_p(r)$.

Theorem 3.11 *The space* $\mathbb{W}_p^m(r)$ *endowed with the norm*

$$\|u\|_{m,p,r} = \left(\int_{\mathbb{R}^d} \sum_{|\alpha|\leq m} (1+|x|^2)^{pr/2}\,|u_\alpha(x)|^p dx \right)^{1/p}$$

is a separable Banach space, and $\mathbb{W}_2^m(r)$ *is a Hilbert space with respect to the scalar product generated by the norm* $\|\cdot\|_{m,2,r}$. *The set* $\mathbb{C}_0^\infty(\mathbb{R}^d)$ *is dense in* $\mathbb{W}_p^m(r)$ *(in the strong topology of* $\mathbb{W}_p^m(r)$*).* □

Warning 3.5 *In what follows, the space* $\mathbb{W}_2^m(r)$ *will be denoted by* $\mathbb{H}^m(r)$. □

The proof of the next result is an easy exercise for the reader. The main point to remember is that

$$\int_{\mathbb{R}^d} \frac{dx}{(1+|x|^2)^{r/2}} < \infty \iff r > \mathrm{d}.$$

Lemma 3.6 (i) *Let* f *be a* $\overline{\mathscr{B}(\mathbb{R}^d)}$*-measurable real-valued function, and, for some* $s, m \in \mathbb{R}_+$,

$$\sum_{|\alpha|\le m} |f_\alpha(x)| \le N(1+|x|^2)^{s/2}.$$

If $p \ge 1$ *and* $rp < -(\mathrm{d}+sp)$, *then* $f \in \mathbb{W}_p^m(r)$.

(ii) *If* $s,\ r \in \mathbb{R}^1$, *then the operator* $\mathcal{M}^s$ *defined by the equality* $\left(\mathcal{M}^s f\right)(x) = (1+|x|^2)^{s/2} f(x)$ *belongs to* $\mathfrak{L}\left(\mathbb{W}_p^n(r), \mathbb{W}_p^m(r-s)\right)$. □

3.5.9. Now we are ready to carry out a detailed analysis of problem (3.5.1), (3.5.2). We make the following assumptions:

- The real-valued functions $a^{\alpha\beta} = a^{\alpha\beta}(t,x,\omega)$, $b^{l\beta} = b^{l\beta}(t,x,\omega)$, and $f^\alpha = f^\alpha(t,x,\omega)$ are $\overline{\mathscr{B}([0,T]\times G)\otimes\mathscr{F}}$-measurable and are predictable for every $x \in G$.
- The real-valued function $\varphi = \varphi(x,\omega)$ is $\overline{\mathscr{B}(G)\otimes\mathscr{F}_0}$-measurable.
- The functions $a^{\alpha\beta}$, $b^{l\beta}$ are uniformly bounded.
- $\mathbb{E}\|\varphi\|_0^2 < \infty$, $\mathbb{E}\int_{[0,T]}\sum_\alpha \|f^\alpha(t)\|_0^2\,dt < \infty$.

Definition 3.6 A function

$$u \in \mathbb{L}_2^\omega([0,T],\mathscr{P};\mathring{\mathbb{H}}^m(G))$$

is called a **generalized solution** of problem *(3.5.1)*, (3.5.2), if, for every $y \in \mathbb{C}_0^\infty(G)$, the following equality holds for $\mathbb{P}$- a.a. $(t,\omega) \in [0,T]\times\Omega$:

$$\begin{aligned}\big(y,u(t)\big)_0 &= (y,\varphi)_0 - \int_{[0,t]}\sum_{|\alpha|\le m}\Big(y_\alpha, \sum_{|\alpha|\le m} a^{\alpha\beta}u_\beta(s) + f^\alpha(s)\Big)_0 ds \\ &+ \sum_{l=1}^{\mathrm{d}_1}\int_{[0,T]}\Big(y, \sum_{|\beta|\le m} b^{l\beta}u_\beta(s)\Big)_0 dw^l(s) + \big(y, M(t)\big)_0.\end{aligned} \tag{3.5.3}$$

□

Note that the integrals in (3.5.3) are well-defined, because, in view of Proposition 3.7(i), the processes $\sum_{|\beta|\le m} a^{\alpha\beta} u_\beta$ and $\sum_{|\beta|\le m} b^{l\beta} u_\beta$ belong to the space $\mathbb{L}_2^\omega([0,T],\mathscr{P};\mathbb{L}_2(G))$.

Remark 3.5 Equality (3.5.3) can be obtained by multiplying (3.5.1) by y and integrating by parts, keeping in mind the boundary conditions.

If the generalized solution of the problem (3.5.1), (3.5.2) belongs to

$$\mathbb{L}_p\big([0,T],\overline{\mathscr{B}([0,T])},\mathfrak{l};\mathbb{W}_p^{2m+1}(G)\big)$$

for a sufficiently large p, then, under some additional regularity of the boundary of G, a suitable Sobolev embedding theorem will imply that the solution has a version in $\mathbb{C}^{2m}(G\cup\Gamma)$ that is continuous in t and such that all its spatial derivatives of order up to $m-1$ are equal to zero on the boundary of G. Such a version satisfies equalities (3.5.1), (3.5.2) for all t,x, and can therefore be called the classical solution of the Dirichlet problem (3.5.1), (3.5.2). Classical solutions and related questions are studied in the next chapter. □

Let us connect the notion of generalized solution with the definition of solution presented earlier in this chapter. Note that, for every $y, v\in \mathring{\mathbb{H}}^m(G)$,

$$\begin{aligned}\Big|\sum_{|\alpha|\le m}\Big(y_\alpha,\sum_{|\beta|\le m}a^{\alpha\beta}(t,\cdot,\omega)v_\beta\Big)_0\Big|\\ \le N\sum_{|\alpha|\le m}\|y_\alpha\|_0\sum_{|\beta|\le m}\|v_\beta\|\le N_1\|y\|_m\cdot\|v\|_m,\end{aligned}\tag{3.5.4}$$

where N_1 does not depend on (t,ω).

Thus, for every fixed $v\in\mathring{\mathbb{H}}^m(G)$ and every $(t,\omega)\in[0,T]\times\Omega$, there exists an operator

$$A(t,\omega)\in\mathfrak{L}\Big(\mathring{\mathbb{H}}^m(G),\mathring{\mathbb{H}}^{-m}(G)\Big)$$

such that

$$-\sum_{|\alpha|\le m}(y_\alpha,\sum_{|\beta|\le m}a^{\alpha\beta}(t,\omega)v_\beta)_0=[y,A(t,\omega)v];\tag{3.5.5}$$

to simplify the notation, throughout the rest of this section we use $[\cdot,\cdot]$ to denote the CBF of the normal triple $\big(\mathring{\mathbb{H}}^m(G),\mathbb{L}_2(G),\mathring{\mathbb{H}}^{-m}(G)\big)$ (see Sect. 3.5.6). By (3.5.4) and (3.5.5), and in view of the isometry between $\mathring{\mathbb{H}}^{-m}(G)$ and $\big(\mathring{\mathbb{H}}^m(G)\big)^*$, there exists a positive number N such that, for all t,ω and every $v\in\mathring{\mathbb{H}}^m(G)$,

$$\|A(t,\omega)v\|_{-m}\le N\|v\|_m.\tag{3.5.6}$$

It follows from the Pettis theorem (Sect. 1.3.9) and equality (3.5.5) that, for every $v \in \mathring{\mathbb{H}}^m(G)$, the process Av is predictable with values in $\mathring{\mathbb{H}}^{-m}(G)$. Next, define the operator

$$B(t,\omega)v := \Big(\sum_{|\beta|\le m} b^{1\beta}(t,\omega)v_\beta, \dots, \sum_{|\beta|\le m} b^{\mathrm{d}_1\beta}(t,\omega)v_\beta\Big). \tag{3.5.7}$$

Then, for every (t,ω), $B(t,\omega) \in \mathfrak{L}\Big(\mathring{\mathbb{H}}^{-m}(G), \mathfrak{L}_2\big(\mathbb{R}^{\mathrm{d}_1}, \mathbb{L}_2(G)\big)\Big)$, and there exists a positive number N that does not depend on t, ω, such that, for all $v \in \mathring{\mathbb{H}}^m(G)$,

$$|||B(t,\omega)v||| \le N\|v\|_m.$$

Also, for every $v \in \mathring{\mathbb{H}}^m(G)$, the process Bv is predictable with values in $\mathbb{L}_2\big(\mathbb{R}^{\mathrm{d}_1}, \mathbb{L}_2(G)\big)$.

Finally, define

$$f := \sum_{|\alpha|\le m} (f^\alpha)_\alpha. \tag{3.5.8}$$

By Proposition 3.7(ii) we find that f is a predictable process with values in $\mathring{\mathbb{H}}^{-m}(G)$.

As a result, we conclude that, if it exists, a generalized solution of problem (3.5.1), (3.5.2) coincides with the solution of the LSES (3.1.1) in the normal triple $\big(\mathring{\mathbb{H}}^m(G), \mathbb{L}_2(G), \mathring{\mathbb{H}}^{-m}(G)\big)$ when the operators A, B and the function f are defined by (3.5.5), (3.5.7) and (3.5.8).

In particular, by Proposition 1.3, a generalized solution of (3.5.1) (3.5.2) has a version in $\mathbb{C}\big([0,T], \mathscr{P}; \mathbb{L}_2(G)\big)$, which we will consider from now on.

Theorem 3.12 *Assume that*

(A_m) *There exist $K \ge 0$ and $\delta > 0$ such that, for every collection of real numbers $\{\xi^\alpha,\ |\alpha| \le m\}$ and all $(t,x,\omega) \in [0,T]\times G\times\Omega$,*

$$2\sum_{\substack{|\alpha|\le m\\ |\beta|\le m}} a^{\alpha\beta}(t,x,\omega)\,\xi^\alpha\xi^\beta - \sum_{l=1}^{\mathrm{d}_1}\Big|\sum_{|\beta|\le m} b^{l\beta}(t,x,\omega)\xi^\beta\Big|^2 \ge \delta\sum_{|\alpha|=m}|\xi^\alpha|^2 - K|\xi^0|^2;$$

ξ^0 is the number ξ^α corresponding to $|\alpha| = 0$.

Then problem (3.5.1), (3.5.2) *has a unique generalized solution, and there exists a positive number C such that*

$$\mathbb{E}\sup_{t\le T}\|u(t)\|_0^2+\mathbb{E}\int_{[0,T]}\|u(s)\|_m^2\,ds$$
$$\le C\,\mathbb{E}\Big(\|\varphi\|_0^2+\int_{[0,T]}\|f(s)\|_{-m}^2\,ds+\|M(t)\|_0^2\Big). \tag{3.5.9}$$

Proof We show that (3.5.1), (3.5.2), when formulated as an LSES in the normal triple

$$\big(\mathring{\mathbb{H}}^m(G), \mathbb{L}_2(G), \mathring{\mathbb{H}}^{-m}(G)\big),$$

satisfies the conditions of Theorem 3.1.

In fact, conditions (B) and (C) of Theorem 3.1 are fulfilled in view of (3.5.6) and the assumptions made in the beginning of this paragraph, and it remains to prove coercivity, that is, to confirm that condition (A_m) of the theorem implies condition (A) from Sect. 3.1.1.

If $v\in\mathring{\mathbb{H}}^m(G)$, then, by (3.5.5) and (3.5.7), we find that, for all t,ω,

$$2[v,Av]+\|\!|Bv|\!\|^2=-2\sum_{\substack{|\alpha|\le m\\|\beta|\le m}}\big(v_\alpha, a^{\alpha\beta}\,v_\beta\big)_0$$
$$+\sum_{l=1}^{d_1}\Big\|\sum_{|\beta|\le m}b^{l\beta}v_\beta\Big\|_0^2=\int_G\Big(-2\sum_{\substack{|\alpha|\le m\\|\beta|\le m}}a^{\alpha\beta}v_\alpha v_\beta(x)$$
$$+\sum_{l=1}^{d_1}\Big|\sum_{|\beta|\le m}b^{l\beta}\,v_\beta(x)\Big|^2\Big)dx.$$

Condition (A_m) of the theorem then implies

$$2[v,Av]+\|\!|Bv|\!\|^2\le-\delta\sum_{|\alpha|=m}\|v_\alpha\|_0^2+K\|v\|_0^2,$$

and then, by the Fridrichs lemma (Sect. 5),

$$2[v,Av]+\|\!|Bv|\!\|^2\le-\delta\,N\|v\|_m^2+K\|v\|_0^2,$$

which is the desired condition (A) from Sect. 3.1.1. □

3.5.10.

Remark 3.6

$$\mathbb{E}\|\varphi\|_0^2 < \infty \text{ and } \mathbb{E}\int_{[0,T]} \sum_{\alpha} \|f^{\alpha}(t)\|_0^2 \, dt < \infty$$

are not necessary for uniqueness, because the difference of two generalized solutions is a generalized solution of the same problem with $\varphi = 0$, $f^{\alpha} = 0$. □

Theorems 3.12 and 3.6 yield the following result.

Corollary 3.3 *A generalized solution of the problem* (3.5.1), (3.5.2) *possesses the Markov property, that is, for every Borel set* $\Gamma \subset \mathbb{L}_2(G)$ *and every* $s,\ t \in [0, T]$ *with* $s \le t$,

$$\mathbb{P}\Big(u(t) \in \Gamma \big| \mathscr{F}_s(u)\Big) = \mathbb{P}\Big(u(t) \in \Gamma \big| u(s)\Big) \ (\mathbb{P}\text{-a.s.}),$$

where $\mathscr{F}_s(u)$ *is the* σ*-algebra generated by* $u(r, x)$ *for* $r \le s$ *and* $x \in G$. □

3.5.11.

Remark 3.7 All the results of this chapter extend to the case of complex-valued operators, initial conditions, and free terms. In this case, we use the real part of the CBF in the corresponding coercivity and dissipativity conditions.

Chapter 4
Itô's Second-Order Parabolic Equations

4.1 Introduction

4.1.1. Let us fix $T_0, T \in \mathbb{R}_+$ with $T_0 \leq T$, and $\mathrm{d}, \mathrm{d}_1 \in \mathbb{N}$. We also fix the stochastic basis $\mathbb{F} = (\Omega, \mathscr{F}, \{\mathscr{F}_t\}_{t \in [0,T]}, \mathbb{P})$ with the usual assumptions, and a standard Wiener process w on $\mathbb{F}$ with values in $\mathbb{R}^{\mathrm{d}_1}$.

Warning 4.1 *Throughout what follows*:

(a) *upper indices, or superscripts, i, i', j, l, k represent coordinates of a vector; the same indices put below as subscripts represent differentiation with respect to the corresponding vector coordinate. For example,*

$$g_{ij} = \frac{\partial^2 g(x)}{\partial x^i \partial x^j};$$

(b) *repeated indices in monomials are summed over. For example,*

$$\int\limits_{[0,t]} h^l(s)\, dw^l(s) = \sum_{l=1}^{\mathrm{d}_1} \int\limits_{[0,t]} h^l(s)\, dw^l(s));$$

(c) *the notation of Sect. 4.4.5 remains in use*;
(d) *in the notations of the spaces $\mathbb{W}_p^m(\mathbb{R}^{\mathrm{d}})$, $\mathbb{H}^p(r, \mathbb{R}^{\mathrm{d}})$, $\mathbb{W}_p^m(r, \mathbb{R}^{\mathrm{d}})$, and $\mathbb{L}_p(r, \mathbb{R}^{\mathrm{d}})$ the argument $\mathbb{R}^{\mathrm{d}}$ will be omitted.*

4.1.2. In this chapter we study the equation

$$\begin{aligned} du(t,x,\omega) = \Big(& a^{ij}(t,x,\omega) u_{ij}(t,x,\omega) + b^i(t,x,\omega) u_i(t,x,\omega) \\ & + c(t,x,\omega) u(t,x,\omega) + f(t,x,\omega) \Big)\, dt \end{aligned} \tag{4.1.1}$$

B. L. Rozovsky, S. V. Lototsky, *Stochastic Evolution Systems*, Probability Theory and Stochastic Modelling 89, https://doi.org/10.1007/978-3-319-94893-5_4

$$+\Big(\sigma^{il}(t,x,\omega)u_i(t,x,\omega)+h^l(t,x,\omega)u(t,x,\omega)+g^l(t,x,\omega)\Big)\,dw^l(t),$$
$$(t,x,\omega)\in(T_0,T]\times\mathbb{R}^{\mathrm{d}}\times\Omega.$$

We investigate the solvability of forward and backward Cauchy problems in generalized and classical settings, that is, in Sobolev spaces and in the space of continuously differentiable functions.

From the formal point of view, it could be natural to consider more general equations than (4.1.1), namely, equations containing a second-order partial differential operator in the dw term, but such equations require a very different approach and are not as common in applications.

Definition 4.1 An equation of the type (4.1.1) is called **parabolic** if, for all $(t,x,\omega)\in[T_0,T]\times\Omega\times\mathbb{R}^{\mathrm{d}}$ and all $\xi\in\mathbb{R}^{\mathrm{d}}$,

$$2a^{ij}(t,x,\omega)\xi^i\xi^j-\sum_{l=1}^{\mathrm{d}_1}|\sigma^{il}(t,x,\omega)\xi^i|^2\ge 0; \tag{4.1.2}$$

the equation is called **super-parabolic** if there exists a positive number δ such that, for t,ω,x,ξ,

$$2a^{ij}(t,x,\omega)\xi^i\xi^j-\sum_{l=1}^{\mathrm{d}_1}\left|\sigma^{il}(t,x,\omega)\xi^i\right|^2\ge\delta\sum_{i=1}^{\mathrm{d}}|\xi^i|^2. \tag{4.1.3}$$

□

In other words, the parabolic condition holds if and only if the matrix $A=(A^{ij})$ with components

$$A^{ij}=2a^{ij}\;-\;\sigma^{il}\sigma^{il}$$

is non-negative definite, and the super-parabolic condition holds if and only if A is uniformly positive definite.

When $\sigma^{il}\equiv 0$, conditions (4.1.2), (4.1.3) turn into the usual conditions from the theory of partial differential equations (see [120, Section 4.1]). For example, if $\sigma^{il}\equiv 0$, then, under some additional regularity conditions on the functions a^{ij},b^i,c, condition (4.1.3) is equivalent to uniform ellipticity of the operator $Lf:=(a^{ij}fi)_j+b^if_i+cf$.

Note that if the coefficients of Eq. (4.1.1) are uniformly bounded, then the condition (4.1.3) is a special case of condition (A) of Theorem 3.12 with $m=1$.

Considering the Cauchy problem for Eq. (4.1.1) as an LSES in some normal triple of Sobolev spaces, we prove below that the super-parabolic condition ensures that the LSES is coercive and the parabolic condition guarantees its dissipativity.

Getting ahead of the story, we may say that, in many aspects of the theory, a super-parabolic Itô equation is similar to a deterministic parabolic equation, whereas a parabolic Itô equation is similar to a deterministic degenerate parabolic equation.

It is important to understand to what extent the parabolic condition is necessary for the solvability of Itô partial differential equations. The following example shows that this condition cannot, in general, be omitted if we want (4.1.1) to be well-posed in the class of square-integrable functions.

Example 4.1 Consider the following special case of Eq. (4.1.1)

$$u(t,x) = \varphi(x) + \int\limits_{[0,t]} a^2 \frac{\partial^2 u(s,x)}{\partial x^2} ds + \int\limits_{[0,t]} \sigma \frac{\partial u(s,x)}{\partial x} dw(s), \tag{4.1.4}$$

where $\varphi \in \mathbb{L}_2$.

We call a function $u = u(t, x, \omega)$ a solution of Eq. (4.1.4) if it is predictable for every x, continuous in t for all (x, ω), twice continuously differentiable in x for all (t, ω), and satisfies (4.1.4) for all $t \in [0, T]$ for $\mathbb{P}$-a.a. ω.

The Fourier transforms of the functions $\varphi(x)$ and $u(t, x)$ will be denoted by $\hat{\varphi}(y)$ and $\hat{u}(t, y)$, respectively.

Taking the Fourier transforms on both sides of (4.1.4),

$$\hat{u}(t,y) = \hat{\varphi}(y) - \int\limits_{[0,t]} (ya)^2 \hat{u}(s,y)\, ds + \mathrm{i} \int\limits_{[0,t]} (\sigma y)\hat{u}(s,y)\, dw(s),$$

where $\mathrm{i} = \sqrt{-1}$ is the imaginary unit.

This equation, considered as an ordinary differential equation for fixed y, has a unique solution

$$\hat{u}(t,y) = \hat{\varphi}(y) \exp\left(-\frac{1}{2}(2a^2 - \sigma^2)y^2 t + \mathrm{i}\sigma y w(t) \right);$$

see [37, Section 4.1]. By Parseval's equality,

$$\|u(t)\|_{\mathbb{L}_2}^2 = \int\limits_{\mathbb{R}^1} |\hat{\varphi}(y)|^2 \exp\Big(-(2a^2 - \sigma^2)y^2 t \Big) dy.$$

If the parabolic condition does not hold, that is, if $2a^2 - \sigma^2 < 0$, then the integral on the right-hand side of the last equality converges only under very special assumptions on φ.

If we drop the requirement of classical solvability of Eq. (4.1.4) and instead interpret the equation as an LSES in $(\mathbb{H}^1, \mathbb{L}_2, \mathbb{H}^{-1})$ in the spirit of Definition 3.6, then the result remains the same: without parabolicity conditions, the equation is not well-posed in $(\mathbb{H}^1, \mathbb{L}_2, \mathbb{H}^{-1})$ and a generalized solution exists only for very special initial conditions φ; see [72, Section III.3].

4.1.3. Below is the list of the main notation to be used extensively in what follows.

$\mathbb{X}$	A separable Banach space
p	A number in the interval $[1, +\infty)$
$\|\cdot\|_{m,p}$	The norm in $\mathbb{W}^m_p$, $p \neq 2$
$\|\cdot\|_m$	The norm in $\mathbb{H}^m = \mathbb{W}^m_2$
$(\cdot,\cdot)_m$	The inner product in $\mathbb{H}^m$
$[\cdot,\cdot]_m$	The CBF of $(\mathbb{H}^{m+1}, \mathbb{H}^m, \mathbb{H}^{m-1})$
$\tilde{\mathscr{F}}$	A sub-σ-algebra of $\mathscr{F}$
$\tilde{\mathbb{P}}$	Restriction of $\mathbb{P}$ to $\tilde{\mathscr{F}}$
$\mathbb{C}([T_0, T], \mathscr{P}; \mathbb{X})$	The set of all strongly continuous $\mathbb{X}$-valued processes
$\mathbb{L}_p([T_0, T]; \mathbb{X})$	$\mathbb{L}_p([T_0, T], \overline{\mathscr{B}([T_0, T])}, \mathfrak{l}; \mathbb{X})$
$\mathbb{L}_p([T_0, T] \times \Omega; \mathbb{X})$	$\mathbb{L}_p([T_0, T] \times \Omega, \overline{\mathscr{B}([T_0, T]) \otimes \mathscr{F}}, \mathfrak{l} \times \mathbb{P}; \mathbb{X})$
$\mathbb{L}_p(\Omega, \tilde{\mathscr{F}}; \mathbb{X})$	$\mathbb{L}_p(\Omega, \tilde{\mathscr{F}}, \tilde{\mathbb{P}}; \mathbb{X})$
$\mathbb{L}_p(\Omega; \mathbb{X})$	$\mathbb{L}_p(\Omega, \mathscr{F}, \mathbb{P}; \mathbb{X})$
$\mathbb{L}_p(\Omega; \mathbb{C}([T_0, T]; \mathbb{X}))$	$\mathbb{L}_p(\Omega, \mathscr{F}, \mathbb{P}; \mathbb{C}([T_0, T]; \mathbb{X}))$
$\mathbb{L}_p([T_0, T] \times \Omega; \mathbb{C}_w\mathbb{X})$	The set of all representatives of $\mathbb{L}_p([T_0, T]) \times \Omega; \mathbb{X})$ that are weakly continuous in t, ($\mathbb{P}$-a.s.) as $\mathbb{X}$-valued functions
$\mathbb{L}^{\omega}_p([T_0, T]; \mathbb{X})$	The set of all $\overline{\mathscr{B}([T_0, T]) \otimes \mathscr{F}}$-measurable functions $f : [T_0, T] \times \Omega \to \mathbb{X}$ such that $f(\cdot, \omega) \in L_p([T_0, T];\ \mathbb{X})$ for $\mathbb{P}$- a.a. ω
$\mathbb{L}_p([T_0, T], \mathscr{P}; \mathbb{X})$	The set of all predictable representatives from $\mathbb{L}_p(T_0, T] \times \Omega; \mathbb{X})$
$\mathbb{L}^{\omega}_p([T_0, T]; \mathscr{P}; \mathbb{X})$	The set of all predictable representatives from $\mathbb{L}^{\omega}_p([T_0, T]; \mathbb{X})$

Warning 4.2 *Similar notations, but with $\overleftarrow{\mathscr{P}}$ instead of $\mathscr{P}$, are used for backward predictable functions.*

4.2 The Cauchy Problem for Super-Parabolic Itô Equations in Divergence Form

4.2.1. In this section we consider the Cauchy problem

$$du(t, x, \omega) = \Big(\big(a^{ij}(t, x, \omega)u_i(t, x, \omega)\big)_j + b^i(t, x, \omega)u_i(t, x, \omega)$$

$$+ c(t, x, \omega)u(t, x, \omega) + f(t, x, \omega)\Big)\, dt$$

$$+\Big(\sigma^{il}(t,x,\omega)u_i(t,x,\omega)+h^l(t,x,\omega)u(t,x,\omega) \tag{4.2.1}$$

$$+g^l(t,x,\omega)\Big)\,dw^l(t,\omega),\quad (t,x,\omega)\in(T_0,T]\times\mathbb{R}^{\mathrm{d}}\times\Omega,$$

$$u(T_0,x,\omega)=\varphi(x,\omega),\quad (x,\omega)\in\mathbb{R}^{\mathrm{d}}\times\Omega. \tag{4.2.2}$$

Equation (4.2.1) differs from (4.1.1) in the term containing the second-order derivatives. If the coefficients a^{il} are differentiable in x, then Eqs. (4.1.1) and (4.2.1) are equivalent, because

$$(a^{ij}u)_i = a^{ij}u_{ij} + a_i^{ij}u_j.$$

In that case, results concerning problem (4.1.1), (4.2.2) can be obtained from the corresponding results for problem (4.2.1), (4.2.2) by reducing Eq. (4.1.1) to the form (4.2.1).

On the other hand, if the super-parabolic condition is fulfilled, then problem (4.2.1), (4.2.2), unlike problem (4.1.1), (4.2.2), is solvable without any assumptions on differentiability or even continuity of the coefficients of the equation.

Throughout this section it will be supposed that the coefficients of Eq. (4.2.1) $a^{ij},\ b^i,\ c,\ \sigma^{il},\ h^l,\ i,\ j = 1,2,\dots d,\ l = 1,2,\dots \mathrm{d}_1$, are real-valued, $\mathscr{B}([T_0,T]\times\mathbb{R}^{\mathrm{d}})\otimes\mathscr{F}$-measurable, uniformly in (t,x,ω) bounded, predictable (for every $x\in\mathbb{R}^{\mathrm{d}}$) functions, and φ is an $\mathscr{F}_{T_0}$-measurable random variable taking values in $\mathbb{L}_2$.

Additionally, it is assumed that

$$f\in\mathbb{L}_2^{\omega}\big([T_0,T],\mathscr{P};\mathbb{H}^{-1}\big),\quad g^l\in\mathbb{L}_2^{\omega}\big([T_0,T],\mathscr{P};\mathbb{L}_2\big),\ l=1,2,\dots,\mathrm{d}_1.$$

In the next paragraph, it will be proved that problem (4.2.1), (4.2.2) has a unique solution in the space $\mathbb{L}_2\big([T_0,T],\mathscr{P};\mathbb{H}^1\big)\cap\mathbb{L}_2\big(\Omega;\mathbb{C}([T_0,T];\mathbb{H}^0)\big)$.

After that, we investigate how regularity of the solution depends on regularity of the input (coefficients of the equation, the initial condition φ and the free terms $f,\ g^l$). In particular, we prove that, under appropriate assumptions about the input, problem (4.2.1), (4.2.2) has a solution in $\mathbb{L}_p\big([T_0,T]\times\Omega;\mathbb{C}_w\mathbb{W}_p^m\big)\cap\mathbb{L}_2\big([T_0,T],\mathscr{P};\mathbb{H}^{m+1}\big)\cap\mathbb{L}_2\big(\Omega;\mathbb{C}([T_0,T];\mathbb{H}^m)\big)$, as well as a classical solution.

4.2.2.

Definition 4.2 A function $u\in\mathbb{L}_2^{\omega}([T_0,T],\mathscr{P};\mathbb{H}^1)$ is called a **generalized solution** of problem (4.2.1), (4.2.2) if, for every $y\in\mathbb{C}_0^{\infty}(\mathbb{R}^{\mathrm{d}})$, the following equality holds for $\ell \times \mathbb{P}$- a.a. (t, ω) :

$$\begin{aligned}
\big(u(t), y\big)_0 = (\varphi, y)_0 + \int\limits_{[T_0,t]} \Big(&- \big(a^{ij} u_i(s), y_i\big)_0 \\
&+ \big(b^i u_i(s) + cu(s), y\big)_0 + [f(s), y]_0 \Big)\, ds \\
&+ \int\limits_{[T_0,T]} \big(\sigma^{il} u_i(s) + h^l(s)u(s) + g^l(s), y\big)_0 \, dw^l(s),
\end{aligned} \tag{4.2.3}$$

where $[\cdot, \cdot]_0$ is the CBF of the normal triple $(\mathbb{H}^1, \mathbb{L}_2, \mathbb{H}^{-1})$. □

All the integrals in (4.2.3) are well-defined by the Cauchy–Schwarz inequality and the first property of a CBF (Sect. 2.5.2).

Our next step is to formulate (4.2.1), (4.2.2) as the LSES (3.1.1). By the Cauchy–Schwarz inequality,

$$\Big|-(a^{ij} v_i, y_j)_0 + (b^i v_i + cv, y)_0\Big| \le N\|v\|_1 \cdot \|y\|_1, \quad v, y \in \mathbb{H}^1, \tag{4.2.4}$$

where the number N does not depend on (t, ω). Hence, for every $v \in \mathbb{H}^1$, the expression on the left-hand side of inequality (4.2.4) is a continuous linear functional on $\mathbb{H}^1$. Because the CBF $[\cdot, \cdot]_0$ establishes an isometric isomorphism between $(\mathbb{H}^1)^*$ and $\mathbb{H}^{-1}$ (see Proposition 3.4), there exists a family of bounded linear operators $A(t, \omega) : \mathbb{H}^1 \to \mathbb{H}^{-1}$ such that, for all $v, y \in \mathbb{H}^1$,

$$[y, A(t, \omega)v]_0 = -\big(a^{ij}(t, \omega)v_i, y_j\big)_0 + \big(b^i(t, \omega)v_i + c(t, \omega)v, y\big)_0, \tag{4.2.5}$$

and

$$\|A(t, \omega)v\|_{-1} \le N\|v\|_1,$$

with N independent of (t, ω).

Define also $B^l(t, \omega)v := \sigma^{il}(t, \omega)v_i + h^l(t, \omega)v$, and

$$B(t, \omega)v := \big(B^1(t, \omega)v, \dots, B^{d_1}(t, \omega)v\big). \tag{4.2.6}$$

Then $B(t, \omega) \in \mathfrak{L}\big(\mathbb{H}^1, \mathfrak{L}_2(\mathbb{R}^{d_1}, \mathbb{L}_2)\big)$ for every (t, ω). By the Pettis theorem, it is easy to verify that Av is a predictable $\mathbb{H}^{-1}$-process and Bv is a predictable $\mathfrak{L}_2\big(\mathbb{R}^{d_1}, \mathbb{L}_2\big)$-process.

Finally, define

$$M(t) := \sum_{l=1}^{d_1} \int_{[T_0,t]} g^l(s)\, dw^l(s).$$

As a result, problem (4.2.1), (4.2.2) becomes a particular case of the LSES (3.1.1), and the definition of a generalized solution of problem (4.2.1), (4.2.2) is equivalent to the definition of a solution of this LSES, as stated in Definition 3.1 (see also Remark 3.2).

Remark 4.1 By Proposition 3.1, a generalized solution of problem (4.2.1), (4.2.2) has a version $u \in \mathbb{C}([T_0, T]; \mathscr{P}; \mathbb{L}_2)$, which we will denote the same way and consider from now on. This, in turn, implies the existence of a set $\Omega' \subset \Omega$ with $\mathbb{P}(\Omega') = 1$ such that u satisfies *(4.2.3)* for all $t \in [T_0, T]$ and $\omega \in \Omega' \subset \Omega$.

□

Theorem 4.1 *If the super-parabolicity condition* (4.1.3) *holds, then there exists at most one generalized solution of problem* (4.2.1), (4.2.2).

If, in addition,

$$\varphi \in \mathbb{L}_2(\Omega; \mathbb{L}_2), \quad f \in \mathbb{L}_2([T_0, T] \times \Omega; \mathbb{H}^{-1}), \quad \textit{and} \quad g^l \in \mathbb{L}_2(T_0, T] \times \Omega; \mathbb{L}_2)$$

for every $l = 1, 2, \ldots, d_1$, *then the problem has a generalized solution* u *in the space* $\mathbb{L}_2([T_0, T]; \mathscr{P}; \mathbb{H}^1) \cap \mathbb{C}([T_0, T]; \mathscr{P}; \mathbb{L}_2)$, *and there exists a positive number* N *depending only on* T_0, T, δ, d, d_1, *and* $\max_{\ell,i,j,t,x,\omega} (|a^{ij}|,\ |b^i,\ |\sigma^{il}|,\ h^l|)$, *such that*

$$\mathbb{E} \sup_{t\in[T_0,T]} \|u(t)\|_0^2 + \mathbb{E} \int_{[T_0,T]} \|u(s)\|_1^2\, ds$$

$$\leq N\mathbb{E}\left(\|\varphi\|_0^2 + \int_{[T_0,T]} \left(\|f(t)\|_{-1}^2 + \sum_{l=1}^{d_1} \|g^l(t)\|_0^2 \right) dt \right).$$

□

Proof To prove the theorem, we verify the coercivity condition (A) in Sect. 3.1.1 for the operators A from (4.2.5) and B from (4.2.6), in the normal triple $(\mathbb{H}^1, \mathbb{L}_2, \mathbb{H}^{-1})$; and then apply Theorems 3.1 and 3.4.

To begin, note that, for every $v \in \mathbb{H}^1$ and all t, ω,

$$2[v, Av]_0 + \|Bv\| = -2(a^{ij} v_i, v_j)_0 + 2(b^i v_i + cv, v)_0$$

$$+ \sum_{l=1}^{d_1} \|\sigma^{il} v_i\|_0^2 + 2\sum_{l=1}^{d_1} (\sigma^{il} v_i, h^l v)_0 + \sum_{l=1}^{d_1} \|h^l v\|_0^2.$$

Next, recall that

$$\|v\|_1^2 = \|v\|_0^2 + \sum_{i=1}^{\mathrm{d}} \|v_i\|_0^2.$$

Then, from the super-parabolicity condition (4.1.3) it follows that

$$\begin{aligned}
&- 2(a^{ij} v_i,\ v_j)_0 + \sum_{l=1}^{\mathrm{d}_1} \|\sigma^{il} v_i\|_0^2 \\
&\qquad = \int_{\mathbb{R}^{\mathrm{d}}} \left(- 2a^{ij} v_i v_j(x) + \sum_{l=1}^{\mathrm{d}_1} |\sigma^{il} v_i(x)|^2 \right) dx \qquad (4.2.7) \\
&\qquad \le -\delta \int_{\mathbb{R}^{\mathrm{d}}} \sum_{i=1}^{\mathrm{d}} |v_i(x)|^2 dx = -\delta \|v\|_1^2 + \delta \|v\|_0^2.
\end{aligned}$$

On the other hand, with the help of the Cauchy–Schwarz inequality we find that

$$\begin{aligned}
U &:= \left| 2(b^i v_i + cv,\, v)_0 + 2 \sum_{l=1}^{\mathrm{d}_1} (\sigma^{il} v_i,\ h^l v)_0 + \sum_{l=1}^{\mathrm{d}_1} \|h^l v\|_0^2 \right| \\
&\le N \left(\|v\|_0^2 + \sum_{i=1}^{\mathrm{d}} \|v_i\|_0 \cdot \|v\|_0 \right).
\end{aligned}$$

Now, making use of $2|ab| \le \varepsilon a^2 + \varepsilon^{-1} b^2$, cf. (3.2.10),

$$U \le \varepsilon N_1 \|v\|_1^2 + N_2(\varepsilon) \|v\|_0^2, \qquad (4.2.8)$$

where the number N_2 depends on ε, but N_1 does not.

Choosing ε sufficiently small, we obtain after combining (4.2.7) and (4.2.8) that

$$2[v,\, Av]_0 + \|\!|Bv|\!\| \le -\delta' \|v\|_1^2 + N_3 \|v\|_0^2,$$

completing the proof of the theorem. □

4.2.3. In this paragraph, we establish better Sobolev space regularity of the generalized solution of (4.2.1), (4.2.2) under additional regularity of the input.

Theorem 4.2 *Suppose that the super-parabolicity condition* (4.1.3) *holds and, for some positive integer m,*

(i) *The functions* $a^{ij}, b^i, c, \sigma^{il}, h^l, \ i_1, j = 1, 2, \ldots, \mathrm{d}, \ l = 1, 2, \ldots, \mathrm{d}_1$ *are m times differentiable in x for all t, x, ω and, together with all the derivatives, are uniformly, in t, x, ω, bounded by a constant K;*
(ii) $\varphi \in \mathbb{L}_2(\Omega; H^m)$, $f \in \mathbb{L}_2([T_0, T] \times \Omega; \mathbb{H}^{m-1})$, $g^l \in \mathbb{L}_2\big([T_0, T]) \times \Omega; \mathbb{H}^m\big)$, *for all* $l = 1, 2, \ldots, \mathrm{d}_1$.

Then the generalized solution u of problem (4.2.1), (4.2.2) *belongs to* $\mathbb{L}_2([T_0, T], \mathscr{P}; \mathbb{H}^{m+1}) \cap C([T_0, T]; \mathscr{P}; \mathbb{H}^m)$, *and there exists a positive number N depending only on* $K, \mathrm{d}, \mathrm{d}_1, m, T_0$, *and T, such that*

$$\mathbb{E} \sup_{t \in [T_0, T]} \|u(t)\|_m^2 + \mathbb{E} \int_{[T_0, T]} \|u(t)\|_{m+1}^2 \, dt$$
$$\le N\, \mathbb{E} \left(\|\varphi\|_m^2 + \int_{[T_0, T]} \left(\|f(t)\|_{m-1}^2 + \sum_{l=1}^{\mathrm{d}_1} \|g^l(t)\|_m^2 \right) dt \right).$$

Proof Since $\mathbb{H}^{m+1} \subset \mathbb{H}^1$, $\mathbb{H}^m \subset \mathbb{H}^0 = \mathbb{L}_2$, *and* $\mathbb{H}^{m-1} \subset \mathbb{H}^{-1}$, and all the embeddings are normal, the hypotheses of Theorem 4.1 are fulfilled and thus problem (4.2.1), (4.2.2) has a unique generalized solution belonging to $\mathbb{L}_2([T_0, T]; \mathscr{P}; \mathbb{H}^1) \cap \mathbb{L}_2(\Omega; \ \mathbb{C}([T_0, T], \mathbb{L}_2))$. Moreover, as was shown in the previous paragraph, this is the unique solution in the normal triple $(\mathbb{H}^1, \mathbb{L}_2, \mathbb{H}^{-1})$ of the LSES (3.1.1) with the operators A and B defined by equalities (4.2.5), (4.2.6), and

$$M(t) := \sum_{i=1}^{\mathrm{d}_1} \int_{[T_0, t]} g^l(s)\, dw^l(s).$$

We now show that Theorem 3.2 is applicable with $\mathbb{V} := \mathbb{H}^{m+1}, \mathbb{U} := \mathbb{H}^m$ and $\mathbb{V}' := \mathbb{H}^{m-1}$.

Define the partial differential operator $L_{\mathrm{div}} = L_{\mathrm{div}}(t, \omega)$ by

$$L_{\mathrm{div}} v := (a^{ij} v_i)_j + b^i v_i + cv, \ v \in \mathbb{H}^{m+1}.$$

By Proposition 3.7(i), there exists a positive number N such that, for all (t, ω) and all $v \in \mathbb{H}^{m+1}$,

$$\|L_{\mathrm{div}} v\|_{m-1} \le N \|v\|_{m+1}.$$

After integration by parts, we conclude that, for every $y \in \mathbb{H}^m$,

$$[y, Av]_0 = (y, L_{\text{div}} v)_0.$$

In other words, the restriction of $A(t, \omega)$ to $\mathbb{H}^{m+1}$ coincides with $L_{\text{div}}(t, \omega)$ and therefore

$$\|Av\|_{m-1} \le N\|v\|_{m+1}, \quad v \in \mathbb{H}^{m+1}. \tag{4.2.9}$$

Next, it follows from Proposition 3.7(i) that there exists a positive number N such that, for all (t, ω) and all $v \in \mathbb{H}^{m+1}$,

$$|\!|\!| B(t, \omega)v |\!|\!| \le N\|v\|_m;$$

this time, $|\!|\!| \cdot |\!|\!|$ is the Hilbert–Schmidt norm in $\mathcal{L}_2(\mathbb{R}^{d_1}, \mathbb{H}^m)$. We now verify condition (A_1) of Theorem 3.2. Let $v \in \mathbb{C}_0^\infty(\mathbb{R}^d)$ and denote by $[\cdot, \cdot]_m$ the CBF of the normal triple $(\mathbb{H}^{m+1}, \mathbb{H}^m, \mathbb{H}^{m-1})$. From the second property of CBF (see Sect. 2.5.2), it follows that $[v, Av]_m = (v, L_{\text{div}} v)_m$. On the other hand, in view of (3.3.2) and self-adjointness of the operator $(\mathbb{I} - \boldsymbol{\Delta})^{1/2}$, we have that $(v, L_{\text{div}} v)_m = ((\mathbb{I} - \boldsymbol{\Delta})^m v, L_{\text{div}} v)_0$. That is,

$$[v, Av]_m = \Big((\mathbb{I} - \boldsymbol{\Delta})^m v, L_{\text{div}} v\Big)_0.$$

By direct computations,

$$(\mathbb{I} - \boldsymbol{\Delta})^m v = \sum_{|\alpha| \le m} (-1)^{|\alpha|} C^\alpha v_{2\alpha}, \tag{4.2.10}$$

where $\alpha = (\alpha^1, \ldots, \alpha^d)$ is a multi-index, $v_{2\alpha} := \partial^{2|\alpha|} v / \partial (x^1)^{2\alpha^1} \ldots \partial (x^d)^{2\alpha^d}$, and $C^\alpha > 0$. In fact, using the binomial and multinomial formulas,

$$\begin{aligned}
(\mathbb{I} - \boldsymbol{\Delta})^m v &= \sum_{k=0}^{m} (-1)^k \frac{m!}{k!(m-k)!} \boldsymbol{\Delta}^k v \\
&= \sum_{k=0}^{m} \frac{m!}{k!(m-k)!} (-1)^k \sum_{|\alpha|=k} \frac{k!}{\alpha_1! \cdots \alpha_d!} v_{2\alpha},
\end{aligned}$$

so that

$$C^\alpha = \frac{m!}{(m - |\alpha|)!\, \alpha_1! \cdots \alpha_d!}.$$

Next, by the product rule for the derivatives and after integration by parts,

$$\begin{aligned}
2[v, Av]_m = &- \sum_{|\alpha|=m} C^\alpha \big(a^{ij}(v_i)_\alpha, (v_j)_\alpha\big)_0 + \sum_{|\alpha|=m} C^\alpha \big(b^i(v_i)_\alpha, v_\alpha\big)_0 \\
&+ \sum_{|\alpha|=m} \sum_{\substack{\beta+\gamma=\alpha \\ |\beta|\ge 1}} C^\alpha \big(a^{ij}_\beta (v_i)_\gamma, (v_j)_\alpha\big)_0 + U_1(v),
\end{aligned} \tag{4.2.11}$$

where $U_1(v)$ denotes all the terms containing derivatives of v of order at most m, so that $|U_1(v)| \le N\|v\|_m^2$.

Similarly,

$$|||Bv|||^2 = \sum_{l=1}^{d_1} \|\sigma^{il} v_i\|_m^2 + 2\sum_{l=1}^{d_1} (\sigma^{il} v_i, h^l v)_m + \sum_{l=1}^{d_1} \|h^l v\|_m^2, \tag{4.2.12}$$

so that

$$\begin{aligned}
\sum_{l=1}^{d_1} \|\sigma^{il} v_i\|_m^2 = &\sum_{l=1}^{d_1} \sum_{|\alpha|=m} C^\alpha \big(\sigma^{il}(v_i)_\alpha, \sigma^{il}(v_i)_\alpha\big)_0 \\
&+ \sum_{l=1}^{d_1} \sum_{|\alpha|=m} \sum_{\substack{\beta+\gamma=\alpha \\ |\beta|\ge 1}} C^\alpha \big(\sigma^{il}_\beta (v_i)_\gamma, \sigma^{il}(v_i)_\alpha\big)_0 + U_2(v)
\end{aligned} \tag{4.2.13}$$

with $|U_2(v)| \le N\|v\|_m^2$, and

$$\begin{aligned}
2\sum_{l=1}^{d_1} (\sigma^{il} v_i, h^l v)_m = &2\sum_{l=1}^{d_1} \sum_{|\alpha|=m} C^\alpha \big(\sigma^{il}(v_i)_\alpha, h^l v_\alpha\big)_0 \\
&+ U_3(v), \quad |U_3(v)| \le N\|v\|_m^2.
\end{aligned} \tag{4.2.14}$$

Combining (4.2.11)–(4.2.14) and (4.1.3),

$$2[v, Av]_m + |||Bv||| \le -\delta \sum_{|\alpha|=m} C^\alpha \sum_{i=1}^{d} \|(v_i)_\alpha\|_0^2 + V(v) + U(v), \tag{4.2.15}$$

where $|U(v)| < N\|v\|_m^2$, and

$$V(v) = \sum_{|\alpha|=m} C^{\alpha}\big(b^i(v_i)_{\alpha}, v_{\alpha}\big)_0 + \sum_{|\alpha|=m}\sum_{\substack{\beta+\gamma=\alpha\\|\beta|\geq 1}} C^{\alpha}\big(a^{ij}_{\beta}(v_i)_{\gamma}, (v_j)_{\alpha}\big)_0$$

$$+\sum_{l=1}^{\mathrm{d}_1}\sum_{|\alpha|=m}\sum_{\substack{\beta+\gamma=\alpha\\|\beta|\geq 1}} C^{\alpha}\big(\sigma^{il}_{\beta}(v_i)_{\gamma}, \sigma^{il}(v_i)_{\alpha}\big)_0$$

$$+2\sum_{l=1}^{\mathrm{d}_1}\sum_{|\alpha|=m} C^{\alpha}\big(\sigma^{il}(v_i)_{\alpha}, h^l v_{\alpha}\big)_0.$$

Keeping in mind that the norm $\|v\|_{m+1}$ is equivalent to $\left(\|v\|_m^2 + \sum_i \|v_i\|_m^2\right)^{1/2}$, whereas $\|v_i\|_m^2 = \big((\mathbb{I} - \boldsymbol{\Delta})^m v_i, v_i\big)_0$, we conclude from (4.2.10) that there exist $N, \delta' > 0$ such that

$$-\delta \sum_{|\alpha|=m} C^{\alpha} \sum_{i=1}^{d} \|(v_i)_{\alpha}\|_0^2 \leq -\delta'\|v\|_{m+1}^2 + N\|v\|_m^2. \tag{4.2.16}$$

On the other hand, in view of the Cauchy–Schwarz inequality and (3.2.10),

$$|V(v)| \leq N\|v\|_{m+1} \cdot \|v\|_m \leq \varepsilon N_1\|v\|_{m+1}^2 + N_2(\varepsilon)\|v\|_m^2, \tag{4.2.17}$$

where N_2 depends on ε and N_1 does not.

Taking ε to be sufficiently small, we obtain from (4.2.15)–(4.2.17) that there exist constants $\delta'' > 0$ and $K \geq 0$ such that, for all (t, ω) and every $v \in \mathbb{C}_0^{\infty}(\mathbb{R}^{\mathrm{d}})$,

$$2[v, A(t,\omega)v]_m + |||B(t,\omega)v||| \leq -\delta''\|v\|_{m+1}^2 + K'\|v\|_m^2. \tag{4.2.18}$$

Since $\mathbb{C}_0^{\infty}(\mathbb{R}^{\mathrm{d}})$ is dense in $\mathbb{H}^{m+1}$ (see Proposition 3.5(ii)), this inequality can be extended to all $v \in \mathbb{H}^{m+1}$. Taking into account (4.2.9) and (4.2.18) we find that the conditions (A_1), (B_1) of Theorem 3.2 are satisfied. The validity of the other conditions of this theorem follows from the assumptions made above and the properties of the Hilbert scale $\{\mathbb{H}^s\}$. To conclude the proof, it remains to apply Theorem 3.2 and Proposition 3.1. □

Suppose that, for $n \in \mathbb{N}$, there is a collection

$$\{a^{ij}_n, b^i_n, c_n, \sigma^{il}_n, h^l_n, f_n, g^l_n, \varphi_n\},\ i, j = 1, 2, \ldots, \mathrm{d},\ l = 1, 2, \ldots, \mathrm{d}_1,$$

satisfying the conditions of Theorem 4.1 (if $m = 0$) or Theorem 4.1 (if $m > 0$), and assume that the conditions are satisfied uniformly in n. Denote by u^n the solution of the corresponding problem (4.2.1), (4.2.2), in which a^{ij} is replaced by a_n^{ij}, b^i by b_n^i etc.

The following result is an immediate consequence of Theorem 3.5.

Corollary 4.1 *Let $a_n^{ij} \to a^{ij}$, $b_n^i \to b$, $c_n \to c$, $\sigma_n^{il} \to \sigma^{il}$, and $h_n^l \to h^l$ as $n \to \infty$ for all i, j, l and $\ell \times \ell_{\mathrm{d}} \times \mathbb{P}$-a.a. (t, x, ω), and*

$$\lim_{n\to\infty} \mathbb{E}\Bigg(\int_{[T_0,T]} \Big(\|f_n(t) - f(t)\|_m^2 + \sum_{l=1}^{\mathrm{d}_1} \|g_n^l(t) - g^l(t)\|_m^2 \Big)\, dt + \|\varphi_n - \varphi\|_m^2 \Bigg) = 0.$$

Then

$$\lim_{n\to\infty} \mathbb{E}\left(\sup_{t\in[T_0,T]} \|u^n(t) - u(t)\|_m^2 + \int_{[T_0,T]} \|u^n(t) - u^2(t)\|_{m+1}^2 dt \right) = 0.$$

□

4.2.4. In this paragraph we derive a corollary of Theorem 4.2 showing that if the coefficients, the initial values, and the external forces of the problem are sufficiently smooth, then the equivalence class of functions representing the generalized solution has a representative from $\mathbb{C}^2(\mathbb{R}^{\mathrm{d}})$ satisfying the equation for all t, x.

We start with an auxiliary result that builds on the Sobolev embedding theorem from Sect. 3.5.5.

Proposition 4.1 *Let (S, Σ) be a measurable space and let ξ be a Σ-measurable mapping of S to $\mathbb{W}_p^m$, $p \geq 1$, $m \in \mathbb{N}$. Assume that there exists a positive integer n such that $m - n > \mathrm{d}/p$. Then there exists a function $\tilde{\xi} : S \times \mathbb{R}^{\mathrm{d}} \to \mathbb{R}^1$ such that*

(a) *the function $\tilde{\xi}$ is $\Sigma \otimes \mathscr{B}(\mathbb{R}^{\mathrm{d}})$-measurable;*
(b) *$\tilde{\xi}(s, \cdot) \in \mathbb{C}_b^n(\mathbb{R}^{\mathrm{d}}) \cap \mathbb{W}_p^m$ and, for every $s \in S$,*

$$\|\tilde{\xi}(s, \cdot)\|_{\mathbb{C}_b^n(\mathbb{R}^{\mathrm{d}})} \leq N \|\xi(s, \cdot)\|_{m,p},$$

where the number N depends only on m, p, d, and n;
(c) *$\|\tilde{\xi}(s, \cdot) - \xi(s, \cdot)\|_{m,p} = 0$ for every $s \in S$.* □

Proof There are two main differences between this result and the corresponding embedding theorem from Sect. 3.5.5: there is an additional parameter s, and x is not confined to a bounded domain. Accordingly, for $k \in \mathbb{N}$, let B_k be the open ball in

$\mathbb{R}^d$ with center at the origin and radius k, and define

$$\xi_k(s,x) = \xi(s,x)1_{\{x\in B_k\}}.$$

Then each ξ_k is a Σ-measurable mapping of S into $\mathbb{W}_p^m(B_k)$. Applying Theorem 3.9 to ξ_k for every s, we conclude that there exists a function $\tilde{\xi}_k : S \to \mathbb{C}_b^n(B_k)$ with the properties:

(i) $\|\tilde{\xi}_k(s,\cdot)\|_{\mathbb{C}_b^n(B_k)} \le N\|\xi_k(s,\cdot)\|_{B_k,m,p} \le N\|\xi(s,\cdot)\|_{m,p}$;
(ii) $\tilde{\xi}_k$ is a Σ-*measurable function in* $\mathbb{C}_b^n(B_k)$;
(iii) $\|\tilde{\xi}_k(s,\cdot) - \xi(s,\cdot)\|_{B_k,m,p} = 0, \quad s \in S.$

In particular, property (1) follows from the continuity of the operator, property (2) implies that $\tilde{\xi}_k(s,x)$ is a Σ-measurable function for every $x \in B_k$, and continuity of each $\tilde{\xi}_k$ and property (3) imply $\tilde{\xi}_k(s,x) = \tilde{\xi}_{k+1}(s,x)$ when $|x| < k$. Using the continuity of $\tilde{\xi}_k$ with respect to x we conclude that each $\tilde{\xi}_k$ is a $\Sigma \otimes \mathscr{B}(B_k)$-measurable mapping from $S \times B_k$ to $\mathbb{R}^1$.

The desired function $\tilde{\xi}$ is then

$$\tilde{\xi}(s,x) := \sum_k \tilde{\xi}_k(s,x)\, 1_{\{k-1\le|x|<k\}}. \qquad \square$$

Now we are ready to state and prove the result about classical solutions.

Theorem 4.3 *Let $n \ge 2$ be an integer and let the assumptions of Theorem 4.2 be satisfied for $m > n + d/2$. Denote by $\tilde{f}$ and $\tilde{g}^l$ the corresponding versions of f and g^l whose existence is ensured by the proposition with $(S,\Sigma) = ((T_0,T]\times\Omega, \mathscr{P})$.*

Then the generalized solution of problem (4.2.1), (4.2.2) *has a version $v = v(t,x,\omega)$ so that*

$$\mathbb{E} \sup_{t\in[T_0,T]} \|u - v\|_m^2(t) = 0.$$

The function v has the following properties:

(a) *For every $x \in \mathbb{R}^d$, the process $v(\cdot,x,\cdot)$ is predictable*;
(b) *For every $(t,\omega) \in [T_0,T]\times\Omega$, the function $v(t,\cdot,\omega)$ is in $\mathbb{C}_b^n(\mathbb{R}^d)$*;
(c) *$v(t,x,\omega)$ as the generalized solution possesses properties enumerated in Theorem 4.2*;
(d) $\mathbb{E} \sup_{t\in[0,T]} \|v(t)\|^2_{\mathbb{C}_f^n(\mathbb{R}^d)} < \infty$;
(e) *If α is a* d*-dimensional multi-index with the length $|\alpha| \le n-2$, then, for every $x \in \mathbb{R}^d$, there exists a set $\tilde{\Omega}_x \subset \Omega$ with $\mathbb{P}(\tilde{\Omega}_x) = 1$, such that, for all $(t,\omega) \in$*

$[T_0, T] \times \tilde{\Omega}_x$,

$$v_\alpha(t,x) = \varphi_\alpha(x) + \int\limits_{[T_0,t]} \Big((a^{ij} v_i)_j + b^i v_i + cv + \tilde{f}(s)\Big)_\alpha (s,x)\, ds$$

$$+ \int\limits_{[T_0,t]} \Big(\sigma^{il} v_i + h^l v + \tilde{g}^l\Big)_\alpha (s,x)\, dw^l(s); \tag{4.2.19}$$

(f) *if* $v_1(t,x,\omega)$ *and* $v_2(t,x,\omega)$ *are two generalized solutions of problem* (4.2.1), (4.2.2) *having the properties* (a), (b)*, then*

$$\mathbb{P}\left(\sup_{\substack{t \in [T_0,T] \\ x \in \mathbb{R}^d}} \Big| v_1(t,x,\omega) - v_2(t,x,\omega) \Big| > 0 \right) = 0. \qquad \square$$

Proof Since by Theorem 4.2 the generalized solution u of problem (4.2.1), (4.2.2) belongs to $\mathbb{H}^{m+1}$ for all $(t,\omega) \in [T_0,T] \times \Omega$, we obtain, in view of the proposition, that there exists a version v of the generalized solution having the properties mentioned in items (a) to (c) of the theorem. The validity of item (d) follows from item (b) of the proposition and the inequality of Theorem 4.2.

In the proof of item (e) we use the family of averaging operators $\mathcal{T}_\varepsilon$, $\varepsilon > 0$, defined in Sect. 1.5.11. In particular, for a function $g = g(t,x)$ that is locally integrable in x,

$$\mathcal{T}_\varepsilon\, g(t,x) = \big(g(t,\cdot), \zeta_{\varepsilon,x}\big)_0,$$

where $\zeta_{\varepsilon,x}(y) = \varepsilon^{-\mathrm{d}}\, \bar{\zeta}(x-y)/\varepsilon)$ is a smooth positive function with compact support and with total mass equal to one. Below, we summarize the properties of $\mathcal{T}_\varepsilon$ to be used in the proof and also later in the book. An interested reader can either verify these properties independently or consult [117, Sections 1.4 and 4.5].

Lemma 4.1

(i) *For every* $\varepsilon > 0$ *and every* $p \geq 1$*, the operator* $\mathcal{T}_\varepsilon$ *maps* $\mathbb{L}_p$ *to* $\bigcap_{m \in \mathbb{N}} \mathbb{W}_p^m$.

(ii) *If* $f \in \mathbb{W}_p^m$*, then*

$$\|\mathcal{T}_\varepsilon f\|_{m,p} \leq \|f\|_{m,p}, \ \lim_{\varepsilon \to 0} \|\mathcal{T}_\varepsilon f - f\|_{m,p} = 0, \ \textit{and} \ \big(\mathcal{T}_\varepsilon f\big)_\alpha = \mathcal{T}_\varepsilon\big(f_\alpha\big),$$

$|\alpha| \leq m$. $\square$

To establish item (e) of the theorem, let α be a d-dimensional multi-index with $|\alpha| \le n-2$. Then, using the definition of the generalized solution with $y = \zeta_{\varepsilon,x}$, we conclude that, for all $(t,\omega) \in [T_0,T] \times \Omega'_x$ with $\mathbb{P}(\Omega'_x) = 1$, the following equality holds:

$$\begin{aligned} \mathcal{T}_\varepsilon v_\alpha(t,x) &= \mathcal{T}_\varepsilon \varphi_\alpha(x) + \int\limits_{[T_0,t]} \mathcal{T}_\varepsilon \Big(L_{\mathrm{div}} v + \tilde{f} \Big)_\alpha (s,x) ds \\ &+ \sum_{l=1}^{\mathrm{d}_1} \int\limits_{[T_0,t]} \mathcal{T}_\varepsilon \Big(B^l v + \tilde{g}^l \Big)_\alpha (s,x)\, dw^l(s). \end{aligned} \tag{4.2.20}$$

From the most recent lemma and Theorem 3.9 it follows that, for every t and ω, we can pass to the limit as $\varepsilon \to 0$ and get $\mathcal{T}_\varepsilon v_\alpha(t,x) \to v_\alpha(t,x)$, $\mathcal{T}_\varepsilon \varphi_\alpha(x) \to \varphi_\alpha(x)$, $\mathcal{T}_\varepsilon \Big(L_{\mathrm{div}} v(t,x) + f(t,x) \Big)_\alpha \to \Big(L_{\mathrm{div}} v(t,x) + f(t,x) \Big)_\alpha$, and $\mathcal{T}_\varepsilon \Big(B^l v(t,x) + \tilde{g}(t,x) \Big)_\alpha \to B^l v(t,x) + \tilde{g}^l(t,x)$, all in the norm of the space $\mathbb{C}^0_b(\mathbb{R}^{\mathrm{d}})$.

Thus, by the dominated convergence theorem, as $\varepsilon \to 0$,

$$\int\limits_{[T_0,t]} \mathcal{T}_\varepsilon \Big(L_{\mathrm{div}} v + \tilde{f} \Big)_\alpha (s,x)\, ds \to \int\limits_{[T_0,t]} \Big(L_{\mathrm{div}} v + \tilde{f} \Big)_\alpha (s,x)\, ds,$$

$$(t,x,\omega) \in [T_0,T] \times \Omega \times \mathbb{R}^{\mathrm{d}};$$

$$\begin{aligned} &\mathbb{E} \left| \sum_{l=1}^{\mathrm{d}_1} \int\limits_{[T_0,t]} \Big(\mathcal{T}_\varepsilon (B^l v + \tilde{g}^l)_\alpha (s,x) - (B^l v + \tilde{g}^l)_\alpha (s,x) \Big)\, dw^l(s) \right|^2 \\ &\le \sum_{l=1}^{\mathrm{d}_1} \int\limits_{[T_0,T]} \mathbb{E} \Big| \mathcal{T}_\varepsilon (B^l v + \tilde{g}^l)_\alpha (s,x) - (B^l v + \tilde{g}^l)_\alpha (s,x) \Big|^2 ds \to 0, \ x \in \mathbb{R}^{\mathrm{d}}. \end{aligned}$$

Making use of these relations and passing in (4.2.19) to the limit over some subsequence $\varepsilon_i \to 0$, we conclude that $v_\alpha(t,x)$ satisfies (4.2.19) for every $(t,x) \in [T_0,T] \times \mathbb{R}^{\mathrm{d}}$ on a suitable set $\tilde{\Omega}_x$ such that $\tilde{\Omega}_x \subset \Omega$ and $\mathbb{P}(\tilde{\Omega}_x) = 1$.

Item (f) is an immediate consequence of the uniqueness of solution. □

Recall (Warning 2.7) that, for every stochastic integral, we always consider the version that is continuous in time. Under the assumptions of the above theorem, the stochastic integral in (4.2.19) has a version that is jointly continuous in t and x. For this version, there exists a set $\tilde{\Omega}$ with $\mathbb{P}(\tilde{\Omega}) = 1$ such that equality (4.2.19) holds on $\tilde{\Omega}$ for all t, x.

Definition 4.3 A function $v : [T_0,T] \times \Omega \times \mathbb{R}^{\mathrm{d}} \to \mathbb{R}^1$ is called a **classical solution** of problem (4.2.1), (4.2.2) if this function belongs to $\mathbb{C}^{0,2}\big([T_0,T] \times \mathbb{R}^{\mathrm{d}}\big)$

with probability one and possesses property *(a)* and property *(d)* with $\alpha = 0$ from Theorem *4.3*. □

The following is an immediate consequence of Theorem 4.3.

Corollary 4.2 *If the conditions of Theorem 4.2 are fulfilled for all* $m \in \mathbb{N}$ *(not necessarily uniformly in m), then the classic solution of problem* (4.2.1), (4.2.2) *exists, is unique, and, with probability one, is infinitely differentiable with respect to x.* □

4.2.5. In the next two paragraphs, we investigate the solvability of system (4.2.1), (4.2.2) in $\mathbb{L}_p([T_0, T] \times \Omega; \mathbb{W}^m_p)$ for $p \geq 2$. We fix the numbers $m \in \mathbb{N} \cup \{0\}$, $p \in [2, \infty)$, and $q = p/(p-1)$. We also assume that $f(t, \cdot, \omega) \in \mathbb{H}^{m-1}$ for all t, ω, and then use Proposition 3.7(ii) to write f as follows:

$$f(t, x, \omega) = f^0(t, x, \omega) + \sum_{i=1}^{d} f^i_i(t, x, \omega),$$

where $f^i \in \mathbb{H}^m$ for all t, ω and $i = 0, 2, \ldots, \mathrm{d}$; recall that f^i_i means $\partial f^i/\partial x^i$. Below we state the assumptions of f in terms of f^i.

Theorem 4.4 *Suppose that the assumptions of Theorem* 4.2 *are fulfilled and, in addition,* $\varphi \in \mathbb{L}_p(\Omega; \mathbb{W}^m_p)$, f^j *and* $g^l \in \mathbb{L}_p(T_0, T] \times \Omega; \mathbb{W}^m_p)$ *for* $j = 0, 1, 2, \ldots \mathrm{d}$, $l = 1, 2, \ldots, \mathrm{d}_1$. *Then the generalized solution u of* (4.2.1), (4.2.2) *belongs to* $\mathbb{L}_p([T_0, T] \times \Omega; \mathbb{C}_w\mathbb{W}^m_p)$ *and*

$$\mathbb{E} \sup_{t\in[T_0,T]} \|u(t)\|^p_{m,p} + \frac{\delta}{2} p(p-1) \sum_{|\alpha|\leq m} \sum_{i=1}^{\mathrm{d}_1} \mathbb{E} \int_{[T_0,T]} |u_\alpha|^{p-2} |(u_i)_\alpha|^2 \, dx ds$$

$$\leq N\,\mathbb{E} \left(\|\varphi\|^p_{m,p} + \int_{[T_0,T]} \left(\sum_{l=0}^{d} \|f^i(t)\|^p_{m,p} + \sum_{l=1}^{\mathrm{d}_1} \|g^l(t)\|^p_{m,p} \right) dt \right), \tag{4.2.21}$$

where the number N depends only on $m, p, \delta, K, T_0, T, \mathrm{d}$, *and* d_1. □

Corollary 4.3 *If there exists a non-negative integer n such that the number m in the assumptions of the theorem satisfies* $(m-n)p > \mathrm{d}$, *then the generalized solution of problem* (4.2.1), (4.2.2) *has a version v that, with probability one, belongs to* $\mathbb{C}^{0,n}_b([T_0, T] \times \mathbb{R}^\mathrm{d})$, *and the left-hand side of* (4.2.21) *can be replaced with* $\mathbb{E} \sup_{t\in[T_0,T]} \|v(t)\|^p_{\mathbb{C}^n_b(\mathbb{R}^\mathrm{d})}$. *If* $n \geq 2$, *then v is a classical solution of problem* (4.2.1), (4.2.2). □

Proof Let B be an open ball in $\mathbb{R}^{\mathrm{d}}$. By Theorem 3.9, $\mathbb{W}_p^m(B)$ is continuously embedded in $\mathbb{C}_b^n(B)$ and therefore the restriction of u on B has a version v_B belonging to $\mathbb{C}_b^n(B)$ for every t. Moreover, $\mathbb{E}\sup_{t\in[T_0,T]}\|v_B(t)\|^p_{\mathbb{C}_b^n(B)}$ is bounded by the right-hand side of (4.2.21).

The embedding of $\mathbb{W}_p^m(B)$ into $\mathbb{C}_b^n(B)$ is compact [88, Section II.2], and then weak continuity of v_B with respect to t in $\mathbb{W}_p^m$ implies $v_B \in \mathbb{C}\big([T_0,T];\mathbb{C}_b^n(B)\big)$. The rest of the proof is identical to the corresponding part of the proof of Theorem 4.3. □

To prove Theorem 4.4 we need the following auxiliary result.

Lemma 4.2 *Suppose that we have* $\overline{\mathscr{B}([T_0,T]\times\mathbb{R}^{\mathrm{d}})\otimes\mathscr{F}}$*-measurable and predictable for every* $x\in\mathbb{R}^{\mathrm{d}}$*, functions* a^{ij}, b^i, c, σ^{il}, *and* h^l, $j,i=1,2,\dots,\mathrm{d}$, $l=1,2,\dots,\mathrm{d}_1$, *satisfying the super-parabolicity condition* (4.1.3) *and assumption* (i) *of Theorem* 4.2.

Given the (possibly random and time-dependent) functions $u\in\mathbb{W}_p^{m+1}$, $g^l\in\mathbb{W}_p^{m+1}$, *and* $f^j\in\mathbb{W}_p^m$, $l=1,2,\dots,\mathrm{d}_1$ *and* $j=0,1,\dots,\mathrm{d}$, *define*

$$
\begin{aligned}
U(m,p,u,f,g) := \sum_{|\alpha|\le m}|u_\alpha|^{p-2}\Big(&-p(p-1)(u_j)_\alpha\big(a^{ij}u_i+f^j\big)_\alpha \\
&+pu_\alpha\big(b^iu_i+cu+f^0\big)_\alpha \\
&+\frac{p(p-1)}{2}\sum_{l=1}^{\mathrm{d}_1}\Big|\sigma^{il}\big(u_i+h^lu+g^l\big)_\alpha\Big|^2\Big).
\end{aligned}
$$

Then, for every $(t,\omega)\in[T_0,T]\times\Omega$,

$$
\begin{aligned}
\int_{\mathbb{R}^{\mathrm{d}}} U(m,p,u,f,g)dx \le& -\frac{\delta}{2}p(p-1)\sum_{|\alpha|\le m}\sum_{i=1}^{\mathrm{d}}\int_{\mathbb{R}^{\mathrm{d}}}|u_\alpha|^{p-2}\left|(u_i)_\alpha\right|^2dx \\
&+N_0\left(\|u\|^p_{m,p}+\sum_{i=0}^{\mathrm{d}}\|f^i\|^p_{m,p}+\sum_{l=1}^{\mathrm{d}_1}\|g^l\|^p_{m,p}\right),
\end{aligned}
$$

where the number N_0 *depends only on* p, d, d_1, δ, m, *and* K.

Proof To simplify the notation, we write

$$
G_\alpha=|u_\alpha|^p,\; G'_\alpha=p|u_\alpha|^{p-2}u_\alpha,\; G''_\alpha=p(p-1)|u_\alpha|^{p-2},\; \sum_{|\alpha|\le m}^{*}=\sum_{\substack{\alpha=\beta+\gamma\\|\alpha|\le m,\ |\beta|\ge1}}.
$$

For two functions $F = F(t, x, \omega)$, $H = H(t, x, \omega)$, the notation

$$F \asymp H$$

means

$$\int_{\mathbb{R}^d} |F - G|\, dx \le N \left(\|u\|_{m,p}^p + \sum_{i=0}^{d} \|f^i\|_{m,p}^p + \sum_{l=1}^{d_1} \|g^l\|_{m,p}^p \right)$$

with N similar to N_0 from the statement of the lemma. In particular, the statement of the lemma becomes

$$U(m, p, u, f, g) \asymp -\frac{\delta}{2} \sum_{|\alpha| \le m} \sum_{i=1}^{d} G''_\alpha \left|(u_i)_\alpha\right|^2.$$

To begin, note that, by the inequality

$$|ab| \le \frac{a^p}{p} + \frac{b^q}{q},\ a, b \ge \mathbb{R}_+,\ q = \frac{p}{p-1}, \tag{4.2.22}$$

we have

$$\sum_{|\alpha| \le m} (cu + f^0)_\alpha G'_\alpha \asymp 0, \quad \sum_{|\alpha| \le m} \sum_{l=1}^{d_1} \left|(h^l u + g^l)_\alpha\right|^2 G''_\alpha \asymp 0.$$

Next, because $(p-1)G'_\alpha = G''_\alpha u_\alpha$,

$$\sum_{|\alpha| \le m} (b^i u_i)_\alpha G'_\alpha \asymp \sum_{|\alpha| \le m} b^i (u_i)_\alpha u_\alpha G''_\alpha.$$

From the inequality (3.2.10) it follows that

$$\sum_{|\alpha| \le m} (u_i)_\alpha G''_\alpha u_\alpha b^i \asymp \frac{\varepsilon N}{2} \sum_{|\alpha| \le m} G''_\alpha \sum_{i=1}^{d} |(u_i)_\alpha|^2$$

and

$$G''_\alpha (u_j)_\alpha f^j_\alpha \le G''_\alpha \left(\frac{\varepsilon}{2} \sum_{j=1}^{d} |(u_j)_\alpha|^2 + \frac{1}{2\varepsilon} \sum_{j=1}^{d} |f^j_\alpha|^2 \right),$$

so that

$$\sum_{|\alpha|\le m} G''_\alpha (u_j)_\alpha f^j_\alpha \asymp \frac{\varepsilon}{2} \sum_{|\alpha|\le m} \sum_{j=1}^{\mathrm{d}} G''_\alpha |(u_j)_\alpha|^2.$$

Making further use of (4.2.22) and (3.2.10), we find

$$\begin{aligned}
\sum_{|\alpha|\le m} G''_\alpha \sum_{l=1}^{\mathrm{d}_1} \left(\sigma^{il} u_i\right)_\alpha \left(h^l u + g^l\right)_\alpha &\asymp \sum_{|\alpha|\le m} G''_\alpha \sum_{l=1}^{\mathrm{d}_1} \sigma^{il} (u_i)_\alpha \left(h^l u_\alpha + g^l_\alpha\right) \\
&\le \frac{\varepsilon N}{2} \sum_{|\alpha|\le m} G''_\alpha \sum_{i=1}^{\mathrm{d}} |(u_i)_\alpha|^2 + \frac{N}{2\varepsilon} \sum_{|\alpha|\le m} G''_\alpha \left(|u_\alpha|^2 + \sum_{l=1}^{\mathrm{d}_1} |g^l_\alpha|^2\right) \\
&\asymp \frac{\varepsilon N}{2} \sum_{|\alpha|\le m} G''_\alpha \sum_{i=1}^{\mathrm{d}} |(u_i)_\alpha|^2.
\end{aligned}$$

Finally, using the super-parabolicity condition (4.1.3),

$$\begin{aligned}
&\sum_{|\alpha|\le m} |u_\alpha|^{p-2} \left(-p(p-1)(u_j)_\alpha \left(a^{ij} u_i\right)_\alpha + \frac{p(p-1)}{2} \sum_{i=1}^{\mathrm{d}_1} \left|\left(\sigma^{il} u_i\right)_\alpha\right|^2 \right) \\
&\quad = \sum_{|\alpha|\le m} |u_\alpha|^{p-2} \Big(- p(p-1)(u_j)_\alpha a^{ij} (u_i)_\alpha + \frac{p(p-1)}{2} \sum_{l=1}^{\mathrm{d}_1} |\sigma^{il} (u_i)_\alpha|^2 \\
&\quad + \sum_{|\alpha|\le m}^{*} |u_\alpha|^{p-2} \Big(- p(p-1)(u_j)_\alpha a^{ij}_\beta (u_i)_\gamma + \frac{p(p-1)}{2} \sum_{l=1}^{\mathrm{d}_1} |\sigma^{ij}_\beta (u_i)_\gamma|^2 \Big) \\
&\quad \le -\delta \sum_{|\alpha|\le m} G''_\alpha \sum_{i=1}^{\mathrm{d}} |(u_i)_\alpha|^2 \\
&+ \sum_{|\alpha|\le m}^{*} |u_\alpha|^{p-2} \Big(- p(p-1)(u_j)_\alpha a^{ij}_\beta (u_i)_\gamma + \frac{p(p-1)}{2} \sum_{l=1}^{\mathrm{d}_1} |\sigma^{il}_\beta (u_i)_\gamma|^2 \Big).
\end{aligned}$$

By (3.2.10) and (4.2.22),

$$\sum_{|\alpha|\le m}^{*} |u_\alpha|^{p-2} \Big(- p(p-1)(u_j)_\alpha a^{ij}_\beta (u_i)_\gamma \Big) \asymp \frac{\varepsilon}{2} \sum_{|\alpha|\le m} G''_\alpha \sum_{j=1}^{\mathrm{d}} |(u_j)_\alpha|^2$$

and

$$\sum_{|\alpha|\le m}^{*} |u_\alpha|^{p-2} \left(\frac{p(p-1)}{2} \sum_{l=1}^{\mathrm{d}_1} |\sigma_\beta^{il}(u_i)_\gamma|^2 \right) \asymp 0.$$

It remains to combine the above inequalities and to choose ε sufficiently small. The result is

$$U(m,p,u,f,g) \asymp -\frac{\delta}{2} \sum_{|\alpha|\le m} \sum_{i=1}^{\mathrm{d}} G_\alpha'' |(u_i)_\alpha|^2,$$

completing the proof of the lemma. □

4.2.6. We are now ready to prove Theorem 4.4. Once again, we use the averaging operator $\mathcal{T}_\varepsilon$, $\varepsilon > 0$, originally defined in Sect. 1.5.11.

To simplify the presentation, we write

$$v_{(\varepsilon)} := \mathcal{T}_\varepsilon v.$$

Consider the problem

$$du^\varepsilon(t) = \left((a_{(\varepsilon)}^{ij} u_i^\varepsilon(t))_j + b_{(\varepsilon)}^i u_i^\varepsilon(t) + c_{(\varepsilon)} u^\varepsilon(t) + \sum_{j=0}^{\mathrm{d}} (f_{(\varepsilon)}^i(t))_j \right) dt$$

$$+ \left(\sigma_{(\varepsilon)}^{il} u_i^\varepsilon(t) + h_{(\varepsilon)}^l u^\varepsilon(t) + g_{(\varepsilon)}^l(t) \right) dw^l(t),$$

$$(t,x,\omega) \in (T_0, T] \times \mathbb{R}^{\mathrm{d}} \times \Omega, \tag{4.2.23}$$

$$u^\varepsilon(T_0) = \varphi_{(\varepsilon)}(x,\omega), \quad (x,\omega) \in \mathbb{R}^{\mathrm{d}} \times \Omega. \tag{4.2.24}$$

The coefficients of this problem as well as functions $\varphi_{(\varepsilon)}$, $g_{(\varepsilon)}^l$, and $f_{(\varepsilon)}^j$ satisfy the assumptions of the theorem for all m. Indeed, the properties of the averaging operator imply that the initial value $\varphi_{(\varepsilon)}$ and the free terms $f_{(\varepsilon)}^j$, $g_{(\varepsilon)}^l$ are infinitely differentiable in x and belong to $\mathbb{L}_{p'}([T_0,T]\times\Omega; \mathbb{W}_{p'}^m)$ for every positive integer m and for both $p' = p$ and $p = 2$, cf. Lemma 4.1(i). The super-parabolicity condition (4.1.3) holds because, by the Cauchy–Schwarz inequality, $\left|v_{(\varepsilon)}\right|^2 \le \left(v^2\right)_{(\varepsilon)}$.

By Theorem 4.3, problem (4.2.23), (4.2.24) has a unique classical solution, which, by Corollary 4.2, is infinitely differentiable in x. Thus, there exists a subset

$\Omega' \subset \Omega$ such that $\mathbb{P}(\Omega') = 1$ and, for all $(t, x, \omega) \in [T_0, T] \times \mathbb{R}^{\mathrm{d}} \times \Omega$ and every d-dimensional multi-index α, the following equality is satisfied (see Theorem 4.3(e)):

$$u_\alpha^\varepsilon(t) = (\varphi_\varepsilon)_\alpha + \int\limits_{[T_0,t]} \left(\left(a^{ij}_{(\varepsilon)} u_i^\varepsilon\right)_j + b^i_{(\varepsilon)} u_i^\varepsilon + c_{(\varepsilon)} u^\varepsilon + \sum_{j=0}^{\mathrm{d}} (f^j_{(\varepsilon)})_j \right)_\alpha (s)\, ds$$
$$+ \int\limits_{[T_0,t]} \left(\sigma^{il}_{(\varepsilon)} u_i^\varepsilon + h^l_{(\varepsilon)} u^\varepsilon + g^l_{(\varepsilon)} \right)_\alpha (s)\, dw^l(s).$$

Similar to the notation G'_α and G''_α introduced in the previous paragraph, we now write $G'_{\alpha,\varepsilon}$ and $G''_{\alpha,\varepsilon}$, so that, for example,

$$G'_{\alpha,\varepsilon} = p|u_\alpha^\varepsilon|^{p-2} u_\alpha^\varepsilon.$$

Then the Itô formula (Sect. 1.5.3) applied to $|u_\alpha^\varepsilon|^p$ shows that, for every $x \in \mathbb{R}^{\mathrm{d}}$, there exists a set Ω_x with $\mathbb{P}(\Omega_x) = 1$ such that, for all $(t, \omega) \in [T_0, T] \times \Omega_x$ and every $N \in \mathbb{R}_+$,

$$\begin{aligned} e^{-Nt}|u_\alpha^\varepsilon(t)|^p &= e^{-NT_0}|\varphi_{(\varepsilon)\alpha}|^p + \int\limits_{[T_0,t]} e^{-Ns}\Bigg(G'_{\alpha,\varepsilon} \Big(\left(a^{ij}_{(\varepsilon)} u_i^\varepsilon\right)_j \\ &\quad + b^i_{(\varepsilon)} u_i^\varepsilon + c_{(\varepsilon)} u^\varepsilon + \sum_{j=0}^{\mathrm{d}} (f^j_{(\varepsilon)})_j \Big)_\alpha \\ &\quad + \frac{G''_{\alpha,\varepsilon}}{2} \sum_{l=1}^{\mathrm{d}_1} \left| \left(\sigma^{il}_{(\varepsilon)} u_i^\varepsilon + h^l_{(\varepsilon)} u^\varepsilon + g^l_{(\varepsilon)}\right)_\alpha \right|^2 \Bigg)(s)\, ds \\ &\quad + \int\limits_{[T_0,t]} e^{-Ns}\, G'_{\alpha,\varepsilon} \left(\sigma^{il}_{(\varepsilon)} u_i^\varepsilon + h^l_{(\varepsilon)} u^\varepsilon + g^l_{(\varepsilon)} \right)_\alpha (s)\, dw^l(s) \\ &\quad - N \int\limits_{[T_0,t]} e^{-Ns}\, |u_\alpha^\varepsilon(s)|^p\, ds. \end{aligned} \tag{4.2.25}$$

In fact, because the stochastic integral in (4.2.25) has a version continuous in (t, x), we will assume that equality (4.2.25) holds for all t, x on the same set of probability one.

Integrating by parts,

$$
\begin{aligned}
&\int_{\mathbb{R}^d} G'_{\alpha,\varepsilon} \left(\left(a^{ij}_{(\varepsilon)} u^{\varepsilon}_i\right)_j + \sum_{j=1}^{d} \left(f^j_{(\varepsilon)}\right)_j \right)_\alpha dx \\
&= -\int_{\mathbb{R}^d} G''_{\alpha,\varepsilon} (u^{\varepsilon}_j)_\alpha \left(a^{ij}_{(\varepsilon)} u^{\varepsilon}_i + \sum_{j=1}^{d} f^j_{\varepsilon}\right)_\alpha dx.
\end{aligned}
\tag{4.2.26}
$$

We now integrate both sides of (4.2.25) over $\mathbb{R}^d$, change the order of integration (cf. Sect. 1.5.9), use (4.2.26), and then sum over all α with $|\alpha| \le m$. Also, recall the function U introduced in Lemma 4.2. Then (4.2.25) becomes

$$
\begin{aligned}
e^{-Nt}\|u^{\varepsilon}(t)\|^p_{m,p} &= e^{-NT_0}\|\varphi_{(\varepsilon)}\|^p_{m,p} + \int_{[T_0,t]} e^{-Ns} \int_{\mathbb{R}^d} U(m,p,u^{\varepsilon}, f_{(\varepsilon)}, g_{(\varepsilon)})\, dx\, ds \\
&+ \int_{[T_0,t]} e^{-Ns} \int_{\mathbb{R}^d} \sum_{|\alpha|\le m} G'_{\alpha,\varepsilon}\left(\sigma^{il}_{(\varepsilon)} u^{\varepsilon}_i + h^l_{(\varepsilon)} u^{\varepsilon} + g^l_{(\varepsilon)}\right)_\alpha dx\, dw^l(s) \\
&- N \int_{[T_0,t]} e^{-Ns}\|u^{\varepsilon}\|^p_{m,p}\, ds.
\end{aligned}
\tag{4.2.27}
$$

For $n \in \mathbb{N}$ define the set $\Gamma_n(\omega) := \{t \in [T_0, T] : \|u^{\varepsilon}(t,\omega)\|_{\mathbb{C}^m_b(\mathbb{R}^d)} > n\}$ and the stopping time

$$
\tau_n(\omega) := \begin{cases} \inf_t \Gamma_n(\omega), & \text{if } \Gamma_r(\omega) \ne \emptyset; \\ T, & \text{if } \Gamma_n(\omega) = \emptyset. \end{cases}
$$

Then, for every $n \in \mathbb{N}$,

$$
\begin{aligned}
\mathbb{E}\left(e^{-Nt\wedge\tau_n}\|u^{\varepsilon}(t\wedge\tau_n)\|^p_{m,p}\right) &= e^{-NT_0}\mathbb{E}\|\varphi_{(\varepsilon)}\|^p_{m,p} \\
&+ \mathbb{E}\int_{[T_0,t\wedge\tau_n]} e^{-Ns}\int_{\mathbb{R}^d} U(m,p,u^{\varepsilon}, f_{(\varepsilon)}, g_{(\varepsilon)})\, dx\, ds \\
&- N\mathbb{E}\int_{[T_0,t\wedge\tau_n]} e^{-Ns}\|u^{\varepsilon}\|^p_{m,p}\, ds.
\end{aligned}
$$

Passing to the limit as $n \to \infty$ and using Lemma 4.2, together with the continuity of $\|u^\varepsilon\|_{m,p}$ in t,

$$\mathbb{E}\left(e^{-Nt}\,\|u^\varepsilon(t)\|_{m,p}^p\right) = e^{-NT_0}\mathbb{E}\|\varphi_{(\varepsilon)}\|_{m,p}^p$$

$$-\frac{\delta}{2}p(p-1)\,\mathbb{E}\sum_{|\alpha|\le m}\sum_{i=1}^{\mathrm{d}}\int_{[T_0,t]}\int_{\mathbb{R}^{\mathrm{d}}}|u_\alpha^\varepsilon|^{p-2}\cdot\left|\left(u_i^\varepsilon\right)_\alpha\right|^2 dx\,ds$$

$$+N_0\,\mathbb{E}\int_{[T_0,t]}e^{-Ns}\left(\|u^\varepsilon\|_{m,p}^p+\sum_{j=0}^{\mathrm{d}}\|f_{(\varepsilon)}^j\|_{m,p}^p\right.$$

$$\left.+\sum_{l=1}^{\mathrm{d}_1}\|g_{(\varepsilon)}^l\|_{m,p}^p\right)ds - N\,\mathbb{E}\int_{[T_0,t]}e^{-Ns}\,\|u^\varepsilon\|_{m,p}^p\,ds.$$

If N is sufficiently large, then

$$\sup_{t\in[T_0,T]}\mathbb{E}\|u^\varepsilon(t)\|_{m,p}^p+\frac{\delta}{2}p(p-1)\,\mathbb{E}\sum_{|\alpha|\le m}\sum_{i=1}^{\mathrm{d}}\int_{[T_0,t]}\int_{\mathbb{R}^{\mathrm{d}}}|u_\alpha^\varepsilon|^{p-2}\cdot\left|\left(u_i^\varepsilon\right)_\alpha\right|^2 dx\,ds$$

$$\le N_1\,\mathbb{E}\left(\|\varphi_{(\varepsilon)}\|_{m,p}^p+\int_{[T_0,T]}\left(\sum_{j=0}^{\mathrm{d}}\|f_{(\varepsilon)}^j\|_{m,p}^p+\sum_{l=1}^{\mathrm{d}_1}\|g_{(\varepsilon)}^l\|_{m,p}^p\right)ds\right)$$

$$\le N_1\mathbb{E}\left(\|\varphi\|_{m,p}^p+\int_{[T_0,T]}\left(\sum_{j=0}^{\mathrm{d}}\|f^j\|_{m,p}^p+\sum_{l=1}^{\mathrm{d}_1}\|g^l(s)\|_{m,p}^p\right)ds\right), \tag{4.2.28}$$

where the second inequality follows from Lemma 4.1.

The next step is to show that (4.2.28) holds with $\mathbb{E}\sup_{t\in[T_0,T]}\|u^\varepsilon(t)\|_{m,p}^p$ instead of $\sup_{t\in[T_0,T]}\mathbb{E}\|u^\varepsilon(t)\|_{m,p}^p$. To this end, put $N = 0$ in equality (4.2.27) and then use (4.2.28) and Lemma 4.2. The result is

$$\mathbb{E}\sup_{t\in[T_0,T]}\|u^\varepsilon(t)\|_{m,p}^p\le N_2\mathbb{E}\left(\|\varphi\|_{m,p}^p+\int_{[T_0,T]}\left(\sum_{j=0}^{\mathrm{d}}\|f^i\|_{m,p}^p+\sum_{l=1}^{\mathrm{d}_1}\|g^l\|_{m,p}^p\right)ds\right)$$

$$+\mathbb{E}\sup_{t\in[T_0,T]}\left|\int_{[T_0,t]}\int_{\mathbb{R}^{\mathrm{d}}}\sum_{|\alpha|\le m}G'_{\alpha,\varepsilon}\left(\sigma_{(\varepsilon)}^{il}u_i^\varepsilon+h_{(\varepsilon)}^l u^\varepsilon+g_{(\varepsilon)}^l\right)_\alpha dx\,dw^l\right|. \tag{4.2.29}$$

By the Burkholder–Davis–Gundy inequality (Sect. 2.2.8), the last term on the right-hand side of (4.2.29) is bounded by

$$\mathbb{E}\int_{[T_0,T]}\left(\sum_{l=1}^{d_1}\left(\int_{\mathbb{R}^d}\sum_{|\alpha|\le m} G'_{\alpha,\varepsilon}\left(\sigma^{il}_{(\varepsilon)}u^\varepsilon_i(s)+h^l_{(\varepsilon)}+g^l_{(\varepsilon)}\right)_\alpha\right)^2 ds\right)^{\frac{1}{2}},$$

which we further bound by

$$N\mathbb{E}\Bigg(\int_{[T_0,T]}\|u^\varepsilon\|^p_{m,p}\sum_{|\alpha|\le m}\int_{\mathbb{R}^d}G''_{\alpha,\varepsilon}\sum_{l=1}^{d_1}\left|\left(\sigma^{il}_{(\varepsilon)}u^\varepsilon_i+h^l_{(\varepsilon)}u^\varepsilon+g^l_{(\varepsilon)}\right)_\alpha\right|^2dxds\Bigg)^{\frac{1}{2}}$$

$$\le\frac{\lambda N}{2}\mathbb{E}\sup_{t\in[T_0,T]}\|u^\varepsilon(t)\|^p_{m,p} \tag{4.2.30}$$

$$+\frac{N}{2\lambda}\mathbb{E}\sum_{l=1}^{d_1}\sum_{|\alpha|\le m}\int_{[T_0,T]}\int_{\mathbb{R}^d}G''_{\alpha,\varepsilon}\left|\left(\sigma^{il}_{(\varepsilon)}u^\varepsilon_i+h^l_{(\varepsilon)}u^\varepsilon+g^l_{(\varepsilon)}\right)_\alpha\right|^2dxds,$$

using the Cauchy–Schwarz inequality together with (3.2.10) [but now it will be λ instead of ε] and the observation that $|G'_{\alpha,\varepsilon}|=|u^\varepsilon_\alpha|^{p/2}\sqrt{qG''_{\alpha,\varepsilon}}$.

Finally, we use (4.2.28) and (4.2.22) to conclude that

$$\mathbb{E}\sup_{t\in[T_0,T]}\|u^\varepsilon(t)\|^p_{m,p}\le\frac{\lambda N}{2}\mathbb{E}\sup_{t\in[T_0,T]}\|u^\varepsilon(t)\|^p_{m,p}$$

$$+N(\lambda)\mathbb{E}\Bigg(\|\varphi\|^p_{m,p}+\int_{[T_0,T]}\left(\sum_{j=0}^{d}\|f^j\|^p_{m,p}+\sum_{l=1}^{d_1}\|g^l\|^p_{m,p}\right)ds\Bigg).$$

Now take λ sufficiently small and combine the last inequality with (4.2.28) to get

$$\mathbb{E}\sup_{t\in[T_0,T]}\|u^\varepsilon(t)\|^p_{m,p}$$

$$+\mathbb{E}\sum_{i=1}^{d}\sum_{|\alpha|\le m}\int_{[T_0,T]}\int_{\mathbb{R}^d}\left|\left(u^\varepsilon_i(s,x)\right)_\alpha\right|^2\cdot|u^\varepsilon_\alpha(s,x)|^{p-2}\,dxds \tag{4.2.31}$$

$$\le N\,\mathbb{E}\left(\|\varphi\|^p_{m,p}+\int_{[T_0,T]}\left(\sum_{j=0}^{d}\|f^j(s)\|^p_{m,p}+\sum_{l=1}^{d_1}\|g^l(s)\|^p_{m,p}\right)ds\right),$$

where the number N depends only on m, p, δ, K, T_0, and T.

As a result, we now have $u^{\varepsilon} \in \mathbb{L}_p\big(\Omega; \mathbb{C}([T_0, T]; \mathbb{W}^m_p)\big)$.

The next step is to pass to the limit as $\varepsilon \to 0$ in problem (4.2.23), (4.2.24) and in inequality (4.2.31).

By Lemma 4.1,

$$\lim_{\varepsilon\to 0} \mathbb{E}\left(\|\varphi_{(\varepsilon)} - \varphi\|^p_{m,p} + \int_{[T_0,T]} \left(\sum_{j=0}^{\mathrm{d}} \|f^i_{(\varepsilon)}(t) - f^i(t)\|^p_{m,p} + \sum_{l=1}^{\mathrm{d}_1} \|g^l_{(\varepsilon)}(t) - g^l(t)\|^p_{m,p} \right) dt \right) = 0.$$

The assumptions of the theorem allow us to use this result with $p = 2$, and then to apply Corollary 4.1 to problems (4.2.1), (4.2.2) and (4.2.23), (4.2.24). We then conclude that there exists a sequence $\{\varepsilon_n\}$, where $\varepsilon_n \to 0$ as $n \to \infty$, such that $\ell \times \ell_{\mathrm{d}} \times \mathbb{P}$-a.s.,

$$\lim_{n\to\infty} \sum_{|\alpha|\le m} |u^{\varepsilon_n}_{\alpha}(t,x,\omega) \; - \; u_{\alpha}(t,x,\omega)| = 0, \tag{4.2.32}$$

where u is the generalized solution of problem (4.2.1), (4.2.2), belonging, in view of Theorem 4.2, to $\mathbb{L}_2(\Omega; \mathbb{C}([T_0, T]; \mathbb{H}^m))$. In particular, for every $|\alpha| \le m$ and every $y \in \mathbb{C}^{\infty}_0(\mathbb{R}^{\mathrm{d}})$, the function $(u_{\alpha}, y)_0$ is $\mathbb{P}$-a.s. continuous in t on $[T_0, T]$, and, because $\mathbb{C}^{\infty}_0(\mathbb{R}^{\mathrm{d}})$ is dense in $\mathbb{L}_q$, we conclude that u is weakly continuous in $\mathbb{W}^m_p$.

It remains to pass to the limit on the left-hand side of (4.2.31). We start with the second term, which is straightforward: by Fatou's lemma,

$$\sum_{i=1}^{\mathrm{d}} \sum_{|\alpha|\le m} \int_{[T_0,T]} \int_{\mathbb{R}^{\mathrm{d}}} |(u_i(s,x))_{\alpha}|^2 \, |u_{\alpha}(s,x)|^{p-2} dx$$
$$\le \liminf_{\varepsilon_n\to 0} \mathbb{E} \sum_{i=1}^{\mathrm{d}} \sum_{|\alpha|\le m} \int_{[T_0,T]} \int_{\mathbb{R}^{\mathrm{d}}} \left|\left(u^{\varepsilon_n}_i(s,x)\right)_{\alpha}\right|^2 \cdot \left|u^{\varepsilon_n}_{\alpha}(s,x)\right|^{p-2} dxds.$$

The first term is more challenging, although, in the end, the argument also relies on the Fatou lemma.

To begin, observe that $\mathbb{E} \sup\limits_{t\in[T_0,T]} \|u(t)\|_{m,p}$ is well-defined because $\|u\|^p_{m,p}$ is a predictable function, which, in turn, follows from predictability of u as an $\mathbb{L}_2$-process and the equality

$$\|u_{\alpha}(t,\omega)\|_{0,p} = \sup_j \left|\big(u_{\alpha}(t,\omega), y^j\big)_0\right|,$$

where $\{y^j\}$ is a countable set of functions from $\mathbb{C}_0^\infty(\mathbb{R}^d)$ that is dense in the unit ball of $\mathbb{L}_q$; $q = p/(p-1)$.

To apply the Fatou lemma, it remains to show that, with probability one,

$$\sup_{t\in[T_0,T]} \|u_\alpha(t)\|_{0,p} \le \liminf_{\varepsilon_n\to 0} \sup_{t\in[T_0,T]} \|u_\alpha^{\varepsilon_n}(t)\|_{0,p}, \ |\alpha| \le m. \tag{4.2.33}$$

By (4.2.32), we can find a countable dense subset $\{t^i\}$ in $[T_0, T]$, a countable set $\{y^j\}$ in $\mathbb{C}_0^\infty(\mathbb{R}^d)$ that is dense in the unit ball of $\mathbb{L}_q$, and a subset $\tilde{\Omega}$ of Ω with $\mathbb{P}(\tilde{\Omega}) = 1$ so that, for all t^i, all y^j, all $|\alpha| \le m$, and all $\omega \in \tilde{\Omega}$,

$$\left|\left(u_\alpha(t^i), y^j\right)_0\right| = \lim_{\varepsilon_n\to 0} \left|\left(u_\alpha^{\varepsilon_n}(t^i), y^j\right)_0\right| \le \liminf_{\varepsilon_n\to 0} \|u_\alpha^{\varepsilon_n}(t^i)\|_{0,p},$$

and, because $u \in \mathbb{L}_2(\Omega; \mathbb{C}([T_0, T]; \mathbb{H}^m))$, each of the functions $(u_\alpha, y^i)_0$ is continuous in t on $[T_0, T]$. As a result,

$$\begin{aligned}\sup_{t\in[T_0,T]} \|u_\alpha(t)\|_{0,p} &= \sup_{t\in[T_0,T]} \sup_j \left|\left(u_\alpha(t), y^j\right)_0\right| \le \sup_j \sup_i \left|\left(u_\alpha(t^i), y^j\right)_0\right| \\ &\le \sup_i \liminf_{\varepsilon_n\to 0} \|u_\alpha^{\varepsilon_n}(t^i)\|_{0,p} \le \liminf_{\varepsilon_n\to 0} \sup_{t\in[T_0,T]} \|u_\alpha^{\varepsilon_n}(t)\|_{0,p},\end{aligned}$$

which confirms (4.2.33) and completes the proof of Theorem 4.4. □

4.2.7.

Remark 4.2 Sometimes it is useful to consider a slightly more general equation than (4.2.1), by including in (4.2.1) an extra term

$$\tilde{c}(t, x, \omega)\left(\tilde{b}^i(t, x, \omega)u(t, x, \omega)\right)_i,$$

where the functions $\tilde{b}^i$, $\tilde{c}$, and $\tilde{c}_i$ satisfy the same assumptions as does the function c in Theorem 4.4. This term will then contribute $\left(\tilde{b}^i u, (\tilde{c}y)_i\right)_0$ to equality (4.2.3). With appropriate changes, the statements of Theorem 4.4 and Corollary 4.3 continue to hold for this modified equation (4.2.1). Working out the details could be a good exercise for the interested reader. □

4.3 The Cauchy Problem for Second-Order Parabolic Itô Equations in Non-divergence Form

4.3.1. In this section we consider the Cauchy problem (4.1.1), (4.2.2), keeping the assumptions from Sect. 4.2.1 on the coefficients, on the initial value, and on the external forces. Additionally, it will be assumed throughout this section that the coefficients a^{ij} $(i, j = 1, 2, \ldots, \mathrm{d})$ are differentiable in x for all $(t, \omega) \in [T_0, T] \times \Omega$ with the derivatives bounded uniformly in t, x and ω, and that $f \in \mathbb{L}_2^{\omega}([T_0, T]; \mathbb{L}_2)$.

Under these assumptions we show that problem (4.1.1), (4.2.2) is analytically very similar to problem (4.2.1), (4.2.2) from the previous section. Recall that $v_i = \partial v/\partial x^i$ and the summation convention is in place.

4.3.2.

Definition 4.4 A function $u \in \mathbb{L}_2^{\omega}\big((T_0, T), \mathscr{P}; \mathbb{H}^1\big)$ is a **generalized solution** of problem (4.1.1), *(4.2.2)* if, for every $y \in \mathbb{C}_0^{\infty}(\mathbb{R}^{\mathrm{d}})$ and $\mathfrak{l} \times \mathbb{P}$- a.a. (t, ω), the following equality holds:

$$\big(u(t), y\big)_0 = (\varphi, y)_0 + \int\limits_{[T_0,t]} \Big(- \big(a^{ij} u_i, y_j\big)_0 + \big((b^i - a^{ij}_j) u_i + cu + f, y\big)_0 \Big)(s)\, ds$$

$$+ \int\limits_{[T_0,t]} \big(\sigma^{il} u_i + h^l u + g^l, y\big)_0(s)\, dw^l(s). \tag{4.3.1}$$

□

Under the above assumptions, all the integrals in (4.3.1) are well-defined.

Warning 4.3 *As in* Sect. 4.2, *we will work with version of the generalized solution of problem (4.1.1),* (4.2.2) *that belongs to* $\mathbb{C}([T_0, T], \mathscr{P}; \mathbb{L}_2)$ *and therefore satisfies equality* (4.3.1) *for all* $t \in [T_0, T]$ *and* $\omega \in \Omega' \subset \Omega$, *where* $\mathbb{P}(\Omega') = 1$. □

In what follows, we fix the numbers $m \in \mathbb{N}$, $p \in [2, \infty)$, and $q = p/(p-1)$.

Theorem 4.5 *Suppose that, for all* $(t, \omega) \in [T_0, T] \times \Omega$, *the coefficients* a^{ij} *are differentiable in* x *up to, and including, the order of* $2 \vee m$, *the functions* σ^{il} *and* h^l *are differentiable in* x *to order* $m + 1$, *and the functions* b^i *and* c *are differentiable in* x *to order* m, *where* $i, j = 1, \ldots, \mathrm{d}$ *and* $l = 1, \ldots, \mathrm{d}_1$. *It is also assumed that all the above functions, as well as their derivatives, are uniformly bounded by the same constant* K.

If parabolicity condition (4.1.2) *holds, then problem (4.1.1),* (4.2.2) *has at most one generalized solution.*

If, in addition,

$$f \in \mathbb{L}_{p'}([T_0, T] \times \Omega; \mathbb{W}^m_{p'}), \; g^l \in \mathbb{L}_{p'}([T_0, T] \times \Omega; \mathbb{W}^{m+1}_{p'}), \; \varphi \in \mathbb{L}_{p'}(\Omega; \mathbb{W}^m_{p'}),$$

for both $p' = 2$ *and* $p' = 2$, *then problem (4.1.1),* (4.2.2) *has a generalized solution* u *such that*

$$u \in \mathbb{L}_2(\Omega; \mathbb{C}([T_0, T]; \mathbb{H}^{m-1})) \bigcap \mathbb{L}_2([T_0, T] \times \Omega; \mathbb{C}_w \mathbb{H}^m) \bigcap \mathbb{L}_p([T_0, T] \times \Omega; \mathbb{C}_w \mathbb{W}^m_p),$$

and

$$\mathbb{E} \sup_{t\in[T_0,T]} \|u(t)\|_{m,p'}^{p'} \le N\mathbb{E}\Big(\|\varphi\|_{m,p'}^{p'} + \int_{[T_0,T]} \Big(\|f(t)\|_{m,p'}^{p'} + \sum_{l=1}^{d_1} \|g^l(t)\|_{m+1,p'}^{p'} \Big) dt \Big),$$

where the number N depends only on p', d, d_1, *K, m, T_0, and T.* □

Compared to the results from the previous section, relaxing the super-parabolicity conditions requires extra smoothness of the coefficients, including at least two derivatives for a^{ij}, and extra regularity of the free terms f, g^l.

Arguing in the same way as in the proofs of Theorem 4.3 and Corollary 4.2, we get

Corollary 4.4 *Suppose that the assumptions of the theorem are fulfilled and $(m-n)p >$ d for some $n \in \mathbb{N} \cup \{0\}$. Then the generalized solution of problem (4.1.1),* (4.2.2) *has a version v with the following properties*:

(a) *For every $x \in \mathbb{R}^d$, $v(\cdot, x, \cdot)$ is predictable;*
(b) *For all $\omega \in \Omega$, $v(\cdot, \cdot, \omega) \in \mathbb{C}_b^{0,n}\big([T_0, T] \times \mathbb{R}^d\big)$;*
(c) *As a generalized solution of problem (4.1.1),* (4.1.2) *it possesses all the properties of u listed in the theorem;*
(d) $\mathbb{E} \sup_{t\in[T_0,T]} \|v(t)\|^p_{\mathbb{C}_b^n(\mathbb{R}^d)}$ *is bounded above by the right-hand of the inequality from the theorem;*
(e) *If $n \ge 2$ and α is a* d*-dimensional multi-index with $|\alpha| \le n-2$ and if $\tilde{f}$ and $\tilde{g}^l$ are the smooth versions of f and g^l according to Proposition 4.1 with $(S, \Sigma) = ([T_0, T] \times \Omega, \mathscr{P})$, then, for every $x \in \mathbb{R}^d$, there exists a set $\Omega_x \subset \Omega$ with $\mathbb{P}(\Omega_x) = 1$ such that, for all $(t, \omega) \in [T_0, t] \times \Omega_x$,*

$$v_\alpha(t) = \varphi_\alpha + \int_{[T_0,t]} \big(a^{ij} v_{ij} + b^i v_i + c + \tilde{f}\big)_\alpha (s, x)\, ds + \int_{[T_0,t]} \big(\sigma^{il} v_i + h^l v + \tilde{g}^l\big)_\alpha (s, x)\, dw^l(s).$$

The stochastic integral in this equality has a continuous in (x, t) version, and, for this version, there exists a set Ω' of probability 1 on which the equality holds for all t, x;

(f) *If $v_1(t,x)$ and $v_2(t,x)$ are generalized solutions of problem (4.1.1),* (4.2.2) *possessing properties* (a), (b), *then*

$$\mathbb{P}\Big(\sup_{\substack{t\in[T_0,T]\\ x\in\mathbb{R}^d}} |v_1(t,x,\omega) - v_2(t,x,\omega)| > 0 \Big) = 0. \qquad \square$$

4.3.3. The classical solution of problem (4.1.1), (4.2.2) is defined similarly to the classical solution of problem (4.2.1), (4.2.2); cf. Definition 4.3. The following result is similar to Corollary 4.2.

Corollary 4.5 *If the assumptions of Theorem* 4.5 *are fulfilled for every $m \in \mathbb{N}$, with constants possibly depending on m, then the classical solution of problem (4.1.1),* (4.2.2) *is infinitely differentiable in x and all the derivatives are continuous in (t,x)* ($\mathbb{P}$-a.s.). □

4.3.4. To prove Theorem 4.5, we need an auxiliary result; cf. Lemma 1.4.

Lemma 4.3 *Suppose that we have functions a^{ij}, b^i, c, σ^{il}, and h^l, $i,j = 1,2,\ldots,$ d, $l = 1,2,\ldots,\mathrm{d}_1$, satisfying all the conditions of Theorem* 4.5 *for some $m = 0,1,2,\ldots$; if $m = 0$, then we additionally assume that each b^i is differentiable with respect to x, with corresponding derivatives uniformly bounded by K.*

Given the (possibly random and time-dependent) functions $u \in \mathbb{W}_p^{m+2}$, $f \in \mathbb{W}_p^m$, and $g^l \in \mathbb{W}_p^{m+1}$, $l = 1,2,\ldots,\mathrm{d}_1$, and strictly positive numbers r^α, $|\alpha| \le m$, define

$$G_\alpha = |u_\alpha|^p,\ G'_\alpha = p|u_\alpha|^{p-2}u_\alpha,\ G''_\alpha = p(p-1)|u_\alpha|^{p-2},$$

and

$$V(m,p,u,f,g,\{r^\alpha\},t) := \int_{\mathbb{R}^d} \sum_{|\alpha|\le m} r^\alpha \Big(G'_\alpha \left(a^{ij}u_{ij} + b^i u_i + cu + f\right)_\alpha + \frac{1}{2} G''_\alpha \sum_{l=1}^{\mathrm{d}_1} |(\sigma^{il}u_i + h^l u + g^l)_\alpha|^2 \Big)(t,x)\,dx.$$

Then, for every $(t,\omega) \in [T_0,T] \times \Omega$,

$$V(m,p,u,f,g,\{r^\alpha\},t) \le N \cdot \Big(\|u\|_{m,p}^p + \|f\|_{m,p}^p + \sum_{l=1}^{\mathrm{d}_1} \|g^l\|_{m+1,p}^p \Big)(t), \tag{4.3.2}$$

where the number N depends only on r^α, p, d, d_1, *K, and m.* □

Proof The first comment is that the numbers r^α are introduced for purely technical reasons: later on, we will need (4.3.2) with $r^\alpha = C^\alpha$, where the numbers C^α are from (4.2.10).

Next, we introduce several notations to make the formulas shorter. For $F, H \in \mathbb{L}_1$ we write $F \sim H$ if

$$\int_{\mathbb{R}^\mathrm{d}} F(x)dx = \int_{\mathbb{R}^\mathrm{d}} H(x)dx,$$

because the proof involves many integration-by-parts steps in the Sobolev spaces, and so writing

$$uv_i \sim u_i v$$

avoids unnecessarily long formulas. Another notation to be used in the proof is

$$F \ll H,$$

meaning that $F \sim H + \Theta$ and

$$|\Theta| \leq N \left(\sum_{|\alpha|\leq m} \left(|u_\alpha|^p + |f_\alpha|^p \right) + \sum_{|\alpha|\leq m+1} \sum_{l=1}^{\mathrm{d}_1} |g_\alpha^l|^p \right).$$

The objective then becomes to show that

$$V \ll 0.$$

From inequality (4.2.22) it follows that $G'_\alpha f_\alpha \ll 0$ and $G'_\alpha (cu)_\alpha \ll 0$. Also, by direct computation,

$$G'_\alpha (b^i u_i)_\alpha \ll (u_i)_\alpha G'_\alpha b^i = \left(G_\alpha b^i \right)_i - G_\alpha b^i_i \sim -G_\alpha b^i_i \ll 0.$$

By inequality (4.2.22),

$$G''_\alpha \sum_{l=1}^{\mathrm{d}_1} |g_\alpha^l|^2 \ll 0, \quad G''_\alpha \sum_{l=1}^{\mathrm{d}_1} |(h^l u)_\alpha|^2 \ll 0,$$

and

$$G''_\alpha (h^l u)_\alpha g_\alpha^l \leq \frac{1}{2}\, G''_\alpha \sum_{l=1}^{\mathrm{d}_1} \left((h^l u_\alpha)^2 + (g_\alpha^l)^2 \right) \ll 0,$$

$$G''_\alpha (\sigma^{il} u_i)_\alpha g_\alpha^l \ll G''_\alpha\, (u_i)_\alpha\, \sigma^{il} g_\alpha^l = (G'_\alpha)_i \sigma^{il} g_\alpha^l.$$

Integrating by parts, we obtain, with the help of inequality (4.2.22), that

$$\begin{aligned}
&\big(G'_\alpha\big)_i \sigma^{il} g^l_\alpha \sim -G'_\alpha(\sigma^{il} g^l_\alpha)_i \ll 0,\\
&G''_\alpha(\sigma^{il}u_i)_\alpha(h^l u)_\alpha \ll G''_\alpha\,(u_i)_\alpha\,\sigma^{il}(h^l u)_\alpha = (G'_\alpha)_i\sigma^{il}(h^l u)_\alpha\\
&\sim -G'_\alpha\big(\sigma^{il}(h^l u)_\alpha\big)_i \ll -G'_\alpha\sigma^{il}(h^i u_i)_\alpha \ll -G'_\alpha\,(u_i)_\alpha\,\sigma^{il}h^l\\
&= -(G_\alpha)_i\sigma^{il}h^l \sim G_\alpha(\sigma^{il}h^l)_i \ll 0.
\end{aligned}$$

The reader should keep in mind that, unlike all other instances, the subscript α in G, G', G'' is not really a derivative.

It remains to prove the inequality

$$\sum_{|\alpha|=k} r^\alpha\left((a^{ij}u_{ij})_\alpha G'_\alpha + \frac{1}{2}G''_\alpha\sum_{l=1}^{d_1}|(\sigma^{il}u_i)_\alpha|^2\right) \ll 0,\ k=0,1,\ldots,m. \tag{4.3.3}$$

If $|\alpha| = 0$, so that $u_\alpha = u$, then (4.3.3) follows from parabolicity condition (4.1.2) after several rounds of integration by parts:

$$\begin{aligned}
a^{ij}u_{ij}G'_\alpha &= p|u|^{p-2}ua^{ij}u_{ij} = p|u|^{p-2}u\Big((a^{ij}u_i)_j - a^{ij}_j u_i\Big)\\
&\sim -p(p-1)|u|^{p-2}a^{ij}u_iu_j - a^{ij}_j\left(|u|^p\right)_i\\
&\sim -p(p-1)|u|^{p-2}a^{ij}u_iu_j + a^{ij}_{ij}|u|^p \ll -p(p-1)|u|^{p-2}a^{ij}u_iu_j \qquad (4.3.4)\\
&= G''_\alpha a^{ij}u_iu_j.
\end{aligned}$$

Now fix a multi-index α with $|\alpha| = k$, $k = 1,2,\ldots,m$. We will use the following notation:

$$\sum\nolimits'_\alpha = \sum_{\substack{\beta,\gamma:\\ \beta+\gamma=\alpha,\\ |\beta|=1}}, \qquad \sum\nolimits''_\alpha = \sum_{\substack{\beta,\gamma:\\ \beta+\gamma=\alpha,\\ |\beta|>1}}.$$

Using the product rule,

$$(a^{ij}u_{ij})_\alpha G'_\alpha \ll a^{ij}(u_{ij})_\alpha G'_\alpha + \sum\nolimits'_\alpha (a^{ij})_\beta\,(u_{ij})_\gamma G'_\alpha.$$

Replacing u with u_α in (4.3.4), we get

$$a^{ij}(u_{ij})_\alpha G'_\alpha = a^{ij}(u_\alpha)_{ij}G'_\alpha \ll G''_\alpha a^{ij}(u_\alpha)_i(u_\alpha)_j,$$

concluding that

$$(a^{ij}u_{ij})_\alpha G' \ll -G''_\alpha a^{ij}(u_\alpha)_i(u_\alpha)_j) + G'_\alpha {\sum}'_\alpha (a^{ij})_\beta (u_{ij})\gamma. \tag{4.3.5}$$

Next,

$$\begin{aligned} G''_\alpha \sum_{l=1}^{\mathrm{d}_1} |(\sigma^{il}u_i)_\alpha|^2 &= G''_\alpha \sum_{l=1}^{\mathrm{d}_1} |(\sigma^{il}u_i)_\alpha - \sigma^{il}(u_i)_\alpha + \sigma^{il}(u_i)_\alpha|^2 \\ &\ll G''_\alpha \cdot \Big((u_j)_\alpha \sigma^{jl}\Big)\cdot\Big((\sigma^{il}u_i)_\alpha - \sigma^{il}(u_i)_\alpha\Big) + G''_\alpha \sum_{l=1}^{\mathrm{d}_1} |\sigma^{il}(u_i)_\alpha|^2 \\ &= 2(G'_\alpha)_j\sigma^{jl}\cdot\Big((\sigma^{il}u_i)_\alpha \;-\; \sigma^{il}(u_i)_\alpha\Big) + G''_\alpha \sum_{l=1}^{\mathrm{d}_1} |\sigma^{il}(u_i)_\alpha|^2. \end{aligned}$$

By the product rule for derivatives,

$$\begin{aligned} &2(G'_\alpha)_j\,\sigma^{jl}\Big((\sigma^{il}u_i)_\alpha - \sigma^{il}(u_i)_\alpha\Big) \\ &\quad = 2(G'_\alpha)_j\,\sigma^{jl} {\sum}'_\alpha \sigma^{il}_\beta (u_i)_\gamma + 2(G'_\alpha)_j\,\sigma^{jl} {\sum}''_\alpha \sigma^{il}_\beta (u_i)_\gamma. \end{aligned}$$

If $|\beta| = 1$, then

$$(\sigma^{il}\sigma^{jl})_\beta = (\sigma^{il})_\beta\sigma^{jl} + \sigma^{il}(\sigma^{jl})_\beta$$

and, because $|\gamma| = m - 1$, integration by parts results in

$$\begin{aligned} 2(G'_\alpha)_j\,\sigma^{jl} {\sum}'_\alpha \sigma^{il}_\beta (u_i)_\gamma &\ll -2G'_\alpha\sigma^{jl} {\sum}_\alpha \sigma^{il}_\beta (u_{ij})_\gamma \\ &= -G'_\alpha {\sum}_\alpha (\sigma^{il}\sigma^{jl})_\beta (u_{ij})_\gamma. \end{aligned}$$

If $|\beta| > 1$, then $|\gamma| \le m - 2$ and integration by parts yields

$$2G'_j\sigma^{jl} {\sum}''_\alpha \sigma^{il}_\beta (u_i)_\gamma \sim 2G'\left(\sigma^{il} \sum_{|\beta|\neq 1} {}^\alpha\sigma^{il}_\beta (u_i)_\gamma\right)_j \ll 0.$$

To summarize,

$$G''_\alpha \sum_{l=1}^{\mathrm{d}_1} |(\sigma^{il}u_i)_\alpha|^2 \ll G''_\alpha \sum_{l=1}^{\mathrm{d}_1} |\sigma^{il}(u_i)_\alpha|^2 - G'_\alpha {\sum}'_\alpha (\sigma^{il}\sigma^{jl})_\beta (u_{ij})_\gamma. \tag{4.3.6}$$

Using the notation

$$A^{ij} = a^{ij} - \frac{1}{2}\sigma^{il}\sigma^{jl},$$

we combine (4.3.5) and (4.3.6) to get

$$\begin{aligned}
G'_\alpha(a^{ij}u_{ij})_\alpha &+ \frac{1}{2}G''_\alpha \sum_{l=1}^{\mathrm{d}_1} |(\sigma^{il}u_i)_\alpha|^2 \\
&\ll G''_\alpha \left(- a^{ij}(u_\alpha)_i(u_\alpha)_j + \frac{1}{2}\sum_{l=1}^{\mathrm{d}_1} |\sigma^{il}(u_\alpha)_i|^2 \right) \\
&\quad + G'_\alpha \sum\nolimits_\alpha' \left((a^{ij})_\beta (u_{ij})_\gamma - \frac{1}{2}(\sigma^{il}\sigma^{jl})_\beta (u_{ij})_\gamma \right) \\
&= G'_\alpha \sum\nolimits_\alpha' A^{ij}_\beta (u_{ij})_\gamma - G''_\alpha A^{ij}(u_\alpha)_i(u_\alpha)_j,
\end{aligned}$$

or, because $0 < r^\circ \le r^\alpha \le r^*$, $G''_\alpha \ge 0$, and $A^{ij}(u_\alpha)_i(u_\alpha)_j \ge 0$,

$$\begin{aligned}
\sum_{|\alpha|=k} r^\alpha &\left(G'_\alpha(a^{ij}u_{ij})_\alpha + \frac{1}{2}G''_\alpha \sum_{l=1}^{\mathrm{d}_1} |(\sigma^{il}u_i)_\alpha|^2 \right) \\
&\ll r^* \sum_{|\alpha|=k} \left| G'_\alpha \sum\nolimits_\alpha' A^{ij}_\beta (u_{ij})_\gamma \right| - r^\circ \sum_{|\alpha|=k} G''_\alpha A^{ij}(u_\alpha)_i(u_\alpha)_j.
\end{aligned}$$

Now define

$$\eta^k := r^* \sum_{|\alpha|=k} \left| G'_\alpha \sum\nolimits_\alpha' A^{ij}_\beta (u_{ij})_\gamma \right| - r^\circ \sum_{|\alpha|=k} G''_\alpha A^{ij}(u_\alpha)_i(u_\alpha)_j; \tag{4.3.7}$$

to complete the proof of the lemma, we need to show that

$$\eta^k \ll 0, \quad k = 1, 2, \ldots, m.$$

By (3.2.10),

$$\begin{aligned}
\left| G'_\alpha \sum\nolimits_\alpha' A^{ij}_\beta (u_{ij})_\gamma \right| &\le \varepsilon N_1 G''_\alpha \sum\nolimits_\alpha' |A^{ij}_\beta (u_{ij})_\gamma|^2 + N_2(\varepsilon) G_\alpha \\
&\ll \varepsilon N_1 G''_\alpha \sum\nolimits_\alpha' |A^{ij}_\beta (u_{ij})_\gamma|^2,
\end{aligned} \tag{4.3.8}$$

where N_1 does not depend on ε.

To handle the term on the right-hand side of (4.3.8), we need the following technical result.

Proposition 4.2 *Suppose that, for* $i, j = 1, 2, \ldots, \mathrm{d}$, *the functions* $P^{ij} = P^{ij}(x)$ *belong to* $\mathbb{C}_b^2(\mathbb{R}^{\mathrm{d}})$, *and* $P^{ij}(x)\xi^i\xi^j \geq 0$ *for every* $x, \xi \in \mathbb{R}^{\mathrm{d}}$.

Then, for every function $v \in \mathbb{C}^2(\mathbb{R}^{\mathrm{d}})$, *every* $x \in \mathbb{R}^{\mathrm{d}}$, *and every* $n = 1, \ldots, \mathrm{d}$,

$$\left(P_n^{ij}(x)v_{ij}(x)\right)^2 \leq N_o\, P^{ij}(x)v_{i\ell}(x)v_{j\ell}(x),$$

where the constant N_o *depends only on the upper bound for the second derivatives of* P^{ij}. □

To prove this, start by noticing that if $F = F(t)$, $t \in \mathbb{R}^1$, is a non-negative function with two bounded derivatives, then, for every $t \in \mathbb{R}^1$,

$$|F'(t)|^2 \leq 2\Big(\sup_{y\in\mathbb{R}^1} |F''(y)|\Big)F(t);$$

if this inequality were to fail at a point t_0 (with $F(t_0) > 0$; when $F(t_0) = 0$, the Taylor formula at t_0 and assumption $F(t) \geq 0$ imply $F'(t_0) = 0$), then the Taylor expansion, at the point t_0, of $F(\tilde{t}_0)$ with $\tilde{t}_0 = t_0 - 2(F(t_0)/F'(t_0))$ would lead to a contradiction with the assumption $F(\tilde{t}_0) \geq 0$. The statement of the proposition at a particular point $x^* \in \mathbb{R}^{\mathrm{d}}$ now follows after a change of variables that diagonalizes the matrix $\left(P^{ij}(x^*)\right)$. The interested reader can fill in the details or look up the complete proof in [120, Lemma 1.7.1]. □

We now use this proposition to conclude that, for every β with $|\beta| = 1$,

$$|A_\beta^{ij}(u_{ij})_\gamma|^2 = |A_\beta^{ij}(u_{\alpha-\beta})_{ij}|^2 \leq N_3 A^{ij}(u_{\alpha-\beta})_{i\ell}(u_{\alpha-\beta})_{j\ell}$$

and therefore

$$|A_\beta^{ij}(u_\gamma)_{ij}|^2 \leq N_3 A^{ij}(u_{\alpha-\beta})_{i\ell}(u_{\alpha-\beta})_{j\ell} = N_3 \sum_{|\alpha|=k} A^{ij}(u_\alpha)_i(u_\alpha)_j.$$

By assumption, for every α,

$$A^{ij}(u_\alpha)_i(u_\alpha)_j \geq 0,$$

which implies

$$\sum\nolimits_\alpha' |A_\beta^{ij}(u_\gamma)_{ij}|^2 \leq N_4 \sum_{|\alpha|=k} A^{ij}(u_\alpha)_i(u_\alpha)_j$$

and

$$\sum_{|\alpha|=k} G''_\alpha \sum\nolimits'_\alpha |A^{ij}_\beta (u_\gamma)_{ij}|^2. \tag{4.3.9}$$

We now combine (4.3.9) and (4.3.8) with sufficiently small ε and put the result in (4.3.7) to deduce that $\eta^k \ll 0$ and complete the proof of (4.3.2). □

4.3.5. We will now complete the proof of the theorem by applying Theorem 3.4 (for uniqueness) and Theorem 3.3 (for everything else). We will work in the Hilbert scale of Sobolev spaces $\{\mathbb{H}^\gamma,\ \gamma \in \mathbb{R}^1\}$, and, when applying Theorem 3.3, use $\lambda = m$.

As was mentioned in Sect. 4.2.1, we can transform Eq. (4.1.1) to the form (4.2.1). Thus, the same arguments as in the proof of Theorem 4.1 demonstrate that a generalized solution of problem (4.1.1), (4.2.2) is also a solution of the LSES of the type (3.1.1) in the normal triple $(\mathbb{H}^1, \mathbb{L}_2, \mathbb{H}^{-1})$ with the operators A, B and the martingale M given by

$$\begin{aligned}
\big[y, A(t,\omega)v\big]_0 &= -\big(a^{ij}(t,\omega)v_i, y_j\big)_0 + \Big(\big(b^i(t,\omega) - a^{ij}_j(t,\omega)\big)v_i + c(t,\omega)v, y\Big)_0,\\
B(t,\omega)v &= \Big(B^1(t,\omega)v, \dots, B^{\mathrm{d}_1}(t,\omega)v\Big),\ B^l v = \sigma^{il} v_i + h^l v,\\
M(t) &= \int_{[0,t]} g^l(s)\, dw^l(s),
\end{aligned}$$

where $v, y \in \mathbb{H}^1$. From the arguments given in the beginning of the proof of Theorem 1.1, it follows that there exists a number K such that, for all t, ω,

$$\|A(t,\omega)v\|_{-1} \le K\|v\|_1,\ v \in \mathbb{H}^1. \tag{4.3.10}$$

Repeating the arguments from the beginning of the proof of Theorem 4.2, we conclude that the operator A, when restricted to $\mathbb{H}^{m+1}$, coincides with the operator

$$L : v \mapsto a^{ij} v_{ij} + b^i v_i + cv$$

and

$$\|A(t,\omega)v\|_{m-1} \le K\|v\|_{m+1},\ v \in \mathbb{H}^{m+1},\ (t,\omega) \in [T_0, T] \times \Omega.$$

By Proposition 3.7 the differentiation operator is bounded from $\mathbb{H}^{m+1}$ to $\mathbb{H}^m$ and from $\mathbb{H}^1$ to $\mathbb{H}^0$, and so

$$|||B(t,\omega)v||| \le K\|v\|_{n+1},\ v \in \mathbb{H}^{n+1},\ n = 0, m,\ (t,\omega) \in [T_0, T] \times \Omega,$$

where $|||\cdot|||$ is the Hilbert–Schmidt norm in either $\mathfrak{L}_2(\mathbb{R}^d, \mathbb{H}^0)$ or $\mathfrak{L}_2(\mathbb{R}^d, \mathbb{H}^m)$ Thus condition (B_2) of Theorem 3.3 is fulfilled, whereas condition (C_2) follows immediately from the assumptions of our theorem. It remains to show that condition (A_2) of Theorem 3.3 is also satisfied.

Take $u \in \mathbb{H}^{m+2}$ so that $Au \in \mathbb{H}^m$ and $[u, Au]_m = (u, Lu)_m$ for every $u \in \mathbb{H}^{m+2}$. From (3.3.2), in view of the self-adjointness of the operator $\Lambda = (\mathbb{I} - \mathbf{\Delta})^{1/2}$, it follows that

$$(u, Lu)_m = \big((\mathbb{I} - \mathbf{\Delta})^m u, Lu\big)_0.$$

Using (4.2.10),

$$\begin{aligned}[u, A(t,\omega)u]_m &+ \frac{1}{2}|||B(t,\omega)u|||^2 \\ &= \sum_{|\alpha|\le m} C^\alpha \left(\big((Lu)_\alpha, u_\alpha\big)_0 + \frac{1}{2}\sum_{l=1}^{d_1} \|(B^l u)_\alpha\|_0^2 \right) \\ &= V\big(m, 2, u, 0, 0, \{C^\alpha\}, t\big),\end{aligned}$$

where V is the function introduced in Lemma 4.1.

From the last inequality and (4.3.2) it follows that condition (A_2) holds for $u \in \mathbb{H}^{m+2}$, and then, by continuity, for $u \in \mathbb{H}^{m+1}$.

Thus our LSES has a solution in the normal triple $(\mathbb{H}^1, \mathbb{L}_2, \mathbb{H}^{-1})$, the solution belongs to $\mathbb{L}_2([T_0, T], \mathscr{P}; \mathbb{H}^m)$, and is a generalized solution of problem (4.1.1), (4.2.2).

From (4.3.10) and (2.3.11) it follows that Theorem 3.4 applies and thus problem (4.1.1), (4.2.2) has a unique generalized solution.

The rest of the proof is identical to the corresponding part of the proof of Theorem 4.4. □

Similar to Corollary 4.1 the following stability result holds.

Corollary 4.6 *Suppose that the assumptions of Theorem* 4.5 *are satisfied. Let*

$$\{a_n^{ij}, b_n^i, c_n, \sigma_n^{il}, h_n^l, f_n, g_n^l, \varphi_n,\ i, j = 1, 2, \dots, d,\ l = 1, 2, \dots, d_1,\ n \in \mathbb{N}\}$$

be a collection of functions satisfying the same assumptions, uniformly in n, as the functions $a^{ij}, b^i, c, \sigma^{il}, h^l, f, g^l, \varphi$. Suppose, as well, that u^n is the generalized solution of problem (4.1.1), (4.2.2), *where a^{ij} is replaced by a_n^{ij}, b^i by b_n^i etc.*

If, for all i, j, l and $\ell \times \ell_d \times \mathbb{P}$- a.a. (t, x, ω),

$$a_n^{ij} \to a^{ij},\ b_n^i \to b^i,\ c_n \to c,\ \sigma_n^{il} \to \sigma^{il},\ h_n^l \to h^l,\ \text{as } n \to \infty,$$

and

$$\lim_{n\to\infty}\mathbb{E}\left(\int_{[T_0,T]}\left(\|f_n(t)-f(t)\|_m^2+\sum_{l=1}^{\mathrm{d}_1}\|g_n^l(t)-g^l(t)\|_{m+1}^2\right)dt+\|\varphi_n-\varphi\|_m^2\right)=0,$$

then

$$\lim_{n\to\infty}\mathbb{E}\sup_{t\in[T_0,T]}\|u^n(t)-u(t)\|_m^2=0.$$

□

4.4 The Forward and Backward Cauchy Problems in Weighted Sobolev Spaces

4.4.1. In this section we study the solvability of the Cauchy problem for Eqs. (4.1.1) and (4.2.1) when the initial value φ and the free terms $f(t)$, $g^l(t)$, $l=1,2,\ldots,\mathrm{d}_1$, belong to weighted Sobolev spaces $\mathbb{W}_p^m(r)$. We fix the numbers $r\in\mathbb{R}^1$, $K\in\mathbb{R}_+$, $p\in[2,\infty)$, and $m\in\mathbb{N}\cup\{0\}$. The case $r<0$ is of special interest because it extends the class of admissible initial conditions and free terms to bounded or mildly growing as $|x|\to\infty$.

We also study the same question for the backward Cauchy problems (cf. Sect. 1.5.13):

$$\begin{aligned}-dv(t,x,\omega)&=\Big(L_{\mathrm{div}}v(t,x,\omega)+f(t,x,\omega)\Big)dt\\&\quad+\Big(B^lv(t,x,\omega)+g^l(t,x,\omega)\Big)*dw^l(t),\\&\qquad(t,x,\omega)\in[T_0,T)\times\mathbb{R}^{\mathrm{d}}\times\Omega;\end{aligned}\tag{4.4.1}$$

$$v(T,x,\omega)=\varphi(x,\omega),\ (x,\omega)\in\mathbb{R}^{\mathrm{d}}\times\Omega,\tag{4.4.2}$$

and

$$\begin{aligned}-dv(t,x,\omega)&=\Big(Lv(t,x,\omega)+f(t,x,\omega)\Big)dt\\&\quad+\Big(B^lv(t,x,\omega)+g^l(t,x,\omega)\Big)*dw^l(t),\\&\qquad(t,x,\omega)\in[T_0,T)\times\mathbb{R}^{\mathrm{d}}\times\Omega;\end{aligned}\tag{4.4.3}$$

$$v(T,x,\omega)=\varphi(x,\omega),\ (x,\omega)\in\mathbb{R}^{\mathrm{d}}\times\Omega.\tag{4.4.4}$$

As before,

$$L_{\text{div}}v = \left(a^{ij}v_i\right)_j + b^i v_i(t,x,\omega) + cv; \ Lv = a^{ij}v_{ij} + b^i v_i + cv;$$

$$B^l v = \sigma^{il}v_i + h^l v; \ i,j = 1,2,\ldots,\mathrm{d},\ l = 1,2,\ldots,\mathrm{d}_1.$$

The coefficients, the initial value and the external forces in problems (4.4.1), (4.4.2) and (4.4.3), (4.4.4) are backward predictable with respect to the standard Brownian motion $W = (w^1,\ldots,w^{\mathrm{d}_1})$; cf. Sect. 1.5.13.

The main part of the section is a collection of existence and uniqueness results for problems ((4.1.1), (4.2.2)), ((4.2.1), (4.2.2)), ((4.4.1), (4.4.2)), and ((4.4.3), (4.4.4)) in the space $\mathbb{L}^p\big([T_0,T]\times\Omega;\mathbb{W}_p^m(r)\big)$ and in the space of differentiable functions.

Throughout this section it is supposed that, for all $i,j = 1,2,\ldots,\mathrm{d}$, $l = 1,2,\ldots,\mathrm{d}_1$, the coefficients $a^{ij}, b^i, c, \sigma^{il}$, and h^l, are $\mathscr{B}([T_0,T]\times\mathbb{R}^{\mathrm{d}})\otimes\mathscr{F}$-measurable functions. Concerning problems (4.1.1), (4.2.2) and (4.4.3), (4.4.4), we also assume that all the functions a^{ij} are in $\mathbb{C}_b^1(\mathbb{R}^{\mathrm{d}})$ for all (t,ω).

4.4.2. In this paragraph we study problem (4.2.1), (4.2.2) under the following assumptions:

- φ is an $\mathscr{F}_{T_0}$-measurable function taking values in $\mathbb{L}_2(r)$;
- $f = f^0 + \sum_{i=1}^{\mathrm{d}} f_i^i(t)$, where $f^j \in \mathbb{L}_2^{\omega}\big([T_0,T],\mathscr{P};\mathbb{L}_2(r)\big)$, $j = 0,1,2,\ldots,\mathrm{d}$;
- $g^l \in \mathbb{L}_2^{\omega}([T_0,T],\mathscr{P};\mathbb{L}_2(r))$, $l = 1,2,\ldots,\mathrm{d}_1$;
- The coefficients of Eq. (4.2.1) are predictable for every $x \in \mathbb{R}^{\mathrm{d}}$.

To simplify the presentation, we write

$$S := (1+|x|^2)^{r/2}, \quad S(i) := \frac{rx^i}{(1+|x|^2)}.$$

With this notation, $S_i = S(i)S$, $(1/S)_i = -S(i)/S$. Note that each $S(i)$ is a bounded function.

Definition 4.1 A function $u \in \mathbb{L}_2^{\omega}([T_0,T],\mathscr{P};\mathbb{H}^1(r))$ is called an r-**generalized solution of problem** (4.2.1), (4.2.2) if u satisfies equality (4.2.3) for every $y \in \mathbb{C}_0^{\infty}(\mathbb{R}^{\mathrm{d}})$ and $\mathfrak{l}\times\mathbb{P}$-a.a. (t,ω). □

If $r = 0$, then the above definition coincides with Definition 4.2. In other words, a 0-generalized solution is the same as a generalized solution.

Theorem 4.6 *Suppose that the following conditions are fulfilled.*

(i) *The functions $a^{ij}, b^i, c, \sigma^{il}$, and h^l, $i,j = 1,2,\ldots,\mathrm{d}$, $l = 1,2,\ldots,\mathrm{d}_1$ are m times differentiable in x. These functions and all their derivatives are bounded by the constant K, uniformly in t,x,ω.*

(ii) *For p' equal to p and 2, $\varphi \in \mathbb{L}_{p'}(\Omega, \mathbb{W}^m_{p'}(r))$, $f^j \in \mathbb{L}_{p'}([T_0, T] \times \Omega; \mathbb{W}^m_{p'}(r))$, where $l = 1, 2, \ldots, \mathrm{d}_1$, and $g^l \in \mathbb{L}_{p'}([T_0, T] \times \Omega; \mathbb{W}^m_{p'}(r))$, where $l = 1, 2, \ldots, \mathrm{d}_1$.*

If the super-parabolicity condition (4.1.3) *holds, then problem* (4.2.1), (4.2.2) *has a unique r-generalized solution. This solution is an element of the space*

$$\mathbb{L}_2\Big(\Omega; \mathbb{C}([T_0, T], \mathbb{H}^{m-1}(r))\Big) \cap \mathbb{L}_2\big([T_0, T] \times \Omega; \mathbb{C}_w \mathbb{H}^m(r)\big) \cap \mathbb{L}_p\big([T_0, T] \times \Omega; \mathbb{C}_w \mathbb{W}^m_p(r)\big)$$

and satisfies

$$\mathbb{E} \sup_{t \in [T_0, T]} \|u(t)\|^{p'}_{m, p', r} + \frac{\delta}{2} p(p-1)\, \mathbb{E} \int_{[T, T_0]} \sum_{|\alpha| \geq m} \sum_{i=1}^{\mathrm{d}} \int_{\mathbb{R}^{\mathrm{d}}} \big|(Su)_\alpha\big|^{p-2}$$

$$\times \big|\big((Su)_i\big)_\alpha\big|^2 dx dt \leq N \mathbb{E}\bigg(\|\varphi\|^{p'}_{m, p', r} + \int_{[T_0, T]} \bigg(\sum_{j=0}^{\mathrm{d}} \|f^i(t)\|^{p'}_{m, p', r} \tag{4.4.5}$$

$$+ \sum_{l=1}^{\mathrm{d}_1} \|g^l(t)\|^{p'}_{m, p', r} \bigg) dt \bigg)$$

with $p' = 2$ and $p' = p$. □

Remark 4.3 Condition (ii) of the theorem is not necessary for uniqueness of r-generalized solutions because the difference of two r-generalized solutions is also an r-generalized solution of the problem with zero initial condition and free terms, so that (ii) automatically holds.

Proof If $r = 0$, then the statement of the theorem coincides with that of Theorem 4.4. We now show how to reduce the case of $r \neq 0$ to $r = 0$.

By Lemma 3.6(ii), $S\varphi \in \mathbb{L}_{p'}(\Omega; \mathbb{W}^m_{p'})$, $Sf^j \in \mathbb{L}_{p'}([T_0, T]; \mathscr{P}; \mathbb{W}^m_{p'})$, and $Sg^l \in \mathbb{L}_{p'}([T_0, T], \mathscr{P}; \mathbb{W}^m_{p'})$ for all $j = 0, 1, 2, \ldots, \mathrm{d}$, $l = 1, 2, \ldots, \mathrm{d}_1$, and p' equal to p and 2.

Now define

$$\tilde{b}^i = b^i - \sum_{j=1}^{\mathrm{d}} a^{ij} S(j), \; \tilde{c} = c - \sum_{i=1}^{\mathrm{d}} S(i) b^i + (1/S)_{ij} a^{ij} S,$$

$$\tilde{f}^0 = Sf^0 - \sum_{i=1}^{\mathrm{d}} S(i) Sf^i, \; \tilde{h}^l = h^l - \sum_{i=1}^{\mathrm{d}} S(i) \sigma^{il},$$

and consider the problem

$$dv = \Big(\big(a^{ij} v_i\big)_j + \tilde{b}^i v_i - \sum_{i=1}^{d} S(i)(a^{ij} v)_j + \tilde{c} v + \tilde{f}^0 - \sum_{i=1}^{d} \big(S f^i\big)_i\Big)\, dt$$
$$+ \Big(\sigma^{il} v_i + \tilde{h}^l + S g^l\Big)\, dw^l,\ (t, x, \omega) \in [T_0, T] \times \mathbb{R}^d \times \Omega, \tag{4.4.6}$$

$$v(T_0) = S\varphi,\ (x, \omega) \in \mathbb{R}^d \times \Omega; \tag{4.4.7}$$

when the coefficients are sufficiently smooth, problem (4.4.6), (4.4.7) is the result of multiplying all terms in (4.2.1), (4.2.2) by S and setting $v = Su$.

The coefficients, the initial condition, and the free terms in problem (4.4.6), (4.4.7) satisfy the assumptions of Theorem 4.4, and so this problem has a generalized solution v belonging to

$$\mathbb{L}_2\Big(\Omega; \mathbb{C}\big([T_0, T]; \mathbb{H}^{m-1}\big)\Big) \cap \mathbb{L}_2\big([T_0, T] \times \Omega; \mathbb{C}_w\mathbb{H}^m\big) \cap \mathbb{L}_p\big([T_0, T] \times \Omega;\ \mathbb{C}_w\mathbb{W}_p^m\big)$$

and satisfying (4.2.21).

We will now go from (4.4.6), (4.4.7) back to (4.2.1), (4.2.2). Take $\eta \in \mathbb{C}_0^\infty(\mathbb{R}^d)$ and use $y = \eta/S$ in the corresponding integral equality (of the form (4.2.3)) for the generalized solution of problem (4.4.6), (4.4.7). Defining $u := v/S$ we conclude that, for every $\eta \in \mathbb{C}_0^\infty(\mathbb{R}^d)$ and all $(t, \omega) \in [T_0, T] \times \Omega'$, where $\Omega' \subset \Omega$ and $\mathbb{P}(\Omega') = 1$,

$$(u(t), \eta)_0 = (\varphi, \eta)_0$$
$$+ \int_{[T_0,t]} \Big(-\big(a^{ij} u_i, \eta_j\big)_0 + (b^i u_i + cu + f^0, \eta)_0 - (f^i, \eta_i)_0\Big)(s)\, ds$$
$$+ \int_{[T_0,t]} \Big(\sigma^{il} u_i + h^l u + g^l(s), \eta\Big)_0 (s)\, dw^l(s).$$

From this equality and Lemma 3.6(ii), it follows that u is an r-generalized solution of (4.2.1), (4.2.2). The required properties of u follows from the corresponding properties of v given by Theorem 4.4.

It remains to prove uniqueness.

Let u_1 and u_2 be two r-generalized solutions of problem (4.2.1), (4.2.2) and define $\overline{u} := u_1 - u_2$. Then $\overline{u}$ satisfies

$$(\overline{u}(t), \eta)_0 = \int\limits_{[T_0,t]} \Big(-(a^{ij}\overline{u}_i, \eta_j)_0 + (b^i u_i + cu, \eta)_0 \Big)(s)\, ds + \int\limits_{[T_0,t]} (\sigma^{il} u_i + h^l u, \eta)_0(s)\, dw^l(s),$$

for $\mathfrak{l}\times\mathbb{P}$-a.a. (t,ω) and every $\eta \in \mathbb{C}_0^\infty(\mathbb{R}^{\mathrm{d}})$. By substituting in the equality Sy instead of η, where $y \in \mathbb{C}_0^\infty(\mathbb{R}^{\mathrm{d}})$, we conclude that $\overline{v} := S\overline{u}$ belongs to $\mathbb{L}_2^\omega\big([T_0, T], \mathscr{P}; \mathbb{H}^1\big)$ and satisfies the following equality

$$(\overline{v}(t), y)_0 = \int\limits_{[T_0,t]} \Big(-(a^{ij}\overline{v}_i - \sum_{i=1}^{\mathrm{d}} S(i)a^{ij}\overline{v}, y_j)_0 + (\tilde{b}^i\overline{v}_i + \tilde{c}\,\overline{v}, y)_0 \Big)(s)\, ds + \int\limits_{[T_0,t]} (\sigma^{il}\overline{v}_i + \tilde{h}^l\overline{v}, y)_0(s)\, dw^l(s),$$

for $\mathfrak{l} \times \mathbb{P}$-a.a. (t,ω) and every $\eta \in \mathbb{C}_0^\infty(\mathbb{R}^{\mathrm{d}})$.

Theorem 4.1 and Remark 4.2 imply $\|\overline{v}(t)\|_m = 0$, ($\mathfrak{l} \times \mathbb{P}$-a.s.). From this and Lemma 3.6(ii), it follows that $\|\overline{u}(t)\|_{m,2,r} = 0$ ($\mathfrak{l} \times \mathbb{P}$-a.s.). □

The next result is an immediate consequence of Proposition 4.1.

Proposition 4.3 *Let (U, Σ) be a measurable space and ξ be a Σ-measurable mapping from U to $\mathbb{W}_p^m(r)$, where $p \geq 1$ and $(m-n)p > \mathrm{d}$ for some $n \in \mathbb{N} \cup \{0\}$. Then there exists a function $\tilde{\xi} : U \times \mathbb{R}^{\mathrm{d}} \to \mathbb{R}$ with the following properties.*

(a) *$\tilde{\xi}$ is $\Sigma \otimes \mathscr{B}(\mathbb{R}^{\mathrm{d}})$-measurable.*
(b) *For every $s \in U, \tilde{\xi}(s) \in \mathbb{W}_p^m(r)$, $S\tilde{\xi}(s) \in \mathbb{C}_b^n(\mathbb{R}^{\mathrm{d}})$ and $\|S\tilde{\xi}(s)\|_{\mathbb{C}_b^n(\mathbb{R}^{\mathrm{d}})} \leq N\|\xi(s)\|_{m,p}$, where N depends only on m, p, d, r, and n.*
(c) *$\|\tilde{\xi}(s) - \xi(s)\|_{m,p,r} = 0$ for every s.* □

After that, similar to Corollary 4.3, we get

Corollary 4.7 *Suppose that the assumptions of Theorem 4.6 are fulfilled and, for some $n \in \mathbb{N}\cup\{0\}$, $(m-n)p > \mathrm{d}$. Then the r-generalized solution of (4.2.1), (4.2.2) has a version (in x) $v = v(t, x, \omega)$ with the following properties:*

(a) *For every $x \in \mathbb{R}^{\mathrm{d}}$, the function $v(\cdot, x, \cdot)$ is a predictable real-valued stochastic process;*
(b) *$Sv(\cdot, \cdot, \omega) \in \mathbb{C}_b^{0,n}([T_0, T] \times \mathbb{R}^{\mathrm{d}})$ for every $\omega \in \Omega$;*
(c) *$v(t, x, \omega)$ is an r-generalized solution of problem (4.2.1), (4.2.2);*

(d) $\mathbb{E} \sup_{t\in[T_0,T]} \|Sv(t)\|^p_{\mathbb{C}^n_b(\mathbb{R}^d)} < \infty$;

(e) *If* $n \geq 2$, $|\alpha| \leq n-2$, *and* $\tilde{\varphi}$, $\tilde{f}^j$, $\tilde{g}^l$ *are the corresponding modifications of* $\tilde{\varphi}$, f^j, g^l *according to the above proposition, then, for every* $x \in \mathbb{R}^d$, *there exists a set* $\Omega_x \subset \Omega$ *with* $\mathbb{P}(\Omega_x) = 1$ *such that, for every* $(t, \omega) \in [T_0, T] \times \Omega_x$,

$$v_\alpha(t) = \varphi_\alpha + \int_{[T_0,t]} \left(L_{\text{div}} v(s) + \tilde{f}(s)\right)_\alpha ds$$

$$+ \int_{[T_0,t]} \left(B^l v(s) + g^l(s)\right)_\alpha dw^l(s);$$

(f) *If* v_1 *and* v_2 *are* r*-generalized solutions of problem* (4.2.1), (4.2.2) *possessing properties* (a), (b), (e), *then*

$$\mathbb{P}\Big(\sup_{\substack{t\in[T_0,t]\\ x\in\mathbb{R}_d}} |v_1(t,x,\omega) - v_2(t,x,\omega)| > 0\Big) = 0. \qquad \square$$

A function with properties (a), (b), (e) is called a **classical solution** of (4.2.1), (4.2.2).

4.4.3. In this paragraph, we study problem (4.1.1), (4.2.2) under the assumptions that φ is an $\mathscr{F}_{T_0}$-measurable random variable taking values in $\mathbb{L}_2(r)$, the processes f, g^l belong to $\mathbb{L}_2^\omega([T_0,T]; \mathscr{P}; \mathbb{L}_2(t))$, and the coefficients of Eq. (4.1.1) are predictable for every x.

Definition 4.2 A function $u \in \mathbb{L}_2^\omega([T_0,T], \mathscr{P}; \mathbb{H}^1(r))$ is called an r-**generalized** solution of problem (4.1.1), (4.2.2) if it satisfies equality (4.3.1) for every $y \in \mathbb{C}_0^\infty(\mathbb{R}^d)$, ($\ell \times \mathbb{P}$-a.s.).

Theorem 4.7 *Given* $m \in \mathbb{N}$ *we suppose that the following assumptions hold.*

(i) *The parabolicity condition* (4.1.2) *is fulfilled;*
(ii) *For all* $i, j = 1, 2, \ldots, d$, $l = 1, 2, \ldots, d_1$, *the functions* a^{ij} *are* $2 \vee m$ *times differentiable in* x, σ^{il} *and* h^l $m+1$ *times differentiable, and* f^i *and* c *are* m *times differentiable. The absolute values of these functions and all their derivatives are bounded by the constant* K;
(iii) *For all* $i = 1, 2, \ldots, d$, *and* $l = 1, 2, \ldots, d_1$, f, g^l, *and* $g^l_i \in \mathbb{L}_{p'}([T_0,T] \times \Omega; \mathbb{W}^m_{p'}(r))$ *and* $\varphi \in \mathbb{L}_{p'}(\Omega; \mathbb{W}^m_{p'}(r))$, $p' = p$ *and* $p' = 2$.

Then problem (4.1.1), (4.2.2) *has a unique r-generalized solution u. This solution belongs to*

$$\mathbb{L}_2\Big(\Omega;\mathbb{C}\big([T_0,T];\mathbb{H}^{m-1}(r)\big)\Big)\cap\mathbb{L}_2\big([T_0,T]\times\Omega;\mathbb{C}_w\mathbb{H}^m(r)\big)\cap\mathbb{L}_p\big([T_0,T]\times\Omega;\mathbb{C}_w\mathbb{W}^m_p(r)\big).$$

Moreover, for $p'=2$ and $p'=p$,

$$\mathbb{E}\sup_{t\in[T_0,T]}\|u(t)\|^{p'}_{m,p',r}\le N\mathbb{E}\Bigg(\|\varphi\|^{p'}_{m,p',r}$$
$$+\int\limits_{[T_0,T]}\Bigg(\|f(t)\|^{p'}_{m,p',r}+\sum_{l=1}^{\mathrm{d}_1}\|g^l(t)\|^{p'}_{m+1,p',r}\Bigg)dt\Bigg), \tag{4.4.8}$$

where the number N depends only on p, d, d_1, K, m, T_0, T, r. □

Remark 4.4 Uniqueness holds under conditions (i) and (ii); cf. Remark 1. □

Corollary 4.8 *Suppose that the assumptions of the theorem are fulfilled and $(m-n)p>\mathrm{d}$ for some $n\in\mathbb{N}\cup\{0\}$. Then the r-generalized solution of problem* (4.1.1), (4.2.2) *has a version (in x) $v=v(t,x,\omega)$ which possess properties* (a) *to* (d) *from Corollary* 4.7. *Moreover,*

(e) *If the assumptions of item* (e) *of Corollary* 4.7 *are satisfied, then*

$$v_\alpha(t,x)=\varphi_\alpha+\int\limits_{[T_0,t]}(Lv(s,x)+\tilde f(s,x))_\alpha ds$$
$$+\int\limits_{[T_0,t]}(B^l v(s,x)+\tilde g^l(s,x))_\alpha dw^l(s);$$

$(t,\omega)\in[T_0,T]\times\Omega_x$, $\mathbb{P}(\Omega_{x,t})=1$.

(f) *If $v_1(t,x)$ and $v_2(t,x)$ are r-generalized solutions of problem* (4.1.1), (4.2.2) *possessing properties* (a), (b), *then*

$$\mathbb{P}\Big(\sup_{\substack{t\in[T_0,T]\\x\in\mathbb{R}^{\mathrm{d}}}}|v_1(t,x,\omega)-v_2(t,x,\omega)|>0\Big)=0.$$ □

Both the theorem and the corollary can be derived from Theorem 4.5 and Corollary 4.4 using the same methods as in the proof of Theorem 4.6 and Corollary 4.7.

4.4.4.

Corollary 4.9 *If the assumptions of Theorem* 4.6 (*or Theorem* 4.7) *are fulfilled for all* $m \in \mathbb{N}$, *not necessarily uniformly in* m, *then the solution of problem* (4.2.1), (4.2.2) (*or* (4.1.1), (4.2.2)) *is infinitely differentiable in* x *and all the derivatives are continuous in* t, x *for* $\mathbb{P}$- a.a. ω. □

4.4.5. In this paragraph, we consider problems (4.4.1), (4.4.2) and (4.4.3), (4.4.4).

Let us recall some constructions related to backward predictability. For fixed $T > 0$, consider a family $\{\mathscr{F}_T^t\}$, $t \in [0, T]$, of sub-σ-algebras from $\mathscr{F}$ such that $\mathscr{F}_T^{t_1} \supset \mathscr{F}_T^{t_2}$ for $t_1 \leq t_2$, $\bigcap_{\varepsilon>0} \mathscr{F}_T^{t-\varepsilon} = \mathscr{F}_T^t$ for $t \leq T$, and $\mathscr{F}_T^T$ is completed with respect to the measure $\mathbb{P}$.

For a standard Wiener process $W = (w^1, \dots, w^{d_1})$, define

$$W_T(t) := W(T) - W(T - t).$$

Then W_T is a standard Wiener process with respect to the family $\{\mathscr{F}_T^{T-t}\}$, $t \in [0, T]$; cf. Sects. 1.5.2 and 1.5.13.

Let a^{ij}, f^i, c, σ^{il}, and h^l, $i, j = 1, \dots, \mathrm{d}$ and $l = 1, \dots, \mathrm{d}_1$ be bounded, $\mathscr{B}([T_0, T] \times \mathbb{R}^{\mathrm{d}}) \otimes \mathscr{F}$-measurable functions on $[T_0, T] \times \Omega \times \mathbb{R}^{\mathrm{d}}$. These functions are also assumed to be backward predictable for every $x \in \mathbb{R}^{\mathrm{d}}$ with respect to the family $\{\mathscr{F}_T^t\}$. When (4.4.3), (4.4.4) is considered we will suppose in addition that the functions $a^{ij} = a^{ij}(t, x, \omega)$ have bounded derivatives of the first order in x.

Warning 4.4 *In what follows, we consider backward predictable functions exclusively with respect to the family* $\{\mathscr{F}_T^t\}$ *and will not mention this family explicitly.* □

The following additional hypotheses are assumed to hold:

(a) φ is an $\mathscr{F}_T^T$-measurable random variable taking values in $\mathbb{L}_2(r)$;
(b) For $l = 1, \dots, \mathrm{d}_1$, $g^l \in \mathbb{L}_2^{\omega}([T_0, T]; \mathbb{L}_2(r))$ and is backward predictable;
(c) In problem (4.4.1), (4.4.3), $f = f^0 + \sum_{i=1}^{\mathrm{d}} f_i^i$, where each $f^j \in \mathbb{L}_2^{\omega}([T_0, T]; \mathbb{L}_2(r))$ and is backward predictable, $j = 0, \dots, \mathrm{d}$. In problem (4.4.3), (4.4.4), $f \in \mathbb{L}_2^{\omega}\big([T_0, T]; \mathbb{L}_2(r)\big)$ and is backward predictable.

Definition 4.3 A backward predictable process $u \in \mathbb{L}_2^{\omega}\big([T_0, T]; \mathbb{H}^1(r)\big)$ is called an r-generalized solution of problem (4.4.1), (4.4.2) if, for every $y \in \mathbb{C}_0^{\infty}(\mathbb{R}^d)$, the following equality holds for $\ell \times \mathbb{P}$- a.a. (t, ω) :

$$\begin{aligned}\big(u(t), y_0\big) &= (\varphi, y)_0 \\ &+ \int\limits_{[t,T]} \Big(- (a^{ij}u_i + f^j, y_j)_0(s) + (b^i u_i + cu + f^0, y)_0\Big)(s)\, ds \\ &+ \int\limits_{[t,T]} (\sigma^{il}u_i + h^l u + g^l, y)_0(s) * dw^l(s).\end{aligned} \tag{4.4.9}$$

For problem (4.4.3), (4.4.4), the corresponding equality is

$$\begin{aligned}\big(u(t), y\big)_0 &= (\varphi, y)_0 \\ &+ \int\limits_{[t,T]} \Big(- (a^{ij}u_i, y_j)_0 + \big((b^i - a^{ij}_j)u_i + cu(s) + f(s), y\big)_0\Big)(s)\, ds \\ &+ \int\limits_{[t,T]} (\sigma^{il}u_i(s) + h^l u + g^l, y)_0(s) * dw^l(s).\end{aligned} \tag{4.4.10}$$

□

A time reversal transforms (4.4.3), (4.4.4) to a problem of the type (4.1.1), (4.2.2). Indeed, consider

$$\begin{aligned}dv(t,x,\omega) &= \Big(a^{ij}(T-t,x,\omega)v_{ij}(t,x,\omega) \\ &+ b^i(T-t,x,\omega)v_i(t,x,\omega) + c(T-t,x,\omega)v(t,x,\omega) \\ &+ f(T-t,x,\omega)\Big)\, dt + \Big(\sigma^{il}(T-t,x,\omega)v_i(t,x,\omega) \\ &+ h^l(T-t,x,\omega)v(t,x,\omega) + g^l(T-t,x,\omega)\Big)\, dw^l_T(t), \\ &(t,x,\omega) \in (0,\ T-T_0] \times \mathbb{R}^d \times \Omega,\end{aligned} \tag{4.4.11}$$

$$v(0,x,\omega) = \varphi(x,\omega),\ (x,\omega) \in \mathbb{R}^d \times \Omega. \tag{4.4.12}$$

It follows that $v = v(t)$ is an r-generalized solution of (4.4.11), (4.4.12) if and only if $u(t) = v(T - t)$ is an r-generalized solution of (4.4.1), (4.4.2). To see this, note first of all that, by the definition of a backward predictable stochastic process, the coefficients and the free terms in (4.4.11) are predictable relative to $\{\mathscr{F}_T^{T-t}\}$, $t \in [0, T - T_0]$, the process $W_T = (w_T^1, \dots, w_T^{d_1})$ is a standard Wiener process with respect to the same family, and the initial condition φ is $\mathscr{F}_T^T = \mathscr{F}_T^{T-t}\big|_{t=0}$-measurable. In other words, (4.4.11), (4.4.12) is a problem of the type (4.1.1), (4.2.2) considered on the stochastic basis $\mathbb{F}_T := (\Omega, \mathscr{F}, \{\mathscr{F}_T^{T-t}\}_{t\in[0,T-T_0]}, \mathbb{P})$ that satisfies the usual assumptions.

Now let v be an r-generalized solution of (4.4.11), (4.4.12). Then, for every $y \in \mathbb{C}_0^\infty(\mathbb{R}^d)$,

$$
\begin{aligned}
\big(v(t), y\big)_0 = (\varphi, y)_0 + \int\limits_{[0,t]} \Big(&- \big(a^{ij}(T-s)v_i(s), y_j\big)_0 \\
&+ \Big(\big(b^i(T-s) - a_j^{ij}(T-s)\big)v_i(s) + c(T-s)v(s) + f(T-s), y\Big)_0\Big)\, ds \\
&+ \int\limits_{[0,t]} \big(\sigma^{il}(T-s)v_i(s) + h^l(T-s)v(s) + g^l(T-s), y_0\big)_0\, dw_T^l(s).
\end{aligned}
\tag{4.4.13}
$$

Define $u = u(t)$ by $u(t) := v(T-t)$. Then $u \in \mathbb{L}_2^\omega([T_0, T], \overleftarrow{\mathscr{P}}; \mathbb{L}_2(r))$, and, after a change of variables $s \mapsto T-s$ in (4.4.13), we find that $u(t)$ satisfies (4.4.10). That is, u is indeed an r-generalized solution of problem (4.4.3), (4.4.4).

Conversely, a change of variables $s \mapsto T - s$ in (4.4.10) shows that $v(t) = u(T - t)$ is an r-generalized solution of (4.4.11), (4.4.12).

The same argument shows that (4.4.1), (4.4.2) is equivalent to a problem of the type (4.2.1), (4.2.2). Accordingly, all the results in this chapter for problems (4.2.1), (4.2.2) and (4.1.1), (4.2.2) are naturally carried over to problems (4.4.1), (4.4.2) and (4.4.3), (4.4.4), respectively.

4.4.6.

Warning 4.5 *In the sequel, when dealing with problems* (4.4.1), (4.4.2) *and* (4.4.3), (4.4.4), *we will use the corresponding results for problems* (4.2.1), (4.2.2) *and* (4.1.1), (4.2.2). □

Remark 4.5 With obvious modifications, all the results in this chapter remain valid when the initial condition, the coefficients, and the free terms are complex-valued.

In particular, conditions (4.1.2), (4.1.3) become, respectively,

$$2\mathrm{Re}(a^{ij}\xi^i\xi^j) - \sum_{l=1}^{\mathrm{d}_1} |\sigma^{il}\xi^i|^2 \geq 0, \tag{4.1.2'}$$

$$2\mathrm{Re}(a^{ij}\xi^i\xi^j) - \sum_{l=1}^{\mathrm{d}_1} |\sigma^{il}\xi^i|^2 \geq \delta|\xi|^2, \tag{4.1.3'}$$

$(t, x, \omega, \xi) \in [T_0, T] \times \mathbb{R}^{\mathrm{d}} \times \Omega \times \mathbb{R}^{\mathrm{d}}$, where $\mathrm{Re}(z)$ is the real part of a complex number z. □

Chapter 5
Itô's Partial Differential Equations and Diffusion Processes

5.1 Introduction

5.1.1. In this chapter we continue the study of the Cauchy problem for second-order parabolic Itô equations, this time concentrating on qualitative, rather than analytical, aspects of the problem. The main objective is to establish various connections between these equations and diffusion processes.

5.1.2. Here is a (partial) list of the notation to be used in this chapter:

- $0 \leq T_0 < T < \infty$, the initial and terminal times.
- s, t, τ, the time variables. As a rule, s will be the backward time variable and t, the forward time variable; τ will usually be a dummy variable in integrals.
- $\mathbb{F} = \big(\Omega, \mathscr{F}, \{\mathscr{F}_t\}_{t\geq 0}, \mathbb{P}\big)$, a stochastic basis with the usual assumptions.
- There will be several standard Wiener processes on $\mathbb{F}$: $W = (w^1, \ldots, w^{\mathrm{d}_1})$ of dimension d_1, $\hat{W} = (\hat{w}^1, \ldots, \hat{w}^{\mathrm{d}_0})$ of dimension d_0 and independent of W, and $\mathcal{W} = (\hat{w}^1, \ldots, \hat{w}^{\mathrm{d}_0}, w^1, \ldots, w^{\mathrm{d}_1})$ of dimension $\mathrm{d}_0 + \mathrm{d}_1$.
- $\tilde{\mathscr{F}}_t^s$, the sigma-algebra generated by $W(t_2) - W(t_1)$, $s \leq t_1 < t_2 \leq t$, and completed with respect to $\mathbb{P}$.
- $(\cdot, \cdot)_0$: the inner product in $\mathbb{L}_2(\mathbb{R}^{\mathrm{d}})$.

Warning 4.1 is in force throughout this chapter. In particular, the summation convention will be used, and subscripts will denote partial derivatives with respect to the corresponding components of $x \in \mathbb{R}^{\mathrm{d}}$.

The focus will be on Eqs. (4.1.1) and (4.4.3), driven by the Wiener process W and with the corresponding coefficients $a^{ij}, b^i, c, \sigma^{il}, h^l$, initial condition φ and free terms f, g^l.

We will use a slightly different version of the parabolicity condition by assuming that, for every $(t, x, \omega) \in [T_0, T] \times \mathbb{R}^d \times \Omega$, the matrix $\Big(a^{ij}(t, x, \omega)\Big)_{i,j=1}^{\mathrm{d}}$ can be

B. L. Rozovsky, S. V. Lototsky, *Stochastic Evolution Systems*, Probability Theory and Stochastic Modelling 89, https://doi.org/10.1007/978-3-319-94893-5_5

represented as

$$a^{ij}(t,x,\omega) = \frac{1}{2}\sigma^{il}(t,x,\omega)\sigma^{jl}(t,x,\omega) + \frac{1}{2}\hat{\sigma}^{ik}(t,x,\omega)\hat{\sigma}^{jk}(t,x,\omega) \tag{5.1.1}$$

with suitable $\hat{\sigma}^{ik}$, $i = 1, 2, \ldots, \mathrm{d}$, $k = 1, 2, \ldots, \mathrm{d}_0$; it is not a coincidence that d_0 is also the dimension of the Wiener process $\hat{W}$.

Remark 5.1 Condition (5.1.1) immediately implies (4.1.2) (the parabolicity condition). Conversely, if the matrix a is symmetric and (4.1.2) holds, then representation (5.1.1) follows by taking $\hat{\sigma}$ as a positive square root of the matrix $\left(2a - \sigma\sigma^*\right)$; see [152, Sections 5.2, 5.3]. □

The diffusion process to be studied in this chapter is $\mathcal{X} = (\mathcal{X}^1, \ldots, \mathcal{X}^{\mathrm{d}})$, with

$$\begin{aligned} \mathcal{X}^i(t,s,x) = x^i &+ \int\limits_{[s,t]} \mathcal{B}^i\big(\tau, \mathcal{X}(\tau,s,x)\big)\, d\tau \\ &+ \int\limits_{[s,t]} \boldsymbol{\Sigma}^{i\ell}\big(\tau, \mathcal{X}(\tau,s,x)\big)\, d\mathcal{W}^{\ell}(\tau), t \in [s,T],\ s \in [T_0,T],\ x \in \mathbb{R}^{\mathrm{d}}, \end{aligned} \tag{5.1.2}$$

where

$$\mathcal{B}^i := b^i - \sigma^{il}h^l$$

and $\boldsymbol{\Sigma}$ is the matrix with d rows and $\mathrm{d}_0 + \mathrm{d}_1$ columns, obtained by combining σ and $\hat{\sigma}$:

$$\boldsymbol{\Sigma}^{i\ell} = \begin{cases} \hat{\sigma}^{il}, & \text{if } \ell = l = 1, \ldots, \mathrm{d}_0, \\ \sigma^{il}, & \text{if } \ell = \mathrm{d}_0 + l = \mathrm{d}_0 + 1, \ldots, \mathrm{d}_0 + \mathrm{d}_1. \end{cases}$$

The process $X = X(t,s,x)$ encountered earlier in Chap. 1 (cf. Eq. (1.5.4)) is a particular case of $\mathcal{X}$, corresponding to $\mathrm{d}_0 = 0$ and $h^l = 0$ so that $\boldsymbol{\Sigma} = \sigma$ and $\mathcal{B} = b$:

$$X^i(t,s,x) = x^i + \int\limits_{[s,t]} b^i\big(\tau, X(\tau,s,x)\big)\, d\tau + \int\limits_{[s,t]} \sigma^{il}\big(\tau, X(\tau,s,x)\big)\, dw^l(\tau).$$

In this chapter, the assumptions about the coefficients will always ensure that both $\mathcal{X}$ and X are well-defined in the sense of Definition 1.16. Note also that the first time argument in $\mathcal{X}$ and X is always bigger than or equal to the second.

In Sect. 5.2.2 it will be shown that if the coefficients, the initial value φ, and the free terms f, g^l are non-random, then the r-generalized solutions of (4.1.1), (4.1.2) and (4.4.3), (4.4.4) have representations similar to the probabilistic representation (1.5.7) of the solution of a deterministic equation.

In particular, if $f \equiv g^l \equiv 0$, then the solution of (4.4.3), (4.4.4) is

$$v(s,x) = \mathbb{E}\big[\varphi(\mathcal{X}(T,s,x)\rho(T,s)|\tilde{\mathscr{F}}_T^s\big], \tag{5.1.3}$$

where

$$\rho(t,s) := \exp\Bigg(\int\limits_{[s,t]} c(\tau, \mathcal{X}(\tau,s,x))d\tau + \int\limits_{[s,t]} h^l\big(\tau, \mathcal{X}(\tau,s,x)\big)\, dw^l(\tau)$$
$$- \frac{1}{2} \int\limits_{[s,t]} \big|h(\tau, \mathcal{X}(\tau,s,x))\big|^2 d\tau \Bigg), \; T_0 \le s \le t \le T; \tag{5.1.4}$$

as usual,

$$|h|^2 = \sum_{l=1}^{\mathrm{d}_1} |h^l|^2 = h^l h^l.$$

If we further assume that $c \equiv \sigma^{il} \equiv h^l \equiv 0$, then (4.4.3) becomes the backward Kolmogorov equation (1.5.8) and representation (5.1.3) turns into (1.5.7).

Representation (5.1.3) implies that (4.4.3), (4.4.4), including the corresponding backward Kolmogorov equation, can be solved by the method of random characteristics, the diffusion process $\mathcal{X}$ being the characteristic, although, unlike (1.5.7), representation (5.1.4) is a conditional averaging over the characteristic, relative to the σ-algebra $\tilde{\mathscr{F}}_T^s$.

The representation of the type (5.1.3) for problem (4.1.1), (4.2.2) or (4.4.3), (4.4.4) will be called the averaging over the characteristics (AOC) formula.

An important corollary of the AOC formula is the maximum principle for the Itô parabolic equation, to be proved in Sect. 5.2.

5.1.3. The connections between diffusion processes and Itô parabolic equations are mutually beneficial. In particular, from AOC formula (5.1.3) with $\hat{\sigma}^{il} \equiv h^l \equiv c \equiv 0$ and $\varphi(x) = x^i$, it follows that, for fixed t, the ith coordinate $X^i(t,s,x) = v(s,x)$ of the process X, as a function of s, x, is an r-generalized solution of

$$-dv(s,x) = L_0 v(s,x)\, ds + B_0^l v(s,x) * dw^l(s), \tag{5.1.5}$$
$$(s,x,\omega) \in [T_0, t) \times \mathbb{R}^{\mathrm{d}} \times \Omega,$$
$$v(t,x) = x^i, \; (x,\omega) \in \mathbb{R}^{\mathrm{d}} \times \Omega, \tag{5.1.6}$$

where

$$L_0 v := \frac{1}{2}\, \sigma^{il}\sigma^{jl} v_{ij} + b^i u_i, \;\; B_0^l v := \sigma^{il}(t,x) v_i, \; l = 1, 2, \ldots, \mathrm{d}_1.$$

In what follows, we refer to problem (5.1.5), (5.1.6) as the **backward diffusion equation**.

If we take the expectation on both sides of the integral form of (5.1.5), (5.1.6), then, after changing the order of integration and assuming that the stochastic integral is a (backward) martingale, we get the backward Kolmogorov equation for the process X.

A slightly more general problem, namely Eq. (5.1.5) with the terminal condition

$$v(t,x) = \varphi(x),\ (x,\omega) \in \mathbb{R}^{\mathrm{d}} \times \Omega, \tag{5.1.7}$$

will be called the **backward Liouville equation** for the diffusion process X.

This terminology is motivated by classical mechanics; cf. [1]. Recall that, given a system of (deterministic) ordinary differential equations $y'(t) = \mathfrak{b}(t, y(t))$, $y \in \mathbb{R}^{\mathrm{d}}$, a first integral of this system is a non-constant function $\Phi = \Phi(t,x)$ such that $d\Phi\big(t, y(t)\big)/dt = 0$, that is, $\Phi\big(t, y(t)\big)$ is (locally) constant on the solutions of the system. The corresponding partial differential equation for the function Φ is often referred to as the Liouville equation. It will be shown in Sect. 5.3 that, for $\mathbb{P}$-a.a. ω, the solution of problem (5.1.5), (5.1.7) is the first integral for the diffusion process $X = X(t,s,x)$, considered as a function of $t \in (s,T]$ for fixed $s \in [T_0, T)$ and fixed $x \in \mathbb{R}^{\mathrm{d}}$.

Unlike deterministic dynamical systems, where time is reversible, stochastic Itô equations of the type (5.1.2) do not allow a direct time reversal because of the Wiener process, and, as a result, have two different Liouville's equations describing two different types of the first integrals. Equation (5.1.6) is backward in time and describes the first integrals that are $\tilde{\mathscr{F}}^t_T$-adapted. There is a similar equation, forward in time and describing an $\tilde{\mathscr{F}}^{T_0}_t$-adapted first integral. In particular, the **forward Liouville equation** for the process $X = X(t, T_0, x)$ is

$$du(t,x) = (B^l_0 B^l_0 - L_0)u(t,x)\,ds - B^l_0 u(t,x)\,dw^l(t), \tag{5.1.8}$$

$$T_0 < t \le T,\ x \in \mathbb{R}^{\mathrm{d}},$$

$$u(T_0, x) = \varphi(x). \tag{5.1.9}$$

With $\varphi(x) = x^i$, we get $u\big(t, X(t,T_0,x)\big) = x^i$ for all $t \in [T_0, T]$. We will see in Sect. 5.3 that, if the coefficients b, σ are smooth enough, then the mapping $x \mapsto X(t, T_0, x)$ is a diffeomorphism for every t, and the ith component X^{-i} of the inverse mapping satisfies (5.1.8), (5.1.9) with $\varphi(x) = x^i$: the equation that is natural to call the **forward equation of the inverse diffusion**. Conversely, the AOC formula for problem (4.1.1), (4.2.2) implies that the **backward equation of the inverse diffusion** is a system of the backward ordinary Itô equations satisfied by $\{X^{-i}(t,\cdot,x)\}$ for fixed t, x; see Sect. 5.3.4.

Section 5.3 also shows how the forward equation of the inverse diffusion leads to stochastic versions of the variation-of-parameters formulas for Itô equations. To be more specific, under suitable regularity conditions on the coefficients,

- given two diffusion processes $X = X(t, s, x)$ and $Y = Y(t, s, x)$, there exists a diffusion process $Z = Z(t, s, x)$ such that $Y(t, s, x) = X\big(t, s, Z(t, s, x)\big)$;
- a change of variables reduces a second-order homogeneous ($f = g = 0$) parabolic Itô equation of the type (4.1.1) to a random parabolic equation (an equation with random coefficients but without the dw part).

In Sect. 5.4 we consider a problem similar to (4.1.1), (4.2.2), but with adjoint operators and random coefficients, and prove another version of the AOC formula, this time for the functional $\int_{\mathbb{R}^d} f(t, x)u(t, x)\,dx$.

5.2 The Method of Stochastic Characteristics

5.2.1. In this section we derive the formulas of averaging over characteristics for problems (4.1.1), (4.2.2) and (4.4.3), (4.4.4); cf. (5.1.3). We then use the result to derive the maximum principle for these problems.

Throughout this section the coefficients $a^{ij}, b^i, c, \sigma^{il}, h^l$ and the functions $\varphi, f, g^l, i, j = 1, 2, \ldots, \mathrm{d}, l = 1, 2, \ldots, \mathrm{d}_1$, are non-random, that is, do not depend on ω.

Let us also recall the notation

$$L = a^{ij}\frac{\partial^2}{\partial x^i \partial x^j} + b^i\frac{\partial}{\partial x^i} + c, \quad B^l = \sigma^{il}\frac{\partial}{\partial x^i} + h^l,$$

and the corresponding Eqs. (4.1.1) and (4.4.3):

$$du = (Lu + f)\,dt + (B^l u + g^l)\,dw^l, \quad -dv = (Lv + f)\,dt + (B^l u + g^l) * dw^l.$$

5.2.2. In addition to (5.1.2), we consider the following system of backward Itô equations:

$$\mathcal{Y}^i(t, s, x) = x^i + \int\limits_{[s,t]} \mathcal{B}^i\big(\tau, \mathcal{Y}(t, \tau, x)\big)d\tau + \int\limits_{[s,t]} \boldsymbol{\Sigma}^{i\ell}\big(\tau, \mathcal{Y}(t, \tau, x)\big) * d\mathcal{W}^\ell(\tau), \tag{5.2.1}$$

$$T_0 < s \le t \le T, \ x \in \mathbb{R}^\mathrm{d},$$

with fixed t and x; the functions $\mathcal{B}^i$, $\boldsymbol{\Sigma}^{i\ell}$ and the Wiener process $\mathcal{W}$ are defined in the previous section.

We assume that both (5.2.1) and (5.1.2) have unique solutions, and each solution has a version that is jointly continuous in t, s, x; cf. Sects. 1.5.6 and 1.5.14. In what follows, we always consider these continuous versions while keeping the same notation $\mathcal{X} = \mathcal{X}(t, s, x)$ and $\mathcal{Y} = \mathcal{Y}(t, s, x)$.

Similar to (5.1.4), define

$$\gamma(t,s) := \exp\Bigg(\int\limits_{[s,t]} c\big(\tau, \mathcal{Y}(t,\tau,x)\big)d\tau + \int\limits_{[s,t]} h^l\big(\tau, \mathcal{Y}(t,\tau,x)\big) * dw^l(\tau) - \frac{1}{2}\int\limits_{[s,t]} (h^l h^l)\big(\tau, \mathcal{Y}(t,\tau,x)\big)\, d\tau \Bigg).$$

Theorem 5.1 *Fix a real number r and assume that the following conditions are satisfied:*

(i) *The functions $a^{ij}, b^i, c, \sigma^{il}, h^l, \sigma_j^{il}, h_j^l$, and a_k^{ij} for $i, j, k = 1, 2, \ldots, \mathrm{d}$ and $l = 1, 2, \ldots, \mathrm{d}_1$, and their first-order derivatives in x are uniformly bounded by the number K.*

(ii) $\varphi \in \mathbb{H}^1(r)$, $f \in \mathbb{L}_2\big([T_0, T]; \mathbb{H}^1(r)\big)$, *and* $g^l \in \mathbb{L}_2\big([T_0, T]; \mathbb{H}^2(r)\big)$.

Let $u = u(t, x)$ be an r-generalized solution of problem (4.1.1), (4.2.2), *and let $v = v(s, x)$ be an r-generalized solution of problem* (4.4.3), (4.4.4). *Then*

$$u(t,x) = \mathbb{E}\Bigg[\int\limits_{[T_0,t]} f\big(s, \mathcal{Y}(t,s,x)\big)\gamma(t,s)\, ds + \int\limits_{[T_0,t]} g^l(s, \mathcal{Y}(t,s,x))\gamma(t,s) * dw^l(s) + \varphi\big(\mathcal{Y}(t,T_0,x)\big)\gamma(t,T_0) \Big| \tilde{\mathscr{F}}_t^{T_0} \Bigg], \quad for\ \mathfrak{l} \times \mathfrak{l}_{\mathrm{d}} \times \mathbb{P}\text{-a.a. } (t,x,\omega), \tag{5.2.2}$$

and

$$v(s,x) = \mathbb{E}\Bigg[\int\limits_{[s,T]} f(t, \mathcal{X}(t,s,x))\rho(t,s)\, dt + \int\limits_{[t,T]} g^l(t, \mathcal{X}(t,s,x))\rho(t,s)\, dw^l(t) + \varphi\big(\mathcal{X}(T,s,x)\big)\rho(T,s) \Big| \tilde{\mathscr{F}}_T^{s} \Bigg], \quad \text{for } \mathfrak{l} \times \mathfrak{l}_{\mathrm{d}} \times \mathbb{P}\text{-a.a. } (s,x,\omega). \tag{5.2.3}$$

□

It is the above formulas that justify calling $\mathcal{X}$ and $\mathcal{Y}$ the **stochastic characteristics** of problems (4.4.3), (4.4.4) and (4.1.1), (4.2.2), respectively.

5.2.3. We will only prove (5.2.3); then (5.2.2) will follow after a time change $t = T - s$; cf. Sect. 4.4.5.

To prove (5.2.3), we first consider sufficiently regular φ, f, g^l, and then pass to the limit. In the case of regular φ, f, g^l, we carry out the proof by reduction to a suitable deterministic problem.

Accordingly, in this and the following two paragraphs, we develop all the technical tools required to implement this plan. The first such tool is a deterministic analog of (5.2.3).

Consider the backward equation

$$-\frac{\partial F(s,x)}{\partial s} = LF(s,x) + f(s,x), \ (s,x) \in [T_0, T) \times \mathbb{R}^d, \tag{5.2.4}$$

$$F(T,x) = \varphi(x), \quad x \in \mathbb{R}^d. \tag{5.2.5}$$

Note that problem (5.2.4), (5.2.5) is a special case of (4.4.3), (4.4.4), and an r-generalized solution of (5.2.4), (5.2.5) could be treated in the sense of Definition 4.3.4. However, it will be helpful to state the definition explicitly.

Definition 5.1 A function $F \in \mathbb{L}_2\big([T_0, T]; \mathbb{H}^1(r)\big)$ is an r**-generalized solution** of the problem (5.2.4), (5.2.5) if, for every $y \in \mathbb{C}_0^\infty(\mathbb{R}^d)$,

$$\begin{aligned}\big(F(s), y\big)_0 &= (\varphi, y)_0 \\ &+ \int\limits_{[s,T]} \Big(-\big(a^{ij}F_j, y_j\big)_0 + \big((b^i - a^{ij}_j)F_i(\tau) + cF + f, y\big)_0\Big)(\tau)\,d\tau.\end{aligned} \tag{5.2.6}$$

□

Define

$$\rho^0(t,s) := \exp\Big(\int\limits_{[s,t]} c\big(\tau, X(r,s,x)\big)d\tau\Big);$$

recall that X is the particular case of the process $\mathcal{X}$ with $\hat{\sigma}^{il} \equiv 0$ and $h^l \equiv 0$.

Theorem 5.2 *Let n be a non-negative integer. Assume that*

(i) *The coefficients σ^{il}, b^i, and c satisfy the conditions of Theorem* 5.1.
(ii) $2a^{ij} = \sigma^{il}\sigma^{jl}$, $i, j = 1, 2, \ldots, \mathrm{d}$.
(iii) *For all $(s,x) \in [T_0, T] \times \mathbb{R}^d$,*

$$|\varphi(x)| + |f(s,x)| \leq K(1 + |x|^2)^{n/2}$$

and, for every $R \in \mathbb{R}_+$ *and all* $t \in [T_0, T]$, z, $z' \in \mathbb{R}^{\mathrm{d}}$ *such that* $|z| \le R$ *and* $|z'| \le R$,

$$|\varphi(z) - \varphi(z')| + |f(s,z) - f(s,z')| \le K(1+R)^n |z - z'|.$$

Then, for every

$$r < -\frac{\mathrm{d}+2n}{2}, \tag{5.2.7}$$

problem (5.2.4), (5.2.5) *has a unique* r*-generalized solution given by*

$$\tilde{F}(s,x) := \mathbb{E}\left(\int_{[s,T]} f\big(\tau, X(\tau,s,x)\big)\rho^0(\tau,s)d\tau + \varphi\big(X(T,s,x)\big)\rho^0(T,s) \right).$$

This solution has the following properties:

- *it is continuous in* (s,x) *on* $[T_0, T] \times \mathbb{R}^{\mathrm{d}}$*;*
- *there exists a number* N *depending only on* K, n, T, *and* T_0 *such that, for all* $(s,x) \in [T_0, T] \times \mathbb{R}^{\mathrm{d}}$,

$$|\tilde{F}(s,x)| \le N(1+|x|^2)^{n/2}, \tag{5.2.8}$$

$$\sum_{i=1}^{\mathrm{d}} |\tilde{F}_i(s,x)| \le N(1+|x|^2)^{n/2}. \tag{5.2.9}$$

□

Proof With Warning 4.5 in mind, uniqueness follows by Theorem 4.7. An application of the Itô formula shows that $\tilde{F}$ satisfies (5.2.6), whereas Lemma 3.11(i), together with (5.2.8) and (5.2.9), implies that

$$\tilde{F} \in \mathbb{L}_2\big([T_0, T]; \mathbb{H}^1(r)\big)$$

as long as r satisfies (5.2.7).

The main part of the proof consists in verifying that the function $\tilde{F}$ is continuous and has properties (5.2.8), and (5.2.9). Some of the arguments are rather technical, and an interested reader can reconstruct them by consulting [65, Sections III.1.5, IV.1.1, and IV.1.4]. □

In fact, the theorem can be extended to unbounded functions c under an additional assumption that $c \le 0$; cf. [65]. In our setting, $|c| \le K$ and the time interval is bounded, so, if necessary, we can always achieve $c \le 0$ by changing the unknown function F in (5.2.4) to $e^{-Ks}F$.

Remark 5.2 Theorem 4.7 also implies that $\tilde{F} \in \mathbb{C}\big([T_0, T]; \mathbb{L}_2(r)\big)$. □

The next technical but useful result shows that if φ and f have compact support, then the function $\tilde{F}$ decays at infinity faster than any power of $|x|$.

Corollary 5.1 *Denote by $B(0, R)$ the open ball in $\mathbb{R}^{\mathrm{d}}$ with the center at the origin and radius R. Suppose that the assumptions of the theorem are fulfilled and, moreover, the functions f, φ, and their first derivatives with respect to x are bounded, uniformly in (s, x), and vanish for $x \notin B(0, R)$. Then*

(i) *Problem* (5.2.4), (5.2.5) *has a unique r-generalized solution for every $r \in \mathbb{R}^1$. This solution is given by $\tilde{F}$ and belongs to $\mathbb{L}_p\big([T_0, T]; \mathbb{W}^1_p(r)\big)$ for every $2 \le p < \infty$.*
(ii) *For every $n \in \mathbb{N}$,*

$$\limsup_{|x|\to\infty} \sup_{s\in[T_0,T]} (1 + |x|^2)^{n/2} \, |\tilde{F}(s, x)| = 0. \qquad \square$$

Proof We begin with the second statement. Suppose for a moment that $f \equiv 0$. Since the drift and diffusion coefficients of the process $X(t, s, x)$ are bounded and the initial condition x is non-random, we use Theorem 1.11 to conclude that, for every $n \in \mathbb{N}$,

$$\mathbb{E} \sup_{s\in[T_0,T]} |X(T, s, x) - x|^{n+1} \le N_0 < \infty, \tag{5.2.10}$$

where the constant N_0 depends only on T, T_0, n, and K. Boundedness of c, compactness of support of φ, and the definition of $\tilde{F}$ imply existence of a constant $\tilde{N}$ such that, for $|x| \ge R$,

$$\begin{aligned}
\sup_{s\in[T_0,T]} |\tilde{F}(s, x)| &\le N_1 \sup_{s\in[T_0,T]} \big|\varphi\big(X(T, s, x)\big)\big| \\
&\le N_2 \sup_{s\in[T_0,T]} \mathbb{P}\big(X(T, s, x) \in B(0, R)\big)
\end{aligned}$$

$$\le N_2 \sup_{s\in[T_0,T]} \mathbb{P}\Big(|X(T, s, x) - x| \ge \mathrm{dist}\big(x, B(0, R)\big)\Big),$$

where $\text{dist}(x, B(0, R))$ is the Euclidean distance between the point x and the ball $B(0, R)$. Then Chebyshev's inequality and (5.2.10) lead to

$$\sup_{s\in[T_0,T]} |\tilde{F}(s, x)| \le N_2 \sup_{s\in[T_0,T]} \frac{\mathbb{E}\big|X(T, s, x) - x\big|^{n+1}}{\Big(\text{dist}\big(x, B(0, R)\big)\Big)^{n+1}}$$

$$\le \frac{N_0 N_2}{\Big(\text{dist}\big(x, B(0, R)\big)\Big)^{n+1}}, \tag{5.2.11}$$

and then (5.2.11) implies part (ii) of the corollary [when $f \equiv 0$].

If f is not identically zero, then

$$\left|\mathbb{E}\int_{[s,T]} f\big(\tau, X(\tau, s, x)\big)\rho^0(\tau, s)\, d\tau\right| \le N_3 \mathbb{E} \sup_{T_0\le s\le t\le T} \big|f\big(t, X(t, s, x)\big)\big|,$$

and a similar application of the Chebyshev inequality completes the proof of part (ii).

To prove part (i), note that, because of compact support, $\varphi \in \mathbb{W}^1_p(r)$ and $f \in \mathbb{L}_p([T_0, T]; \mathbb{W}^1_p(r))$ for every $2 \le p < \infty$ and $r \in \mathbb{R}^1$. Thus, we obtain from Theorem 4.7 that problem (5.2.4), (5.2.5) has a unique r-generalized solution F belonging to $\mathbb{L}_p\big([T_0, T]; \mathbb{W}^1_p(r)\big)$ for every $2 \le p < \infty$ and $r \in \mathbb{R}^1$.

On the other hand, from part (ii) and the theorem of this paragraph we find that $\tilde{F}$ is also an r-generalized solution of problem (5.2.4), (5.2.5) for every $r < -(\mathrm{d} + 2n)/2$. In view of the uniqueness of r-generalized solutions of the problem, $F(s, x) = \tilde{F}(s, x)$, $\ell \times \ell_{\mathrm{d}}$-a.s., which completes the proof of the corollary. □

5.2.4. The next technical tool needed to prove Theorem 5.1 is of independent interest. It provides a precise L_p bound on the function $\tilde{F}$ from the previous paragraph and is known as the Krylov–Fichera inequalities.

Fix $r \in \mathbb{R}^1$ as in Theorem 5.1 and introduce the following notation:

$$S(x) = (1 + |x|^2)^{r/2},\ S(i, x) = rx^i(1 + |x|^2)^{-1},$$

$$\delta_{ij} = \begin{cases} 1, & i = j, \\ 0, & i \ne j, \end{cases}$$

$$S(i, j, x) = S(i, x)S(j, x) + r\delta_{ij}(1 + |x|^2)^{-1} - 2rx^i x^j(1 + |x|^2)^{-2}.$$

Then, for all $i, j = 1, 2, \dots, \mathrm{d}$, $S_i(x) = S(i, x,)S(x)$, $S_{ij}(x) = S(i, j, x)S(x)$, $|S(i, x)| \le |r|$, $|S(i, j, x)| \le r^2 + 3|r|$. Because of this close connection between $S(i)$, $S(i, j)$ and the corresponding partial derivatives, we extend the summation convention to $S(i)$ and $S(i,j)$, as in

$$a^{ij}(s,x)S(i,j,x) = \sum_{i,j=1}^{\mathrm{d}} a^{ij}(s,x)S(i,j,x).$$

Next, some more notation:

$$\begin{aligned}
c^*(s,x) &= a^{il}_{ij}(s,x) + 2a^{ij}_{i}(s,x)S(j,x) + a^{ij}(s,x)S(i,j,x) - b^i(s,x)S(i,x)\\
&\quad - b^i_i(s,x) + c(s,x),\\
H_t &= [T_0,t]\times\mathbb{R}^{\mathrm{d}},\ \mathbb{L}_p(H_t,r) = \mathbb{L}_p\big([T_0,t];\mathbb{L}_p(r)\big),\\
\lambda &= \sup_{(s,x)\in H_t} c^*(s,x),\ \tilde{\lambda} = (p-1)\sup_{(s,x)\in H_t}|c(s,x)|.
\end{aligned}$$

Assumptions of Theorem 5.2 imply $|\lambda| + |\tilde{\lambda}| < \infty$.

Theorem 5.3 (The Krylov–Fichera Inequalities) *Suppose that the conditions of Theorem 5.2 are satisfied and fix the numbers $s \le t \in [T_0,T]$, $p \in [1,\infty)$, and $r \in \mathbb{R}^1$. Then, for every $\varphi \in \mathbb{L}_p(r)$ and $\psi \in \mathbb{L}_p(H_t,r)$,*

$$\Big\|\mathbb{E}\Big(\varphi\big(X(t,s,\cdot)\big)\,\rho^0(t,s)\Big)\Big\|^p_{\mathbb{L}_p(r)} \le e^{(\lambda+\tilde{\lambda})(t-s)}\|\varphi\|^p_{\mathbb{L}_p(r)}, \tag{5.2.12}$$

$$\left\|\mathbb{E}\int_{[s,t]}\psi\big(\tau,X(\tau,s,\cdot)\big)\,\rho^0(\tau,s)\,d\tau\right\|^p_{\mathbb{L}_p(H_t,r)} \le |t-T_0|^p e^{(\lambda\vee 0+\tilde{\lambda})(t-T_0)}\,\|\psi\|^p_{\mathbb{L}_p(H_t,r)}. \tag{5.2.13}$$

□

Proof We start by making the following observations:

- It is enough to prove inequality (5.2.12) for $p=1$; the general case follows by the Hölder inequality applied to

$$\mathbb{E}\big(\varphi\rho^0\big) = \mathbb{E}\Big(\big(\varphi(\rho^0)^{1/p}\big)\cdot(\rho^0)^{(p-1)/p}\Big);$$

  note also that $\rho^0(t,s) \le e^{\tilde{\lambda}(t-s)}$.
- Inequality (5.2.13) follows from (5.2.12) after similar applications of the Hölder inequality, first to the expectation and then to the resulting $\int_{[s,t]}$ integral.

- Because $X(t,t,x) = x$, inequality (5.2.12) is equivalent to the claim that the function

$$s \mapsto e^{\lambda s} \left\| \mathbb{E}\Big(\varphi\big(X(t,s,\cdot)\big)\, \rho^0(t,s)\Big) \right\|_{\mathbb{L}_p(r)},\ T_0 \le s \le t,$$

is non-decreasing.

- To prove (5.2.12), it is enough to consider $\varphi \ge 0$ and $\varphi \in \mathbb{C}_0^\infty(\mathbb{R}^{\mathrm{d}})$; the general case then follows by considering positive and negative parts of φ and an approximation.

Accordingly, we will prove (5.2.12) for $p = 1$ and for non-negative smooth φ with compact support.

Let $\eta^{(1)} = \eta^{(1)}(s)$ be a non-negative function from $\mathbb{C}_0^\infty((T_0,t))$ and let $\eta^{(2)} = \eta^{(2)}(x)$ be a non-negative function from $\mathbb{C}_0^\infty(\mathbb{R}^{\mathrm{d}})$ with $\eta^{(2)}(0) = 1$. For $R > 0$, define

$$\eta^{(R)}(s,x) = \eta^{(1)}(s)\eta^{(2)}(x/R),\ y^{(R)}(s,x) = S(x)\eta^{(R)}(s,x).$$

Also, for fixed $t \in (T_0, T]$, define

$$F^0(s,x) = \mathbb{E}\Big(\varphi\big(X(t,s,\cdot)\big)\, \rho^0(t,s)\Big),\ T_0 \le s \le t.$$

By Theorem 5.2, using $y^{(R)}$ as a test function and moving all the derivatives, including time, to y^R,

$$\int\limits_{H_t} F^0(s,x)\left(-\frac{\partial y^{(R)}}{\partial s} + \big(a^{ij}y^{(R)}\big)_{ij} - \big(b^i y^{(R)}\big)_i + cy^{(R)}\right)(s,x)\,dxds = 0,$$

or, after computing the derivatives and moving terms around,

$$\int\limits_{H_t} S(x)F^0(s,x)\left(\frac{\partial\eta^{(R)}(s,x)}{\partial s} - c^*(s,x)\eta^{(R)}(s,x)\right)dxds$$

$$= \int\limits_{H_t} S(x)F^0(s,x)\left(2a_i^{ij}\eta_j^{(R)} + a^{ij}\eta_{ij}^{(R)} + 2a^{ij}\eta_i^{(R)}S(j) - b^i\eta_i^{(R)}\right)(s,x)\,dxds.$$

Now we pass to the limit as $R \to \infty$ and use the dominated convergence theorem, which is justified by Theorem 5.2 and Corollary 5.1(ii). The right-hand side of the above equality becomes zero because each of $\eta_i^{(R)}$, $\eta_{ij}^{(R)}$ tends to zero as $R \to \infty$. As a result,

$$\int\limits_{H_t} S(x)F^0(s,x)\left(\frac{d\eta^{(1)}(s)}{ds} - c^*(s,x)\,\eta^{(1)}(s)\right)dsdx = 0,$$

and consequently, because $c^*(s,x) - \lambda \leq 0$,

$$\int_{H_t} S(x)F^0(s,x)\left(\frac{d\eta^{(1)}(s)}{ds} - \lambda\,\eta^{(1)}(s)\right) dsdx \leq 0.$$

Let us write

$$N(s) = e^{\lambda s}\|F^0(s)\|_{\mathbb{L}_1(r)},\ M(s) = \eta^{(1)}(s)\,e^{-\lambda s}.$$

Then the last inequality becomes

$$\int_{[T_0,t]} N(s)M'(s)\,ds \leq 0, \tag{5.2.14}$$

which indeed implies that the function N is non-decreasing, as desired. If N were continuously differentiable, then, after integration by parts, we would conclude that $N'(s) \geq 0$. The reader is welcome to confirm that, while N is not differentiable but only continuous (by the results from the previous paragraph), inequality (5.2.14) is enough to argue that N is non-decreasing, thus completing the proof of Theorem 5.3. □

5.2.5. The last auxiliary result is about completeness of a particular system of random variables and will be necessary to move between deterministic and stochastic partial differential equations. The result is well-known and is essentially equivalent to completeness of Hermite polynomials in $\mathbb{L}_2(\mathbb{R}^1)$. For details, see, for example [45, §4.4].

For a (non-random) function

$$q \in \mathbb{L}_\infty([T_0,T];\mathbb{R}^{d_1}) := \mathbb{L}_\infty\big([T_0,T],\overline{\mathscr{B}([T_0,T])},\ell;\mathbb{R}^{d_1}\big)$$

and $T_0 \leq s < t \leq T$, define the random variable

$$\mathfrak{q}_t(s) := \exp\Big(\int_{[s,t]} q^l(\tau)\,dw^l(\tau) - \frac{1}{2}\int_{[s,t]} (q^l q^l)(\tau)d\tau\Big). \tag{5.2.15}$$

Lemma 5.1 *If* $\xi \in \mathbb{L}_2(\Omega,\tilde{\mathscr{F}}^s_T;\mathbb{R}^1)$ *and* $\mathbb{E}\Big(\xi\mathfrak{q}_T(s)\Big) = 0$ *for every* $q \in \mathbb{L}_\infty([T_0,T],\mathbb{R}^{d_1})$, *then* $\mathbb{P}\big(\xi = 0\big) = 1$. □

5.2.6. Now we are ready to prove Theorem 5.1. As was mentioned earlier, we only consider problem (4.4.3), (4.4.4).

To begin, assume that f, g^l, and φ as well as their first derivatives in x are uniformly bounded and vanish when $|x| \geq R$ for some $R > 0$.

Let $v = v(s, x)$ be the r-generalized solution of the corresponding problem (4.4.3), (4.4.4) and, for $q \in \mathbb{L}_\infty([T_0, T], \mathbb{R}^{d_1})$, define

$$\tilde{v}(s, x) = v(s, x)\mathfrak{q}_T(s).$$

Fixing $y \in \mathbb{C}_0^\infty(\mathbb{R}^d)$ and applying the Itô formula (Sect. 1.5.3) to the process $\big(v(s), y\big)_0 \mathfrak{q}_T(s)$, $s \in [T_0, T)$, we conclude that $\tilde{v}$ is an r-generalized solution of

$$- d\tilde{v}(s, x) = \Big(a^{ij}\tilde{v}_{ij} + \tilde{b}^i \tilde{v}_i + \tilde{c}\tilde{v} + \tilde{f}\Big)(s, x)\, ds \tag{5.2.16}$$

$$+ \Big(\sigma^{il}\tilde{v}_i + \tilde{h}^l \tilde{v} + \tilde{g}^l\Big)(s, x) * dw^l(s),\ (s, x) \in [T_0, T) \times \mathbb{R}^d,$$

$$\tilde{v}(T, x) = \varphi(x),\ x \in \mathbb{R}^d, \tag{5.2.17}$$

where

$$\tilde{b}^i = b^i + q^l\sigma^{il},\ \tilde{c} = c + q^l h^l,\ \tilde{f} = \mathfrak{q}_T(f + q^l g^l),\ \tilde{h}^l = h^l + q^l,\ \tilde{g}^l = \mathfrak{q}_T g^l.$$

Note that $\tilde{f}$ and $\tilde{g}^l$ are backward predictable relative to the family $\{\tilde{\mathscr{F}}_T^s\}$; also recall that q^l is not random.

Under our current assumptions,

$$\varphi \in \mathbb{L}_p\big(\Omega; \mathbb{W}_p^1(r)\big),\ \tilde{f} \in \mathbb{L}_p\big([T_0, T]\times\Omega; \mathbb{W}^1{}_p(r)\big),\ g^l \in \mathbb{L}_p\big([T_0, T]\times\Omega; \mathbb{W}_p^2(r)\big)$$

for every $p \ge 2$. From Theorem 4.7 and Corollary 4.8, and in view of uniqueness of r-generalized solutions to a problem of the type (4.4.3), (4.4.4) and Warning 4.5, it then follows that $\tilde{v} \in \mathbb{L}_p\big([T_0, T] \times \Omega, \overleftarrow{\mathscr{P}}; \mathbb{W}_p^1(r)\big)$ for every $2 \le p < \infty$, has a continuous in t, x version (also denoted by $\tilde{v}$), and

$$\mathbb{E} \sup_{\substack{t \in [T_0, T] \\ x \in \mathbb{R}^d}} |\tilde{v}(t, x)|^p < \infty.$$

It also follows that $\mathbb{E}\tilde{v} \in \mathbb{L}_2\big([T_0, T]; \mathbb{H}^1(r)\big)$. Let us now take the expectation on both sides of the integral form of (5.2.16), (5.2.17). Since $\tilde{v} \in \mathbb{L}_2\big([T_0, T] \times \Omega, \overleftarrow{\mathscr{P}}; \mathbb{H}^1(r)\big)$, the expectation of the stochastic integral is zero. Changing the order of integration in the obtained equality, we conclude that the (non-random) function $\mathbb{E}\tilde{v}$ is an r-generalized solution, in the sense of Definition 5.1, of the problem

$$-\frac{\partial F}{\partial s} = a^{ij}F_{ij} + \tilde{b}^i F_i + \tilde{c} + \tilde{f},\ (s, x) \in [T_0, T) \times \mathbb{R}^d, \tag{5.2.18}$$

$$F(T, x) = \varphi(x), x \in \mathbb{R}^d. \tag{5.2.19}$$

On the other hand, by Theorem 5.2 and Corollary 5.1, the following equality holds on $[T_0, T] \times \mathbb{R}^d$, $\mathfrak{l} \times \mathfrak{l}_d$-a.s.:

$$\mathbb{E}\tilde{v}(s,x) = \mathbb{E}\left(\int\limits_{[s,T]} \tilde{f}\big(\tau, \tilde{\mathscr{X}}(\tau,s,x)\big) \exp\left(\int\limits_{[s,\tau]} \tilde{c}\big(\tau_1, \tilde{\mathscr{X}}(\tau_1,s,x)\big)d\tau_1 \right) d\tau \right.$$

$$\left. + \varphi(\tilde{\mathscr{X}}(T,s,x)) \exp\left(\int\limits_{[s,T]} \tilde{c}\big(\tau, \tilde{\mathscr{X}}(\tau,s,x)\big)d\tau \right) \right) \equiv \mathbb{E}\big(v(s,x)\mathfrak{q}_T(s)\big), \tag{5.2.20}$$

where $\tilde{\mathscr{X}} = \tilde{\mathscr{X}}(t,s,x)$ is the diffusion process evolving in forward time t:

$$\tilde{\mathscr{X}}^i(t,s,x) = x^i + \int\limits_{[s,t]} \tilde{b}^i\big(\tau, \tilde{\mathscr{X}}(\tau,s,x)\big)\, d\tau$$

$$+ \int\limits_{[s,t]} \boldsymbol{\Sigma}^{i\ell}\big(\tau, \tilde{\mathscr{X}}(\tau,s,x)\big)\, d\mathcal{W}^\ell(\tau), \quad t \in [s,T],\ i = 1, 2, \ldots, \mathrm{d}.$$

Define

$$\tilde{\rho}(t,s) = \exp\left(\int\limits_{[s,t]} \tilde{c}\big(\tau, \mathscr{X}(\tau,s,x)\big)d\tau \right.$$

$$\left. + \int\limits_{[s,t]} \tilde{h}^l\big(\tau, \mathscr{X}(\tau,s,x)\big)\, dw^l(\tau) - \frac{1}{2} \int\limits_{[t,s]} (\tilde{h}^l\tilde{h}^l)\big(\tau, \mathscr{X}(\tau,s,x)\big)\, d\tau \right).$$

Now we use Girsanov's theorem (Sect. 1.5.7) to rewrite equality (5.2.20) as

$$\mathbb{E}\big(v(s,x)\mathfrak{q}_T(s)\big)$$

$$= \mathbb{E}\left(\int\limits_{[s,T]} \tilde{f}\big(\tau, \mathscr{X}(\tau,s,x)\big)\tilde{\rho}(\tau,s)d\tau + \varphi\big(\mathscr{X}(T,s,x)\big)\tilde{\rho}(T,s) \right)$$

$$= \mathbb{E}\Bigg(\int\limits_{[s,T]} \Big(f\big(\tau, \mathscr{X}(\tau,s,x)\big) + q^l(\tau)g^l\big(\tau, \mathscr{X}(\tau,s,x)\big)\Big)\rho(\tau,s)\mathfrak{q}_\tau(s)\, d\tau \tag{5.2.21}$$

$$+ \varphi\big(\mathscr{X}(T,s,x)\big)\rho(T,s)\mathfrak{q}_T(s) \Bigg).$$

We now apply the Itô formula to the product

$$\mathfrak{q}_t(s)\Big(\int\limits_{[s,t]} f\big(\tau,\mathscr{X}(\tau,s,x)\big)\rho(\tau,s)d\tau+\int\limits_{[s,t]} g^l\big(\tau,\mathscr{X}(\tau,s,x)\big)\rho(\tau,s)\,dw^l(\tau)\Big),$$

considered as a function of $t \in [s,T]$, integrate in t over the interval $[s,T]$, and take the expectation of both sides of the result:

$$\begin{aligned}\mathbb{E}\Big(\int\limits_{[s,T]}\Big(f\big(\tau,\mathscr{X}(\tau,s,x)\big)+q^l(\tau)g^l\big(\tau,\mathscr{X}(\tau,s,x)\big)\Big)\rho(\tau,s)\mathfrak{q}_\tau(s)d\tau\Big)\\ =\mathbb{E}\Big(\mathfrak{q}_T(t)\Big(\int\limits_{[s,T]} f\big(\tau,\mathscr{X}(\tau,s,x)\big)\rho(\tau,s)d\tau\\ +\int\limits_{[s,T]} g^l\big(s,\mathscr{X}(\tau,s,x\big)\rho(\tau,s)\,dw^l(s)\Big)\Big).\end{aligned} \tag{5.2.22}$$

From (5.2.21), (5.2.22) it follows that, for all $t \in [T_0,T]$ and $\mathfrak{l}_\mathrm{d}$-a.a. x,

$$\begin{aligned}\mathbb{E}\Big(v(s,x)\mathfrak{q}_T(s)\Big)=\mathbb{E}\Big(\mathfrak{q}_T(s)\,\mathbb{E}\Big[\int\limits_{[s,T]} f\big(\tau,\mathscr{X}(\tau,s,x)\big)\rho(\tau,s)d\tau\\ +\int\limits_{[s,T]} g^l\big(\tau,\mathscr{X}(\tau,s,x)\big)\rho(\tau,s)\,dw^l(\tau)+\varphi\big(\mathscr{X}(T,s,x)\big)\rho(T,s)\big|\tilde{\mathscr{F}}_T^s\Big]\Big).\end{aligned}$$

Together with Lemma 5.1, this yields equality (5.2.3).

We now remove the additional assumptions (related to smoothness and compact support) on φ, f, and g^l. To this end, choose sequences $\{\varphi^n\}$, $\{f^n\}$ and $\{g^{l,n}\}$ whose elements are smooth with compact support and such that, as $n\to\infty$, $\varphi^n\to\varphi$ in $\mathbb{H}^1(r)$, $f^n\to f$ in $\mathbb{L}_2\big([T_0,T];\,\mathbb{H}^1(\mathrm{r})\big)$, and $g^{l,n}\to g^l$ in $\mathbb{L}_2\big([T_0,T];\,\mathbb{H}^2(r)\big)$ for every $l=1,2,\ldots,\mathrm{d}_1$.

Denote by v^n the r-generalized solution of problem (4.4.3), (4.4.4) corresponding to φ^n, f^n, $g^{l,n}$, for which (5.2.3) holds. To complete the proof, we need to pass to the limit $n\to\infty$ in (5.2.3).

From inequality (4.4.8) it follows that, for every $s \in [T_0, T]$, $v^n(s)$ converges in $\mathbb{L}_2(\mathbb{R}^d \times \Omega, \mathscr{B}(\mathbb{R}^d) \otimes \mathscr{F}, \ell_d \times \mathbb{P}; \mathbb{R}^1)$ to the solution $v = v(s)$ of problem (4.4.3), (4.4.4) corresponding to φ, f, g^l, as $n \to \infty$.

Convergence of the right-hand side of (5.2.3) follows from the Krylov–Fichera inequalities (Theorem 5.3).

For example, consider

$$\mathbb{E}\Big[\int\limits_{[s,T]} g^{l,n}\big(\tau, \mathscr{X}(\tau, s, x)\big)\rho(\tau, s)\, dw^l(\tau)\big|\tilde{\mathscr{F}}^s_T\Big].$$

By Girsanov's theorem, we obtain

$$\begin{aligned}
&U^n \\
&:= \int\limits_{H_T} S^2(x)\,\mathbb{E}\Big(\mathbb{E}\Big[\int\limits_{[s,T]} (g^{l,n} - g^l)\big(\tau, \mathscr{X}(\tau, s, x)\big)\rho(\tau, s)\, dw^l(\tau)\Big|\tilde{\mathscr{F}}^s_T\Big]\Big)^2 dxds \\
&\le \int\limits_{H_T} S^2(x)\,\mathbb{E}\int\limits_{[t,T]} \sum_{l=1}^{d_1} \Big|g^{l,n} - g^l\Big|^2\big(s, \mathscr{X}(\tau, s, x)\big)\rho^2(\tau, s)\, d\tau dxds \qquad (5.2.23) \\
&= \int\limits_{H_T} S^2(x)\,\mathbb{E}\int\limits_{[t,T]} \sum_{l=1}^{d_1} \Big|g^{l,n} - g^l\Big|^2\big(s, \hat{\mathscr{X}}(\tau, s, x)\big)\hat{\rho}(\tau, s)\, d\tau dxds,
\end{aligned}$$

where $\hat{\mathscr{X}} = \hat{\mathscr{X}}(t, s, x)$ is the diffusion process

$$\begin{aligned}
\hat{\mathscr{X}}^i(t, s, x) &= x^i + \int\limits_{[s,t]} (b^i - 2h^l\sigma^{il})\big(\tau, \hat{\mathscr{X}}(\tau, s, x)\big)\, d\tau \\
&\quad + \int\limits_{[s,t]} \Sigma^{i\ell}\big(\tau, \hat{\mathscr{X}}(\tau, s, x)\big) dW^\ell(\tau),\ i = 1, 2, \ldots, d,\ t \in [s, T],
\end{aligned}$$

and

$$\hat{\rho}(t, s) := \exp\Big(\int\limits_{[s,t]} (2c + h^l h^l)\big(\tau, \hat{\mathscr{X}}(\tau, s, x)\big)\, d\tau\Big).$$

By (5.2.13),

$$U^n \leq N \int\limits_{H_T} S^2(x) \sum_{l=1}^{\mathrm{d}_1} \left|g^{l,n} - g^l\right|^2(t,x)\,dtdx,$$

with N independent of n, and therefore $\lim\limits_{n\to\infty} U^n = 0$.

We can pass to the limit in the remaining terms in the same way, completing the proof of Theorem 5.1. □

Some applications of the AOC formulas (5.2.2), (5.2.3) will be considered in the following sections of this chapter and also in Chap. 6. Here we discuss one especially important result, namely, the maximum principle, which is an immediate consequence of formulas (5.2.2), (5.2.3).

Corollary 5.2 (The Maximum Principle) *Suppose that conditions of Theorem* 5.1 *are satisfied.*

(i) *If* $g^l = 0$ *for all* $l = 1, 2, \ldots, \mathrm{d}$, $f \geq 0$ ($\ell \times \ell_{\mathrm{d}}$- a.s.), *and* $\varphi \geq 0$ (ℓ_{d}- a.s.), *then the* r*-generalized solutions of problems* (4.1.1), (4.2.2) *and* (4.4.3), (4.4.4) *are non-negative* ($\ell \times \ell_{\mathrm{d}} \times \mathbb{P}$- a.s.).

(ii) *If* $h^l = g^l = 0$ *for all* $l = 1, 2, \ldots d$, $f \leq 0$, $c \leq 0$ ($\ell \times \ell_{\mathrm{d}}$- a.s.), *and* $\varphi \leq 1$ (ℓ_{d}- a.s.), *then the* r*-generalized solutions of problems* (4.1.1), (4.2.2) *and* (4.4.3), (4.4.4) *satisfy* $u(t,x) \leq 1$ *and* $v(s,x) \leq 1$, ($\ell \times \ell_{\mathrm{d}} \times \mathbb{P}$- a.s.). □

In the proof of this corollary we can make use of Theorem 5.3 to show that a change of c, f, h^l, and g^l on a set of zero $\ell \times \ell_{\mathrm{d}}$-measure and φ on a set of zero ℓ_d measure does not change the right-hand side of formulas (5.2.2) and (5.2.3) (up to $\ell \times \ell_{\mathrm{d}} \times \mathbb{P}$-equivalence).

Remark 5.3 If the matrix (a^{ij}) satisfies super-parabolicity condition (4.1.3), then it is possible to relax the regularity conditions on the coefficients and the functions φ, f, g^l can be made less restrictive.

Indeed, for Theorem 5.1 and the maximum principle to hold in the super-parabolic case, it suffices to assume that all the coefficients are bounded, a^{ij} are differentiable in x, with bounded derivatives, f, $g^l \in \mathbb{L}_2([T_0, T]; \mathbb{L}_2(r))$, and $\varphi \in \mathbb{L}_2(r)$. From Theorem 4.6 it follows that problems (4.2.1), (4.2.2) and (4.4.1), (4.4.2) have unique r-generalized solutions. □

5.2.7.

Remark 5.4 The maximum principle can hold even when the coefficients, the initial (terminal) conditions and the free terms in problems (4.1.1), (4.2.2) and (4.4.3), (4.4.4) are random; see reference [74] or Sect. 5.4.2 below. □

5.3 Inverse Diffusion Processes, Variation of Parameters and the Liouville Equations

5.3.1. In this section we consider the diffusion process $X = X(t, s, x)$ given by

$$X^i(t,s,x) = x^i + \int_{[s,t]} b^i\big(\tau, X(\tau,s,x), \omega\big)\, d\tau + \int_{[s,t]} \sigma^{il}\big(\tau, X(\tau,s,x), \omega\big)\, dw^l(\tau)\,, \tag{5.3.1}$$

$$t \in [s,T],\ i = 1,2,\ldots,\mathrm{d},\ s \in [T_0,T).$$

Unless explicitly stated otherwise, in this section the coefficients b and σ are random and

- the functions $b^i = b^i(t,x,\omega)$ and $\sigma^{il} = \sigma^{il}(t,x,\omega)$ are continuous in (t,x), bounded, and predictable (relative to $\{\tilde{\mathscr{F}}_t^{T_0}\}$) for every x;
- the functions b^i are differentiable in x and σ^{il} are twice differentiable in x, with all the derivatives bounded.

From Sects. 1.5.4 and 1.5.6 we know that, under these assumptions, system (5.3.1) has a unique predictable solution, and this solution has a version that is jointly continuous in t, s, x. From now on, we always consider this continuous version, still denoted by X.

The main objective of this section is to show that, for all t, s and $\mathbb{P}$-a.a. ω, the mapping $x \mapsto X(t,s,x)$ is a diffeomorphism of $\mathbb{R}^{\mathrm{d}}$, and to derive the equations for the inverse mapping X^{-1}. Other results of this section include the Liouville equation for $X(t,x,T_0)$ and various stochastic versions of the variation-of-parameters formula.

Warning 5.1 *Throughout what follows, when there is no danger of confusion, the starting time s of X and other diffusion processes of this type will be omitted if $s = T_0$, that is, $X(t, T_0, x,) \equiv X(t,x)$.* □

5.3.2. Let us fix an integer $m \geq 3$. We assume that $\sigma^{il}(\cdot,\cdot,\omega) \in \mathbb{C}_b^{0,m+1}([T_0,T] \times \mathbb{R}^d)$ and $b^i(\cdot,\cdot,\omega) \in \mathbb{C}_b^{0,m}([T_0,T] \times \mathbb{R}^{\mathrm{d}})$ for all $\omega \in \Omega$, $i = 1,2,\ldots,\mathrm{d}$, $l = 1,2,\ldots,\mathrm{d}_1$.

Definition 5.2 For a non-negative integer k, a family of mappings $\Psi(t,\cdot,\omega)$: $\mathbb{R}^{\mathrm{d}} \to \mathbb{R}^{\mathrm{d}}$ for $t \in [T_0,T]$, $\omega \in \Omega$, is called a **stochastic flow** of $\mathbb{C}^{0,k}$-diffeomorphisms of $\mathbb{R}^{\mathrm{d}}$ if

(i) For every $t \in [T_0,T]$ and $\mathbb{P}$- a.a. ω, the mapping $x \mapsto \Psi(t,x,\omega)$ is a bijection of $\mathbb{R}^{\mathrm{d}}$, and both Ψ and its inverse Ψ^{-1} belong to $\mathbb{C}^{0,k}([T_0,T] \times \mathbb{R}^{\mathrm{d}})$.
(ii) The processes $\Psi(\cdot,x,\cdot)$ and $\Psi^{-1}(\cdot,x,\omega)$ are predictable for every x.

(iii) For $\mathbb{P}$- a.a. ω, $\Psi(t, \cdot, \omega)$ is a one-parameter semigroup of transformations of $\mathbb{R}^{\mathrm{d}}$, that is, $\Psi\Big(t_1, \Psi(t_2, \cdot, \omega), \omega\Big) = \Psi(t_1 + t_2, \cdot, \omega)$, ($\mathbb{P}$- a.s.), for all $t_1, t_2 \in [T_0, T]$ such that $t_1, t_2 \leq T$; cf. *[1]*. □

Since we only consider stochastic flows of diffeomorphisms, the attribute "stochastic" will usually be dropped.

Recall the partial differential operators L_0 and B_0^l defined by

$$L_0 u = \frac{1}{2}\sigma^{il}\sigma^{jl}u_{ij} + b^i u_i, \;\; B_0^l u = \sigma^{il}u_i.$$

Proposition 5.1 *Under the assumptions of this paragraph, in particular, existence of $m \geq 3$ spatial derivatives for b and $m + 1$ derivatives for σ, the family of mappings $x \mapsto X(t, x)$ is a flow of $\mathbb{C}^{0,m-1}$-diffeomorphisms of $\mathbb{R}^{\mathrm{d}}$, and, for every $r < -(\mathrm{d}/2 + 1)$, the ith coordinate ($i = 1, 2, \ldots, \mathrm{d}$) of the inverse mapping X^{-1} is the unique r-generalized solution of the problem*

$$du(t,x) = \big(B_0^l B_0^l - L_0\big)u(t,x)\,dt + B_0^l u(t,x)\,dw^l(t), \qquad (5.3.2)$$

$$(t, x, \omega) \in (T_0, T] \times \mathbb{R}^{\mathrm{d}} \times \Omega;$$

$$u(T_0, x) = x^i, \;\; (x, \omega) \in \mathbb{R}^{\mathrm{d}} \times \Omega. \qquad (5.3.3)$$

This solution belongs to

$$\mathbb{L}_2\Big(\Omega; \mathbb{C}\big([T_0, T]; \mathbb{H}^{m-1}(r)\big)\Big) \bigcap \mathbb{L}_p\big(\Omega; \mathbb{C}_w\big([T_0, T]; \mathbb{W}_p^m(r)\big)\big)$$

for every $p \in [2, \infty)$. □

Proof Note that problem (5.3.2), (5.3.3) is a particular case of (4.1.1), (4.2.2). From Lemma 3.6 it follows that the function φ, defined by $\varphi(x) \equiv x^i$, belongs to $\mathbb{W}_p^m(r)$ for every $r < -(\mathrm{d}/2 + 1)$ and $p \in [2, \infty)$. The theorem and the corollary of Sect. 4.4.3 then imply that, for such p and r, problem (5.3.2), (5.3.3) has a unique r-generalized solution in the required space, and this solution has a version in $\mathbb{C}^{0,m-1}([T_0, T] \times \mathbb{R}^{\mathrm{d}})$. The Itô–Ventcel formula (Sect. 1.5.10) applied to $u\big(t, X(t,x)\big)$ shows that $du\big(t, X(t,x)\big) = 0$ or

$$u\big(t, X(t,x)\big) = u\big(T_0, X(T_0, x)\big) = u(T_0, x) = x^i$$

for all x, t, ($\mathbb{P}$-a.s.). In view of the continuity of u and X in t, x, we can choose a single ω-set of probability 1 on which the equality $u\big(t, X(t,x)\big) = x^i$ holds for all x, t simultaneously. To complete the proof it suffices to observe that, in view of the results of Sect. 1.5.5, the function X belongs to $\mathbb{C}^{0,m-1}([T_0, T] \times \mathbb{R}^{\mathrm{d}})$, ($\mathbb{P}$- a.s.), and, by uniqueness of solution, is a one-parameter semigroup of transformations of $\mathbb{R}^{\mathrm{d}}$. □

Warning 5.2 *Throughout what follows we denote the ith component of $X^{-1}(t,s,x)$ by $X^{-i}(t,s,x)$ and, when $s = T_0$, by $X^{-i}(t,x)$.* □

Now we are in a position to introduce the method of **variation of parameters** (cf. [1]) for an ordinary Itô differential equation. Beside the process X, we will consider other similar processes, for which the corresponding coefficients will be identified by the sub-script to the left of the letter, as in ${}_{\mathrm{Y}}b^i$. The same notations will be used for the corresponding operators L_0 and M_0^l, for example,

$$ {}_{\mathrm{Y}}L_0 u = \frac{1}{2}\, {}_{\mathrm{Y}}\sigma^{il}\, {}_{\mathrm{Y}}\sigma^{il}\, u_{ij} + {}_{\mathrm{Y}}b^i\, u_i . $$

Consider the equation

$$ Y^i(t,x) = x^i + \int\limits_{[T_0,t]} {}_{\mathrm{Y}}b^i\big(\tau, Y(\tau,x)\big)\,d\tau + \int\limits_{[T_0,t]} {}_{\mathrm{Y}}\sigma^{il}\big(\tau, Y(\tau,x)\big)\,dw^l(\tau), \quad t \in [T_0,T],\ i = 1,2,\ldots,\mathrm{d}. \tag{5.3.4} $$

It is assumed that the functions ${}_{\mathrm{Y}}b^i = {}_{\mathrm{Y}}b^i(t,x,\omega)$ and ${}_{\mathrm{Y}}\sigma^{il} = {}_{\mathrm{Y}}\sigma^{il}(t,x,\omega)$ are $\mathscr{B}([T_0,T]\times\mathbb{R}^{\mathrm{d}})\otimes\mathscr{F}$-measurable and predictable for every x. We do not make any additional regularity assumptions, in particular, to guarantee existence and uniqueness of solution of (5.3.4), and the following theorem explains why.

Theorem 5.4

(i) *System* (5.3.4) *has a solution if and only if the following system has a solution:*

$$ Z^i(t,x) = x^i + \int\limits_{[T_0,t]} {}_{\mathrm{Z}}b^i\big(\tau, Z(\tau,x)\big)\,d\tau + \int\limits_{[T_0,t]} {}_{\mathrm{Z}}\sigma^{il}\big(\tau, Z(\tau,x)\big)\,dw^l(\tau), \quad t \in [T_0,T],\ i = 1,2,\ldots,\mathrm{d}, \tag{5.3.5} $$

where

$$ {}_{\mathrm{Z}}b^i(t,x) := \Big(\big({}_{\mathrm{Y}}L_0 - L_0 - ({}_{\mathrm{Y}}B_0^l - B_0^l)\,{}_{\mathrm{Y}}B_0^l\big)u\Big)\big(t, X(t,x)\big), $$

$$ {}_{\mathrm{Z}}\sigma^{il}(t,x) = \big(({}_{\mathrm{Y}}B_0^l - B_0^l)u\big)\big(t, X(t,x)\big), $$

and u is the solution of (5.3.2), (5.3.3)*; in particular, there is a different u for different i. Moreover, for every pair Y, Z of solutions, there exists a set $\Omega' \subset \Omega$*

with $\mathbb{P}(\Omega') = 1$ *such that, for all* $(t, x, \omega) \in [T_0, T] \times \mathbb{R}^{\mathrm{d}} \times \Omega'$,

$$Y(t, x) = X\big(t, Z(t, x)\big). \tag{5.3.6}$$

(ii) *The solution of system* (5.3.4) *is unique if and only if the solution of system* (5.3.5) *is unique.* □

Proof Differentiating the identities

$$X^{-i}\big(t, X(t, x)\big) = x^i = X^i\big(t, X^{-1}(t, x)\big)$$

with respect to x^j, we get

$$X_l^{-i}\big(t, X(t, x)\big)X_j^l(t, x) = X_l^i\big(t, X^{-1}(t, x)\big)X_j^{-l}(t, x) = \begin{cases} 1, & i = j \\ 0, & i \neq j \end{cases} \tag{5.3.7}$$

and, differentiating the first one in (5.3.7) with respect to $x^{i'}$,

$$X_{lk}^{-i}\big(t, X(t, x)\big)\, X_{i'}^k(t, x)X_j^l(t, x) = -X_l^{-i}\big(t, X(t, x)\big)X_{ji'}^l. \tag{5.3.8}$$

Now apply the Itô–Ventcel formula to the right-hand side of (5.3.6), to get (5.3.4) and the "if" part of the statement.

The "only if" part follows after applying the Itô–Ventcel formula to $X^{-i}\big(t, Y(t, x)\big)$, $i = 1, 2, \ldots, \mathrm{d}$, making use of the fact that $X^{-i}(t, x)$ is the solution of problem (5.3.2), (5.3.3).

Now suppose that the solution of (5.3.4) is unique and let Z, Z' be two solutions of (5.3.5). Then both $X\big(t, Z(t, x)\big)$ and $X\big(t, Z'(t, x)\big)$ satisfy (5.3.4), meaning that, by uniqueness and the diffeomorphic properties of X,

$$\begin{aligned} 1 &= \mathbb{P}\Big(X\big(t, Z(t, x)\big) = X\big(t, Z'(t, x)\big)\Big) = \mathbb{P}\Big(X^{-1}\big(t, X\big(t, Z(t, x)\big)\big) \\ &= X^{-1}\big(t, X\big(t, Z'(t, x)\big)\big)\Big) = \mathbb{P}\Big(Z(t, x) = Z'(t, x)\Big), \end{aligned}$$

for all t, x, that is, uniqueness for (5.3.4) implies uniqueness for (5.3.5). To get the proof in the other direction, just read the above chain of equalities from right to left. □

Corollary 5.3 *Let* ${}_{\mathrm{Y}}b^i \equiv 0$ *and* ${}_{\mathrm{Y}}\sigma^{il} \equiv 0$, *and let* Z *be the corresponding process defined by* (5.3.5)*. Then there exists a set* $\Omega' \subset \Omega$ *with* $\mathbb{P}(\Omega') = 1$ *such that, for all* $(t, x, \omega) \in [T_0, T] \times \mathbb{R}^{\mathrm{d}} \times \Omega'$,

$$X^{-1}(t, x, \omega) = Z(t, x, \omega).$$

Here are two immediate applications of the theorem. In particular, these applications confirm that it is natural to call equality (5.3.6) the **variation of parameters** formula for diffusion processes.

Transformation killing a drift. If we take

$$ {}_{Z}\sigma \equiv 0,\ {}_{Z}b^i(t,x) = -b^j X_j^{-i}\big(t, X(t,x)\big), $$

or, in the matrix-vector notation,

$$ {}_{Z}b(t,x) = \left(\frac{\partial X(t,x)}{\partial x}\right)^{-1} b\big(t, x(t,x)\big), $$

where $\partial X(t,x)/\partial x$ is the Jacobian of the transformation $x \mapsto X(t,x)$, then ${}_{Y}b \equiv 0$ and ${}_{Y}\sigma \equiv \sigma$. In other words, we went from process X with non-zero drift b to process $Y = X\big(t, Z(t,x)\big)$ with zero drift ${}_{Y}b$ and the same diffusion coefficient as in X.

Transformation generating a drift and diffusion. This time, we will start with X such that $b^i \equiv 0$ and the first $\mathrm{d}_2 \le \mathrm{d}_1$ columns of the matrix σ are zero: $\sigma^{il} \equiv 0$ for $l \le \mathrm{d}_2$, $\mathrm{d}_2 \le \mathrm{d}_1$. The goal is to construct the process Z so that the corresponding process Y has prescribed drift ${}_{Y}b$ and the diffusion matrix ${}_{Y}\sigma$ such that ${}_{Y}\sigma^{il} = \sigma^{il}$ for $l = \mathrm{d}_2 + 1, \dots, \mathrm{d}_1$. This goal is achieved by taking

$$ {}_{Z}b^i(t,x) = \left(\frac{1}{2}\sum_{l=1}^{d_2} {}_{Y}\sigma^{jl}\, {}_{Y}\sigma^{kl}\, X_{jk}^{-l} + {}_{Y}b^j X_j^{-i}\right)\big(t, X(t,x)\big); $$

$$ {}_{Z}\sigma^{il}(t,x) = \begin{cases} \left({}_{Y}\sigma^{jl} X_j^{-i}\right)\big(t, X(t,x)\big), & \text{if } l \le \mathrm{d}_2, \\ 0, & \text{if } l > \mathrm{d}_2. \end{cases} $$

5.3.3. In this paragraph, we use Theorem 5.4 to derive a variation of parameters formula for second-order parabolic Itô equations.

Consider problem (4.1.1), (4.2.2). We assume that

- the coefficients are $\mathscr{B}([T_0, T] \times \mathbb{R}^{\mathrm{d}}) \otimes \overline{\mathscr{F}}$-measurable and predictable for every x;
- $f \equiv g^l \equiv 0$;
- $\sigma^{il}(\cdot,\cdot,\omega) \in \mathbb{C}^{0,4}([T_0, T] \times \mathbb{R}^{\mathrm{d}})$ and $h^l(\cdot,\cdot,\omega) \in \mathbb{C}^{0,2}([T_0, T] \times \mathbb{R}^{\mathrm{d}})$ for all $\omega \in \Omega$, $i = 1, 2, \dots, \mathrm{d}$, and $l = 1, 2, \dots, \mathrm{d}_1$.

Recall that, in this chapter, we always assume that the matrix (a^{ij}) has the form

$$ a^{ij} = \sigma^{il}\sigma^{jl} + \hat{\sigma}^{ik}\hat{\sigma}^{jk}. $$

Let $\eta = \eta(t, x)$ be the diffusion process

$$\eta^i(t, x) = x^i - \int_{[T_0,t]} \sigma^{il}\big(t, \eta(t, x)\big)\, dw^l(t), \ t \in [T_0, T], \ i = 1, 2, \ldots, \mathrm{d}.$$

By Corollary 5.3, η has a version that is a flow of $\mathbb{C}^{0,2}$-diffeomorphisms of $\mathbb{R}^{\mathrm{d}}$. Denote by $\eta^{-i}(t, x)$ the ith coordinate of the inverse mapping η^{-1}.

Proposition 5.1 and Corollary 4.8 yield that, for every i, the function $(t, x) \mapsto \eta^{-i}(t, x)$ has a version that is the classical solution of problem (5.3.2), (5.3.3) with $b^i \equiv 0$ and σ^{il} replaced by $-\sigma^{il}$.

In the sequel we consider only these versions of η and η^{-1} while keeping the same notation.

Next, we define several random functions:

$$\psi(t, x) = \exp\left(- \int_{[T_0,t]} h^l\big(\tau, \eta(\tau, x)\big)\, dw^l(\tau) + \frac{1}{2} \int_{[T_0,t]} (h^l h^l)\big(\tau, \eta(\tau, x)\big)\, d\tau \right),$$

$$\widetilde{\psi}(t, x) = \exp\left(\int_{[T_0,t]} h^l(\tau, x)\, dw^l(\tau) - \frac{1}{2} \int_{[T_0,t]} (h^l h^l)(\tau, x)\, d\tau \right),$$

$$\tilde{a}^{ij}(t, x) = \frac{1}{2}\Big(\hat{\sigma}^{mk}\hat{\sigma}^{nk}\eta_m^{-i}\eta_n^{-j}\Big)\big(t, \eta(t, x)\big),$$

$$\begin{aligned}\tilde{b}^i(t, x) &= \Big(\big(b^j - h^l\sigma^{jl} - \sigma_m^{jl}\sigma^{ml}\big)\eta_j^{-i} + \frac{1}{2}\hat{\sigma}^{mk}\hat{\sigma}^{nk}\eta_{mn}^{-i}\Big)\big(t, \eta(t, x)\big)\\ &\quad + \psi(t, x)\big(\hat{\sigma}^{mk}\hat{\sigma}^{nk}\widetilde{\psi}_m\eta_n^{-i}\big)\big(t, \eta(t, x)\big),\end{aligned}$$

$$\begin{aligned}\tilde{c}(t, x) &= \big(c - h_i^l\sigma^{il}\big)\big(t, \eta(t, x)\big)\\ &+\psi(t, x)\widetilde{\psi}_i\big(t, \eta(t, x)\big)\Big(b^i - \sigma^{il}h^l - \sigma_j^{il}\sigma^{jl} + \frac{1}{2}\hat{\sigma}^{mk}\sigma^{nk}\eta_{mn}^{-i}\Big)\big(t, \eta(t, x)\big)\\ &\quad + \frac{1}{2}\psi(t, x)\big(\hat{\sigma}^{mk}\sigma^{nk}\widetilde{\psi}_{mn}\big)\big(t, \eta(t, x)\big),\end{aligned}$$

and a partial differential operator

$$\tilde{L}v = \tilde{a}^{ij}v_{ij} + \tilde{b}^i v_i + \tilde{c}v.$$

Note that if $\hat{\sigma} \equiv 0$, which is a possibility, then $\tilde{L}$ is a first-order operator. Also, assumptions on h^l and σ^{il} imply

$$\psi, \tilde{\psi} \in \mathbb{C}^{0,2}\big([T_0, T] \times \mathbb{R}^{\mathrm{d}}\big).$$

In addition to problem (4.1.1), (4.2.2), consider

$$\frac{\partial v(t,x)}{\partial t} = \tilde{L}v(t,x)\,dt,\ (t,x,\omega) \in (T_0,T] \times \mathbb{R}^d \times \Omega; \tag{5.3.9}$$

$$v(T_0,x) = \varphi(x),\ (x,\omega) \in \mathbb{R}^{\mathrm{d}} \times \Omega. \tag{5.3.10}$$

Denote by $\mathfrak{U}$ the set of classical solutions of problem (4.1.1), (4.2.2) and by $\widetilde{\mathfrak{U}}$, the set of classical solutions of (5.3.9), (5.3.10); each of these sets can be empty.

Theorem 5.5 *Suppose that, for* $\mathbb{P}$- a.a. $\omega \in \Omega$,

$$L \in \mathfrak{L}\Big(\mathbb{C}^{0,2}([T_0,T] \times \mathbb{R}^{\mathrm{d}}),\, \mathbb{C}^{0,0}([T_0,T] \times \mathbb{R}^{\mathrm{d}})\Big),$$

$$B^l \in \mathfrak{L}\Big(\mathbb{C}^{0,2}([T_0,T] \times \mathbb{R}^{\mathrm{d}}),\, \mathbb{C}^{0,1}([T_0,T] \times \mathbb{R}^{\mathrm{d}})\Big),\ l = 1,2,\ldots,\mathrm{d}_1,$$

$$\tilde{L} \in \mathfrak{L}\Big(\mathbb{C}^{0,2}([T_0,T) \times \mathbb{R}^{\mathrm{d}}),\, \mathbb{C}^{0,0}([T_0,T] \times \mathbb{R}^{\mathrm{d}})\Big).$$

Then, for $\mathbb{P}$- a.a. $\omega \in \Omega$, *the mapping*

$$\Phi : u(t,x) \mapsto \psi(t,x)\,u\big(t,\eta(t,x)\big) \tag{5.3.11}$$

belongs to $\mathfrak{L}\Big(\mathbb{C}^{0,2}([T_0,T] \times \mathbb{R}^{\mathrm{d}})\Big)$ *and defines a bijection between* $\mathfrak{U}$ *and* $\widetilde{\mathfrak{U}}$, *with inverse*

$$\Phi^{-1} : v(t,x) \mapsto \widetilde{\psi}(t,x)v\big(t,\eta^{-1}(t,x)\big). \tag{5.3.12}$$

□

Proof Note that

$$d\psi = -\psi h^l dw^l + \psi h^l h^l dt,\ \ d\widetilde{\psi} = \widetilde{\psi} h^l dw^l,$$

implying, in particular, $d(\psi\widetilde{\psi}) = 0$ or, more precisely, $\psi(t,x)\widetilde{\psi}\big(t,\eta(t,x)\big) = 1$.

Then, taking u, a classical solution of (4.1.1), (4.2.2) and v, a classical solution of (5.3.9), (5.3.10), the proof consists in the application of the Itô–Ventcel formula to the right-hand side of equalities (5.3.11), (5.3.12) using (5.3.7) and (5.3.8). The computations are long but straightforward. Below are some of the highlights of the transition from (4.1.1) to (5.3.9).

The transformation

$$u(t,x) \mapsto u\big(t,\eta(t,x)\big)$$

eliminates the $\sigma^{il}u_i dw^l$ term in (4.1.1), together with $\frac{1}{2}\sigma^{il}\sigma^{jl}u_{ij}$; after that, multiplication by ψ eliminates the $h^l u dw^l$ term. Indeed, the application of the Itô–Ventcel formula, followed by the usual Itô formula, yields

$$du\big(t, \eta(t, x)\big) = Lu\,dt + B^l u\,dw^l - B_0^l u\,dw^l + \frac{1}{2}\sigma^{il}\sigma^{jl}u_{ij}\,dt - \sigma^{il}(M^l u)_i\,dt$$

$$= \frac{1}{2}\hat{\sigma}^{ik}\hat{\sigma}^{jk}u_{ij}\,dt + b^i u_i\,dt + cu\,dt + h^l u\,dw^l - \sigma^{il}\sigma_i^{jl}u_j\,dt - \sigma^{il}(h^l u)_i\,dt,$$

$$d\Big(\psi(t, x)u\big(t, \eta(t, x)\big)\Big) = \psi\,du + u\,d\psi - \psi u h^l h^l\,dt;$$

after further substitutions, we see that both $\psi h^l u\,dw^l$ and $\psi u h^l h^l\,dt$ cancel out from the right-hand side of the last equality.

The formulas for the coefficients $\tilde{a}^{ij}, \tilde{b}^i, \tilde{c}$ in the equation satisfied by v follow by the chain rule. Namely, with

$$u(t, x) = \widetilde{\psi}(t, x)v\big(t, \eta^{-1}(t, x)\big)$$

we find

$$u_i = \widetilde{\psi}_i v + \widetilde{\psi} v_m \eta_i^{-m},\ u_{ij} = \widetilde{\psi}_{ij} v + 2\widetilde{\psi}_i v_m \eta_j^{-m} + \widetilde{\psi} v_{mn} \eta_i^{-m}\eta_j^{-n} + \widetilde{\psi} v_m \eta_{ij}^{-m},$$

so that, for example,

$$\hat{\sigma}^{mk}\hat{\sigma}^{nk}u_{mn} = \widetilde{\psi}\hat{\sigma}^{mk}\hat{\sigma}^{nk}\eta_m^{-i}\eta_n^{-j}v_{ij} + \cdots,$$

from which the expression for $\tilde{a}^{ij}$ follows. Lower-order derivatives will have additional terms coming from the Itô–Ventcel formula.

The details are left as an exercise for the interested reader. □

5.3.4. As was mentioned in the introduction, we call problem (5.3.2), (5.3.3) the forward equation of inverse diffusion. This equation describes the dynamics of $X^{-1}(t, s, x)$ with respect to t for a fixed s. Now we will derive the backward equation of inverse diffusion, describing the dynamics of $X^{-1}(t, s, x)$ with respect to s for fixed t. If the coefficients of system (5.3.1) are non-random, then this equation immediately follows from the AOC formula (5.2.2) and the forward equation of inverse diffusion.

Corollary 5.4 *Suppose that the assumptions of* Sect. 5.2 *are satisfied with* $m = 3$ *and the functions* b^i, σ^{il}, $i = 1, 2, \ldots, \mathrm{d}$, $l = 1, 2, \ldots, \mathrm{d}_1$, *do not depend on* ω.

Denote by $Y = Y(t,s,x)$ a continuous in (t,s,x) version of the solution

$$Y^i(t,s,x) = x^i + \int_{[s,t]} (\sigma^{il}_j \sigma^{jl} - b^i)\big(\tau, Y(t,\tau,x)\big)d\tau$$

$$- \int_{[s,t]} \sigma^{il}\big(\tau, Y(t,\tau,x)\big) * dw^l(\tau),\ s \in [T_0,t],\ i = 1,2,\dots,\mathrm{d}. \tag{5.3.13}$$

Then there exists a set $\Omega' \subset \Omega$ with $\mathbb{P}(\Omega') = 1$ such that, for all $(t,x,\omega) \in [T_0,T] \times \mathbb{R}^{\mathrm{d}} \times \Omega'$,

$$X^{-i}(t,x) = Y^i(t,T_0,x). \tag{5.3.14}$$

□

Proof Let us apply Theorem 5.4 to problem (5.3.2), (5.3.3). Comparing (4.1.1) and (5.3.2), we conclude that $\hat{\sigma} = 0$, so that, in Eq. (5.2.1) for the stochastic characteristic $\mathcal{Y}$, we have

$$\mathcal{B}^i = \sigma^{il}_j \sigma^{jl} - b^i,\ \boldsymbol{\Sigma} = -\sigma,\ \mathcal{W} = W,$$

and, in (5.2.2),

$$\varphi(x) = x^i,\ f = h^l = g^l = 0,\ \gamma(t,s) = 1.$$

It follows that $\mathcal{Y}^i(t,T_0,x) \equiv Y^i(t,T_0,x)$ is $\tilde{\mathcal{F}}^t_{T_0}$-measurable and so

$$X^{-i}(t,x) \equiv X^{-i}(t,T_0,x) = u(t,x) = \mathcal{Y}^i(t,T_0,x) \equiv Y^i(t,T_0,x)$$

for $\mathfrak{l} \times \mathfrak{l}_{\mathrm{d}}$- a.a. t,x. Since both parts of equality (5.3.14) are continuous in t,x ($\mathbb{P}$-a.s.), we conclude that equality (5.3.14) holds for all t,x on the same set of probability one. □

5.3.5. In this paragraph we derive the **backward diffusion equation**. Here, as well as in the previous paragraph, we suppose that the coefficients b^i, σ^{il} for all i,l do not depend on ω, but we do not assume these coefficients are continuous in t.

Given $r < -(d/2+1)$ and $i = 1,2,\dots,\mathrm{d}$ consider the problem

$$-dv(s,x) = L_0 v(s,x)\,ds + B^l_0 v(s,x) * dw^l(s)\,, \tag{5.3.15}$$

$$(s,x,\omega) \in [T_0,T) \times \mathbb{R}^{\mathrm{d}} \times \Omega,$$

$$v(T,x) = x^i,\ \ (x,\omega) \in \mathbb{R}^{\mathrm{d}} \times \Omega. \tag{5.3.16}$$

By the theorem and corollary of Sect. 4.3.2 this problem has a unique r-generalized solution, with a continuous in t, x version, which we consider below using the same notation.

Problem (5.3.15), (5.3.16) will be called the backward diffusion equation, or backward equation of the forward diffusion, as it describes the dynamics of the diffusion process $X = X(t, s, x)$ in the backward time variable s. The precise result is as follows.

Corollary 5.5 *Let* $X = X(T, s, x)$ *be the solution of* (5.3.1) *and let* $v = v(s, x)$ *be an* r*-generalized solution of problem* (5.3.15), (5.3.16)*. Then there exists a set* $\Omega' \subset \Omega$ *with* $\mathbb{P}(\Omega') = 1$ *such that* $X(T, s, x) = v(s, x)$ *for all* $(s, x, \omega) \in [T_0, T] \times \mathbb{R}^{\mathrm{d}} \times \Omega'$. □

The argument is identical to the proof of Corollary 5.4.

To summarize, there are two diffusion processes:

- Direct, denoted by $X = X(t, s, x)$, given by (5.3.1) with s, x fixed, $t \geq s$, and usually coming without the "direct" modifier;
- Inverse, denoted by $Y = Y(t, s, x)$ and given by (5.3.13) (backward equation of the inverse diffusion) with t, x fixed, $s \leq t$.

There are two related stochastic parabolic equations:

- Equation (5.3.2), describing the evolution of the components of $X^{-1}(t, s, x)$ forward in time and known as the **forward equation of the inverse diffusion**,
- Equation (5.3.15), describing the evolution of the components of $X(t, s, x)$ backward in time and known as the backward equation of the direct diffusion or simply **backward diffusion equation**.

In particular,

- If $u = u(t, x)$ solves (5.3.2) and $u(T_0, x) = x^i$, then

$$u(t, x) = X^{-i}(t, T_0, x) = Y^i(t, T_0 x),$$

  where X^{-i} is the i-the coordinate of the mapping that is inverse to $x \mapsto X(t, T_0, x)$;
- If $v = v(s, x)$ solves (5.3.15) and $v(T, x) = x^i$, then

$$v(s, x) = X^i(T, s, x).$$

Note also that if we start with a diffusion process X defined by (5.3.1) and want to make sense out of all the other three diffusion equations related to X, then it is necessary to have the coefficients in (5.3.1) both forward and backward predictable, essentially forcing the assumption that the coefficients are non-random.

5.3.6. Now we are ready to settle the questions related to first integrals and Liouville equations.

Definition 5.3 A mapping $\xi : [T_0, T] \times \mathbb{R}^d \times \Omega \to \mathbb{R}^1$ is called a **first integral** of system (5.3.1) on $[T_0, T]$ if, for every $t \in [T_0, T]$, this mapping is $\mathscr{B}(\mathbb{R}^d) \otimes \mathscr{F}$-measurable and, for all $t, \tau \in [T_0, T]$ such that $t + \tau \leq T$,

$$\xi\big(t, X(t, T_0, x)\big) = \xi(t + \tau, X(t + \tau, T_0, x)\big) \ (\mathbb{P}\text{- a.s.}).$$

A first integral $\xi = \xi(t, x)$ of system (5.3.1) will be called **direct** if it is predictable in t for every x, and **inverse** if it is backward predictable in t for every x. □

The following result is an immediate consequence of the AOC formula (5.2.2), Proposition 5.1, and Corollary 4.4.

Proposition 5.2 *If*

$$b^i(\cdot, \cdot, \omega) \in \mathbb{C}_b^{0,3}([T_0, T] \times \mathbb{R}^d), \ \sigma^{il}(\cdot, \cdot, \omega) \in \mathbb{C}_b^{0,4}([T_0, T] \times \mathbb{R}^d)$$

for all $\omega \in \Omega$, $i = 1, 2, \dots, \mathrm{d}$, *and* $l = 1, 2, \dots, \mathrm{d}_1$, *then, for every function* $\varphi \in \mathbb{C}_b^2(\mathbb{R}^d)$, *the function* $\xi(t, x) = \varphi(X^{-1}(t, x, T_0))$ *is a direct first integral of* (5.3.1) *on* $[T_0, T]$. *This function is the unique classical solution of*

$$du(t, x) = (B_0^l B_0^l - L_0)u(t, x)\, dt - B_0^l u(t, x)\, dw^l(t),$$

$$(t, x, \omega) \in (T_0, T] \times \mathbb{R}^d \times \Omega,$$

$$u(T_0, x) = \varphi(x), \ x \in \mathbb{R}^d.$$ □

Note that, by construction, we have $\xi\big(t, X(t, T_0, x)\big) = \varphi(x)$.

5.3.7. The following result concerning inverse first integrals is an immediate consequence of AOC formula (5.2.3) and Corollary 4.4.

Proposition 5.3 *Suppose that the coefficients of system* (5.3.1) *do not depend on* ω *and, for all* $t \in [T_0, T]$, $i = 1, 2, \dots, \mathrm{d}$ *and* $l = 1, 2, \dots, \mathrm{d}_1$, $b^i(t, \cdot)$ *and* $\sigma^{il}(t, \cdot)$ *belong to* $\mathbb{C}_b^3(\mathbb{R}^d)$. *Then, for every* $\varphi \in \mathbb{C}_b^2(\mathbb{R}^d)$, *the function* $\xi(s, x) = \varphi\big(X(T, s, x)\big)$ *is the inverse first integral of system* (5.3.1) *on* $[T_0, T]$. *This function is the unique classical solution of*

$$-dv(s, x) = L_0 v(s, x)\, dt + B_0^l v(s, x) * dw^l(s), \tag{5.3.17}$$

$$(s, x, \omega) \in [T_0, T) \times \mathbb{R}^d \times \Omega,$$

$$v(T, x) = \varphi(x), \ x \in \mathbb{R}^d. \tag{5.3.18}$$

□

This time, we have $\xi\big(t, X(t, T_0, x)\big) = \varphi\big(X(T, T_0, x)\big)$, which is a consequence of the uniqueness of solution of (5.3.1).

Remark 5.5 As was mentioned in the introduction to this chapter, the **backward Liouville equation** (5.3.17), (5.3.18) is a generalization of the backward Kolmogorov equation. The latter can be obtained from the former by taking expectation on both sides of (5.3.17).

5.4 Representation of Measure-Valued Solutions

5.4.1. Consider the problem

$$du(t,x,\omega) = L^* u(t,x,\omega)\,dt + B^{l*}u(t,x,\omega)\,dw^l(t), \tag{5.4.1}$$

$$(t,x,\omega) \in (0,T] \times \mathbb{R}^{\mathrm{d}} \times \Omega,$$

$$u(0,x,\omega) = \varphi(x,\omega), \quad (x,\omega) \in \mathbb{R}^{\mathrm{d}} \times \Omega, \tag{5.4.2}$$

where

$$L^*u = L^*(t,x,\omega)u := \big(a^{ij}(t,x,\omega)u\big)_{ij} - \big(b^i(t,x,\omega)u\big)_i + c(t,x,\omega)u,$$

$$B^{l*}u = B^{l*}(t,x,\omega)u := -\big(\sigma^{il}(t,x,\omega)u\big)_i + h^l(t,x,\omega)u.$$

The operators L^* and B^{l*} are sometimes called formal adjoints of L and B^l because $(Lf,g)_0 = (f, L^*g)$ and $(B^l f, g) = (f, B^{l*}g)$ for all $f, g \in \mathbb{C}_0^\infty(\mathbb{R}^{\mathrm{d}})$, as long as the functions a^{ij}, b^i, and σ^{il} are sufficiently smooth in x.

Let $\mathfrak{F}_0$ be a sub-σ-algebra of $\mathscr{F}_0$, completed with respect to the measure $\mathbb{P}$, and define $\mathfrak{F}_t := \mathfrak{F}_0 \vee \tilde{\mathscr{F}}_t^0$ for every $t \in [0,T]$. Write $\mathbb{F}(\mathfrak{F}) := (\Omega, \mathscr{F}, \{\mathfrak{F}_t\}_{t\in[0,T]}, \mathbb{P})$ and denote by $\mathscr{P}(\mathfrak{F})$ the σ-algebra of predictable sets on $\mathbb{F}(\mathfrak{F})$.

By construction, $W = (w^1, \ldots, w^{\mathrm{d}_1})$ is a standard Wiener process on $\mathbb{F}(\mathfrak{F})$; recall that $\tilde{\mathscr{F}}_t^0$ is the filtration generated by W.

Definition 5.4 A function $u : [0,T] \times \Omega \to \mathbb{L}_2$ will be called measure-valued if, for every $t \in [0,T]$, it belongs, $\mathbb{P}$- a.s., to the cone of non-negative functions from $\mathbb{L}_1$. □

Given a sub-σ-algebra $\mathscr{G} \subset \mathscr{F}$ and an $\mathbb{R}^{\mathrm{d}}$-valued random variable ξ, recall that the **regular conditional probability distribution** of ξ with respect to $\mathscr{G}$ is a function $P_{\mathscr{G}}^{\xi} : \mathscr{B}(\mathbb{R}^{\mathrm{d}}) \times \Omega \to \mathbb{R}^1$ such that $P_{\mathscr{G}}^{\xi}(\cdot,\omega)$ is a probability measure on $(\mathbb{R}^1, \mathscr{B}(\mathbb{R}^{\mathrm{d}})$ for $\mathbb{P}$-a.a ω, and, for every $\Gamma \in \mathscr{B}(\mathbb{R}^{\mathrm{d}})$, the random variable $P_{\mathscr{G}}^{\xi}(\Gamma,\cdot)$ is $\mathscr{G}$-measurable and $P_{\mathscr{G}}^{\xi}(\Gamma,\omega) = \mathbb{P}[\xi \in \Gamma|\mathscr{G}]$, $\mathbb{P}$-a.s.

Throughout this section it is assumed that

$$P_{\mathfrak{F}_0}^{x_0}(\Gamma) = \int_\Gamma \varphi(x)dx, \quad \Gamma \in \mathscr{B}(\mathbb{R}^{\mathrm{d}})$$

and $\varphi \in \mathbb{L}_2(\Omega, \mathfrak{F}_0; \mathbb{H}^1)$.

As far as the coefficients in (5.4.1), the assumptions are as follows:

- all the coefficients are $\mathscr{P}(\mathfrak{F})$-measurable for every $x \in \mathbb{R}^{\mathrm{d}}$;
- for all $(t, \omega) \in [0, T] \times \Omega$, $i, j = 1, 2, \dots \mathrm{d}$, and $l = 1, 2, \dots, \mathrm{d}_1$, a^{ij} and σ^{il} are three times differentiable in x, b^i and h^l are two times differentiable in x, and c is differentiable in x. All the coefficients and their derivatives are bounded, uniformly in (t, x, ω), by the number K.

Remark 5.6 Under the above assumptions, $L^* u = a^{ij} u_{ij} + \tilde{b}^i u_i + \tilde{c} u$ and $B^{l*} u = -\sigma^{il} u_i + \tilde{h}^l$, where $\tilde{b}^i = 2a^{ij}_j - b^i$, $\tilde{c} = c + a^{ij}_{ij} - b^i_i$, and $\tilde{h}^l = h^l - \sigma^{il}_i$. Then, by Theorem 4.5 problem (5.4.1), (5.4.2) has a unique solution $u \in \mathbb{L}_2([0, T], \mathscr{P}(\mathfrak{F}); \mathbb{H}^1) \cap \mathbb{L}_2\big(\Omega; \mathbb{C}([0, T]; \mathbb{L}_2)\big)$.

5.4.2. The following is the main result of this section.

Theorem 5.6 *A generalized solution u of problem* (5.4.1), (5.4.2) *is measure-valued and, for every $\psi \in \mathbb{L}_\infty$ and all $t \in [0, T]$,*

$$\int_{\mathbb{R}^{\mathrm{d}}} \psi(x)\, u(t, x)\, dx = \mathbb{E}\Big[\psi\big(\mathcal{X}(t)\big)\rho(t)\big|\mathfrak{F}_t\Big], \quad \mathbb{P}\text{-a.s.}, \tag{5.4.3}$$

where $\mathcal{X} = \mathcal{X}(t)$ is the solution of

$$\begin{aligned} \mathcal{X}^i(t) = x^i_0 + & \int_{[0,t]} \mathcal{B}^i\big(s, \mathcal{X}(s)\big)\, ds \\ & + \int_{[0,t]} \boldsymbol{\Sigma}^{i\ell}\big(s, \mathcal{X}(s)\big)\, d\mathcal{W}^\ell(s), \quad i = 1, 2, \dots, \mathrm{d},\ t \in [0, T], \end{aligned} \tag{5.4.4}$$

with functions $\mathcal{B}^l$, $\boldsymbol{\Sigma}^{i\ell}$ and Wiener process $\mathcal{W}$ defined in Sect. 5.1.2, and

$$\begin{aligned} \rho(t) := \exp\Bigg(& \int_{[0,t]} c\big(s, \mathcal{X}(s)\big)\, ds + \int_{[0,t]} h^l\big(s, \mathcal{X}(s)\big)\, dw^l(s) \\ & - \frac{1}{2} \int_{[0,t]} (h^l h^l)\big(s, \mathcal{X}(s)\big)\, ds \Bigg). \end{aligned}$$

□

Remark 5.7 Formula (5.4.3) is an AOC formula for the integral functional of the solution of system (5.4.1), (5.4.2), with $\mathcal{X}$ playing the role of the **stochastic characteristic.**

The special form of the operators and absence of the external forces in (5.4.1) are not essential, although these assumptions do simplify the presentation and make the results immediately applicable to the problem of optimal non-linear filtering in Sect. 6.3.

Corollary 5.6 *There exists a $\mathscr{P}(\mathfrak{F})$-measurable function $u \in \mathbb{L}_1([0,T] \times \Omega; \mathbb{C}_w\mathbb{L}_1)$ which is a generalized solution of problem (5.4.1), (5.4.2) and, for all $(t,\omega) \in [0,T] \times \Omega$, belongs to the cone of non-negative functions in $\mathbb{L}_1$.* □

5.4.3. To prove the theorem, it is enough to verify (5.4.3) for $\psi \in \mathbb{C}_0^\infty(\mathbb{R}^d)$. After that, the case of general ψ follows by the dominated convergence theorem, and then, by considering non-negative ψ, we conclude that u is indeed measure-valued; in particular, taking $\psi \equiv 1$ confirms that $u(t,\cdot) \in \mathbb{L}_1$.

We will need the following result, which is also of independent interest.

Theorem 5.7 *For every $f \in \mathbb{C}_b^{1,2}([0,T] \times \mathbb{R}^d)$, the conditional expectation $\mathbb{E}\big[f\big(t, \mathcal{X}(t)\big)\rho(t)\big|\mathfrak{F}_t\big]$ has a version, denoted by $\Phi_t[f]$, such that*

$$d\Phi_t[f] = \Phi_t\left[Lf + \frac{\partial f}{\partial t}\right]dt + \Phi_t[B^l f]\,dw^l(t),\ t \in [0,T]. \tag{5.4.5}$$

□

Proof We begin with the following auxiliary statement.

Lemma 5.2 *Let $g \in \mathbb{L}_2([0,T], \mathscr{P}; \mathbb{R}^{\mathrm{d}_0})$, where $\mathrm{d}_0 \in \mathbb{N}$, and let $\hat{w}$ be a standard Wiener process in $\mathbb{R}^{\mathrm{d}_0}$, relative to the family $\{\mathscr{F}_t\}$. Assume that $(w^1,\dots,w^{\mathrm{d}_1})$ and $\hat{w}$ are independent. Then*

$$\mathbb{E}\left[\int_{[0,t]} g^k(s)d\hat{w}^k(s)\Big|\mathfrak{F}_t\right] = 0 \ \ (\mathbb{P}\text{- a.s.}).$$ □

Proof First, assume that

$$g^k(s) = g_0^k 1_{\{0\}}(s) + \sum_{i=0}^{n-1} g_i^k 1_{\{(t_i,t_{i+1}]\}}(s),$$

where $0 = t_0 < t_1 < \ldots < t_n = t$ and each g_i^k is an $\mathscr{F}_{t_i}$-measurable random variable. Then

$$\mathbb{E}\left[\int_{[0,t]} g^k(s)d\hat{w}^k(s)\Big|\mathfrak{F}_t\right] = \sum_{i=0}^{n-1}\mathbb{E}\big[g_i^k\big(\hat{w}^k(t_{i+1}) - \hat{w}^k(t_i)\big)\big|\mathfrak{F}_t\big].$$

Let ζ be an $\mathfrak{F}_t$-measurable random variable. We need to show that

$$\mathbb{E}\big(\zeta g_i^k\big(\hat{w}^k(t_{i+1}) - \hat{w}^k(t_i)\big)\big) = 0$$

for all i, k, which is true because g_i^k is $\mathscr{F}_{t_i}$-measurable and $\hat{w}^k(t_{i+1}) - \hat{w}^k(t_i)$ is independent of both $\mathscr{F}_{t_i}$ and ζ. As a result,

$$\mathbb{E}\left[\int\limits_{[0,t]} g^k(s)d\hat{w}^k(s)\Big|\mathfrak{F}_t\right] = \sum_{i=0}^{n-1}\mathbb{E}\big[g_i^k\big(\hat{w}^k(t_{i+1}) - \hat{w}^k(t_i)\big)\big|\mathfrak{F}_t\big] = 0.$$

The general case follows after passing to the limit. □

To prove equality (5.4.5), apply Itô's formula to the product $f\big(t, \mathcal{X}(t)\big)\rho(t)$ and then take the conditional expectation, with respect to $\mathfrak{F}_t$, of both sides of the resulting integral equality. Then use the above lemma, together with Theorem 1.15 and the equality

$$\mathbb{E}\big[Lf\big(s, \mathcal{X}(s)\big)\rho(s)\big|\mathfrak{F}_t\big] = \mathbb{E}[Lf\big(s, \mathcal{X}(s)\big)\rho(s)|\mathfrak{F}_s] \quad (\mathbb{P}\text{- a.s.})$$

(by independence of the increments of w^l). Equality (5.4.5) follows.

5.4.4. Next, we need to resolve a certain technical issue. By construction, the mapping

$$h \mapsto \Phi_t[h] = \mathbb{E}\Big[h\big(\mathcal{X}(t)\big)\rho(t)\big|\mathfrak{F}_t\Big]$$

is well-defined for every $t \in [0, T]$, $h \in \mathbb{C}_b^0(\mathbb{R}^d)$ and is linear: if $h_1, h_2 \in \mathbb{C}_b^0$ and $\alpha, \beta \in \mathbb{R}^1$, then, with probability one,

$$\Phi_t[\alpha h_1 + \beta h_2] = \alpha\Phi_i[h_1] + \beta\Phi_t[h_2]. \tag{5.4.6}$$

Under additional regularity of h, we have representation (5.4.5), which, in particular, implies the continuity of $\Phi_t[h]$ in time and consequently the existence of a single probability-one set on which (5.4.6) holds for all $t \in [0, T]$.

However, the exceptional subset of Ω, on which (5.4.6) fails, can depend on h_1, h_2, α, and β, whereas our proof requires a version of $\Phi_t[\cdot]$ which is a linear functional on $\mathbb{C}_b^0(\mathbb{R}^d)$ and, for sufficiently regular f, possesses stochastic differential (5.4.5). In particular, we need (5.4.6) to hold on the same set of full probability for all h_1, h_2, α, β. This technical issue, and the need to prove that (a version of) $\Phi_t[\psi]$ can indeed be written in the form $\int\limits_{\mathbb{R}^d} v(t,x)\psi(x)dx$ for some function v, are the reasons why we cannot simply claim the conclusion of Theorem 5.6 from Theorem 5.7.

This point is worth emphasizing: the only problem-specific computation in the proof is the Itô formula in the derivation of (5.4.5), which is effectively left to the reader as an exercise. The rest of the rather long argument resolves the above measurability and existence issues using tools from functional analysis and various existence/uniqueness/regularity results for LSESs from the previous chapters of the book.

We begin with the construction of the required version of Φ using the following corollary to the Hahn–Banach theorem; cf. [26, Section II.5].

Proposition 5.4 *Let $\mathbb{X}$ be a Banach space with topological dual $\mathbb{X}^*$. Let $\{x_n\}$ be a sequence of elements of $\mathbb{X}$ and $\{c_n\}$ be a sequence of real numbers. Then the following two conditions are equivalent:*

1. *There exist an $x^* \in \mathbb{X}^*$ and $N > 0$ such that $\|x^*\|_{\mathbb{X}^*} \le N$ and $x^* x_n = c_n$ for every $n \in \mathbb{N}$;*
2. *For every finite set of real numbers $\{\alpha_i\}$,*

$$\left|\sum_i \alpha_i c_i\right| \le N \left\|\sum_i \alpha_i x_i\right\|_{\mathbb{X}}. \qquad \square$$

Using Proposition 5.4, we prove

Theorem 5.8 *There exists a set $\Omega' \subset \Omega$ with $\mathbb{P}(\Omega') = 1$ such that, for every $(t, \omega) \in [0, T] \times \Omega'$, there exists a continuous linear functional $\tilde{\Phi}_t[\cdot](\omega)$ on $\mathbb{C}^0_b(\mathbb{R}^{\mathrm{d}})$ and*

(a) *For every $f \in \mathbb{C}^0_b([0, T] \times \mathbb{R}^{\mathrm{d}})$, $\tilde{\Phi}_t[\psi]$ is a continuous (in t), $\mathscr{P}(\mathfrak{F})$-measurable version of $\Phi_t[\psi]$, and*

$$|\tilde{\Phi}_t[f](\omega)| \le \left(\sup_{x \in \mathbb{R}^{\mathrm{d}}} |f(t, x)|\right) \tilde{\Phi}_t[1](\omega) < \infty, \quad (t, \omega) \in [0, T] \times \Omega'. \tag{5.4.7}$$

(b) *For every $f \in \mathbb{C}^0_b([0, T] \times \mathbb{R}^{\mathrm{d}})$, $\tilde{\Phi}_t[\psi]$ satisfies* (5.4.5). $\square$

Proof Take a countable set $\{\eta_i\}$, $i \in \mathbb{N}$, dense in $\mathbb{C}^0_b([0, T] \times \mathbb{R}^{\mathrm{d}})$ in the sup norm and with $\eta_i \in \mathbb{C}^0_b([0, T]) \times \mathbb{R}^{\mathrm{d}}$. Given a $t \in [0, T]$ and a finite collection of rational numbers r_i, the definition of Φ implies

$$\left|\sum_i r_i \Phi_t[\eta_i]\right| \le \max_{x \in \mathbb{R}^{\mathrm{d}}} \left|\sum_i r_i \eta_i(t, x)\right| \Phi_t[1], \quad (\mathbb{P}\text{- a.s.}). \tag{5.4.8}$$

By (5.4.5), every $\Phi_t[\eta_i]$ is continuous in t, which implies the existence of an $\Omega' \subset \Omega$ with $\mathbb{P}(\Omega') = 1$ such that (5.4.8) holds for all $\omega \in \Omega$, and all finite collections of r_i, η_i; for $\omega \notin \Omega'$, re-define Φ by setting $\Phi_t[\eta_i] = 0$, $\Phi_t[1] = 0$, $t \in [0, T]$, $i \in \mathbb{N}$. Proposition 5.4 then implies the existence of a family of linear bounded functionals $\tilde{\Phi}_t$, $t \in [0, T]$, on $\mathbb{C}^0_b([0, T])$ such that

$$\tilde{\Phi}_t[\eta_i] = \Phi_t[\eta_i], \tag{5.4.9}$$

for all $(t, \omega) \in [0, T] \times \Omega'$ and all $i \in \mathbb{N}$, and

$$\left|\tilde{\Phi}_t[f]\right| \leq \max_{x \in \mathbb{R}^{\mathrm{d}}} |f(t, x)| \; \Phi_t[1] \tag{5.4.10}$$

for all $(t, \omega) \in [0, T] \times \Omega'$ and $f \in \mathbb{C}_b^0([0, T] \times \mathbb{R}^{\mathrm{d}})$.

Now, given $f \in \mathbb{C}_b^0([0, T] \times \mathbb{R}^{\mathrm{d}})$, let $\{\eta_{i_n}\}$ be a subsequence of $\{\eta_i\}$ such that

$$\lim_{n \to \infty} \sup_{t,x} |f(t, x) - \eta_{i_n}(t, x)| = 0.$$

By (5.4.9) and continuity of $\tilde{\Phi}_t$,

$$\tilde{\Phi}_t[\psi] = \lim_{n \to \infty} \tilde{\Phi}_t[\eta_{i_n}] = \lim_{n \to \infty} \Phi_t[\eta_j] = \Phi_t[f]; \tag{5.4.11}$$

the last equality in (5.4.11) holds on a probability one subset of Ω' that might depend on f, and this is the main reason to consider $\tilde{\Phi}$ instead of Φ.

Then, together with (5.4.9) and (5.4.10), representation (5.4.5) of $\Phi_t[\eta_i]$ implies continuity of the function $t \mapsto \tilde{\Phi}_t[f]$ for every $f \in \mathbb{C}_b^0([0, T] \times \mathbb{R}^{\mathrm{d}})$, as well as the remaining statements of the theorem. □

Remark 5.8 Analysis of the proofs of Theorems 5.7 and 5.8 shows that both theorems hold if the functions $\mathcal{B}^i$ and $\boldsymbol{\Sigma}^{i\ell}$ are uniformly Lipschitz in x, which is far less restrictive than the conditions imposed in Theorem 5.6. □

5.4.5. Next, we will prove Theorem 5.6 under additional assumptions. For some portion of the proof, we will work in the Hilbert scale $\{\mathbb{H}^\alpha, \alpha \in \mathbb{R}^1\}$; cf. Sect. 3.5.3. In particular, $(\cdot, \cdot)_\alpha$ and $\|\cdot\|_\alpha$ are the inner product and the norm in $\mathbb{H}^\alpha$, and $[\cdot, \cdot]_\alpha$ is the canonical bilinear functional of the normal triple $(\mathbb{H}^{\alpha+1}, \mathbb{H}^\alpha, \mathbb{H}^{\alpha-1})$.

Fix an integer $\mathfrak{n}$ such that $\mathfrak{n} - 1$ that is bigger than $\mathrm{d}/2$. The number $\mathfrak{n}$ plays an auxiliary role: it is necessary to have $\mathbb{H}^{\mathfrak{n}-1}$ embedded in the space of continuous functions on $\mathbb{R}^{\mathrm{d}}$; the reason for writing $\mathfrak{n} - 1$ now is that, in most of the subsequent computations, the number of interest will be $\mathfrak{n}$.

Lemma 5.3 *In addition to the assumptions of Theorem 5.6 let the following conditions be satisfied.*

(i) *For all* $(t, \omega) \in [0, T] \times \Omega$, $i = 1, 2, \ldots, \mathrm{d}$, $\ell = 1, 2, \ldots, \mathrm{d} + \mathrm{d}_1$, *and* $l = 1, 2, \ldots, \mathrm{d}_1$, *the functions* $b^i(t, \cdot, \omega)$, $c(t, \cdot, \omega)$ $\boldsymbol{\Sigma}^{i\ell}(t, \cdot, \omega)$, $h^l(t, \cdot, \omega)$ *belong to* $\mathbb{C}_b^{2\mathfrak{n}+1}(\mathbb{R}^{\mathrm{d}})$, *and* $\psi \in \mathbb{C}_0^\infty(\mathbb{R}^{\mathrm{d}})$.
(ii) *Equation* (5.4.1) *is super-parabolic.*

Then formula (5.4.3) *holds.* □

Proof We begin with the following result.

Proposition 5.5 *Under the assumptions of the lemma there exists a set* $\Omega' \subset \Omega$ *with* $\mathbb{P}(\Omega') = 1$ *and a function* $\tilde{v} : [0, T] \times \Omega \to \mathbb{H}^{n-1}$ *belonging to* $\mathbb{L}_2([0, T] \times \Omega, \mathscr{P}(\mathfrak{F}); \mathbb{H}^n)$ *such that, for every* $\psi \in \mathbb{C}_0^\infty(\mathbb{R}^d)$ *and all* $(t, \omega) \in [0, T] \times \Omega'$,

$$\tilde{\Phi}_t[\psi] = \left(\psi, \tilde{v}\right)_{n-1}(t). \tag{5.4.12}$$

□

Proof From Theorem 5.8 and Theorem 3.9 (Sobolev embedding) we conclude that there exist a number $N \in \mathbb{R}_+$ and a set Ω' with $\mathbb{P}(\Omega') = 1$ such that, for all $(t, \omega) \in [0, T] \times \Omega'$ and $\psi \in \mathbb{C}_0^\infty(\mathbb{R}^d)$,

$$\left|\tilde{\Phi}_t[\psi]\right| \le \sup_{x \in \mathbb{R}^d} \left|\psi(x)\right| \Phi_t[1] \le N\|\psi\|_m \, \Phi_t[1], \; m > d/2. \tag{5.4.13}$$

Using (5.4.13) with $m = n - 1$ and the Riesz representation theorem in $\mathbb{H}^{n-1}$, we get the existence of a $\tilde{v} = \tilde{v}(t, x, \omega)$ such that $\tilde{v} \in \mathbb{H}^{n-1}$ and (5.4.12) holds for all $(t, \omega) \in [0, T] \times \Omega'$. The $\mathscr{P}(\mathfrak{F})$-measurability of $\tilde{v}$ follows by Theorem 5.8 and the Pettis theorem in Sect. 1.3.9.

To complete the proof, observe that

$$\begin{aligned}\mathbb{E} \int_{[0,T]} \|\tilde{v}(t)\|_{n-1}^2 \, dt &= \mathbb{E} \int_{[0,T]} \sup_{\psi \in \mathbb{C}_0^\infty(\mathbb{R}^d)} \frac{\left|\tilde{\Phi}_t[\psi]\right|^2}{\|\psi\|_{n-1}^2} \, dt \\ &\le N \int_{[0,T]} \mathbb{E}|\Phi_t[1]|^2 \, dt < \infty,\end{aligned}$$

because

$$\mathbb{E}\left|\Phi_t[1]\right|^2 \le \mathbb{E}\rho^2(t) \le e^{(2K+K^2)T}.$$

□

To prove the lemma, it remains to show that

$$\tilde{v} \in \mathbb{L}_2\Big([0, T] \times \Omega; \mathbb{H}^{2n}\Big) \bigcap \mathbb{L}_2\Big(\Omega; \mathbb{C}([0, T]; \mathbb{H}^{2n-1})\Big), \tag{5.4.14}$$

and $\Lambda^{2n-2}\tilde{v} = u$, where $\Lambda = (\mathbb{I} - \boldsymbol{\Delta})^{1/2}$, $\boldsymbol{\Delta}$ is the Laplace operator (cf. Sect. 3.5.3), and u is the solution of (5.4.1), (5.4.2). Then (5.4.12) will imply (5.4.3), because

$$\left(\psi, \tilde{v}(t)\right)_{n-1} = \left(\Lambda^{n-1}\psi, \Lambda^{n-1}\tilde{v}(t)\right)_0 = \left(\psi, \Lambda^{2n-2}\tilde{v}(t)\right)_0.$$

Here is an outline of the argument. Combining Proposition 5.5 and Theorem 5.8, we conclude that, for all $t \in [0, T]$ and $\psi \in \mathbb{C}_0^\infty(\mathbb{R}^d)$,

$$\begin{aligned}\big(\tilde{v}(t), \psi\big)_{n-1} = \big(\tilde{v}(0), \psi\big)_{n-1} + \int_{[0,t]} \big(\tilde{v}, L\psi\big)_{n-1}(s)\, ds \\ + \int_{[0,t]} \big(\tilde{v}, B^l\psi\big)_{n-1}(s)\, dw^l(s)\end{aligned} \tag{5.4.15}$$

on the same set of probability one. If we had $\tilde{v} \in \mathbb{L}_2([0, T] \times \Omega; \mathbb{H}^n)$, then we would argue that $\tilde{v}$ is a generalized solution of the equation

$$d\tilde{v} = \boldsymbol{A}\tilde{v}\, dt + \boldsymbol{B}\tilde{v}\, dW$$

in the normal triple $(\mathbb{H}^n, \mathbb{H}^{n-1}, \mathbb{H}^{n-2})$, with suitable operators $\boldsymbol{A}$, $\boldsymbol{B}$. Also, by (5.4.12),

$$(\tilde{v}(0), \psi)_{n-1} = \tilde{\Phi}_0[\psi] = (\varphi, \psi)_0,$$

which implies $\tilde{v}(0) = \Lambda^{-2(n-1)}\varphi \in \mathbb{H}^{2n-1}$. Then Theorem 3.2 would give the required regularity of $\tilde{v}$, and equality $\Lambda^{2n}\tilde{v} = u$ follows by uniqueness of solution. A (minor) technical complication, namely, that we only have $\tilde{v} \in \mathbb{L}_2([0, T] \times \Omega; \mathbb{H}^{n-1})$, is resolved by switching to $v = \Lambda^{-2}\tilde{v}$.

The details are as follows. If $v = \Lambda^{-2}\tilde{v}$, then $v \in \mathbb{L}_2([0, T] \times \Omega; \mathbb{H}^{n+1})$ and

$$(v, \psi)_n = (\Lambda v, \Lambda\psi)_{n-1} = (\Lambda^2 v, \psi)_{n-1} = (\tilde{v}, \psi)_{n-1},$$

so that, instead of (5.4.16), we now need

$$v \in \mathbb{L}_2\Big([0, T] \times \Omega; \mathbb{H}^{2n+2}\Big) \bigcap \mathbb{L}_2\Big(\Omega; \mathbb{C}\big([0, T]; \mathbb{H}^{2n+1}\big)\Big), \tag{5.4.16}$$

and $u = \Lambda^{2n} v$.

Define the operators $\boldsymbol{A}$ and $\boldsymbol{B}^l$ by

$$[\boldsymbol{A}\eta, \psi]_n = -\big(\eta_j, a^{ij}\psi_i\big)_n + \big(\eta, (b^i - a^{ij}_j)\psi_i\big)_n + \big(\eta, c\psi\big)_n, \tag{5.4.17}$$

and

$$\big(\boldsymbol{B}^l\eta, \psi\big)_n = -(\eta_i, \sigma^{il}\psi)_n + \big(\eta, (h^l - \sigma^{il}_i)\psi\big)_n, \tag{5.4.18}$$

for every $\eta, \psi \in \mathbb{H}^{n+1}$. Direct computations, very similar to Sects. 4.2.2 and 4.2.3, show that the operators $\boldsymbol{A} = \boldsymbol{A}(t, \omega)$ and $\boldsymbol{B}^l = \boldsymbol{B}^l(t, \omega)$ are well-defined for all

$(t, \omega) \in [0, T] \times \Omega$ and have the following properties:

- For $m = \mathfrak{n}$ and $m = 2\mathfrak{n}$, $\boldsymbol{A} : \mathbb{H}^{m+1} \to \mathbb{H}^{m-1}$ and $\boldsymbol{B}^l : \mathbb{H}^{m+1} \to \mathbb{H}^m$ are uniformly bounded;
- If $\psi \in \mathbb{C}_0^\infty(\mathbb{R}^{\mathrm{d}})$, then

$$[\boldsymbol{A}\eta, \psi]_{\mathfrak{n}} = (\eta, L\psi)_{\mathfrak{n}}, \ \left(\boldsymbol{B}^l\eta, \psi\right)_{\mathfrak{n}} = (\eta, B^l\psi)_{\mathfrak{n}};$$

- With $\boldsymbol{B} = (\boldsymbol{B}^1, \ldots, \boldsymbol{B}^{\mathrm{d}_1})$, there exists a $\delta > 0$ and an $N \in \mathbb{R}^1$ such that, for all $\eta \in \mathbb{H}^{\mathfrak{n}+1}$ and $(t, \omega) \in [0, T] \times \Omega$,

$$2\,[\boldsymbol{A}\eta, \eta]_m + \|\boldsymbol{B}\eta\|^2 \le -\delta\,\|\eta\|_{m+1}^2 + N\|\eta\|_m^2, \ m = \mathfrak{n}, 2\mathfrak{n}.$$

As a result, (5.4.15) becomes the definition of the generalized solution of

$$v(t) = v(0) + \int_{[0,t]} \boldsymbol{A}u(s)\,ds + \int_{[0,t]} \boldsymbol{B}v(s)\,dW(s), \ t \in [0, T], \tag{5.4.19}$$

in the normal triple $(\mathbb{H}^{\mathfrak{n}+1}, \mathbb{H}^{\mathfrak{n}}, \mathbb{H}^{\mathfrak{n}-1})$. We now apply Theorem 3.2 to (5.4.19) and conclude that (5.4.19) has a unique generalized solution, and the solution satisfies (5.4.16). On the other hand, equality (5.4.15), when written in terms of the inner product $(\cdot, \cdot)_0$, implies that $\Lambda^{2\mathfrak{n}}v$ is a generalized solution of (5.4.1), (5.4.2), which, by uniqueness, must coincide with u.

The completes the proof of Lemma 5.3. □

5.4.6. Now we can finish the proof of Theorem 5.6. As was mentioned earlier, it is enough to consider $\psi \in \mathbb{C}_0^\infty(\mathbb{R}^{\mathrm{d}})$.

By Lemma 5.3, we have the result for a super-parabolic equation with smooth coefficients. Accordingly, we construct a suitable family of super-parabolic equations with smooth coefficients approximating (5.4.1), (5.4.2) and then pass to the limit.

Let $\tilde{w}(t)$ be a d-dimensional Wiener process on $\mathbb{F}(\mathfrak{F})$ independent of $\mathcal{W}$ and let $\tilde{\mathcal{W}}$ be a $(\mathrm{d} + \mathrm{d}_0 + \mathrm{d}_1)$-dimensional Wiener process whose first d components are $\tilde{w}$, followed by $\mathrm{d}_0 + \mathrm{d}_1$ components of $\mathcal{W}$.

In this paragraph we use the notation $\xi_{(\varepsilon)}(s, x, \omega) := (\mathcal{T}_\varepsilon \xi(s, \cdot, \omega))(x)$, first introduced in Sect. 4.2.6, where $\mathcal{T}_\varepsilon$ is the Sobolev averaging operator, first introduced in Sect. 1.5.11.

Given $\varepsilon > 0$ denote by ${}_\varepsilon\boldsymbol{\Sigma}$ the matrix $({}_\varepsilon\boldsymbol{\Sigma}^{i\ell})$, $i = 1, 2, \ldots, \mathrm{d}$, $\ell = 1, 2, \ldots, \mathrm{d} + \mathrm{d}_0 + \mathrm{d}_1)$ obtained by putting together, left to right, the $\mathrm{d} \times \mathrm{d}$ identity matrix times $\sqrt{2\varepsilon}$, the $\hat{\sigma}_{(\varepsilon)}$ matrix, and the $\sigma_{(\varepsilon)}$ matrix:

$${}_\varepsilon\boldsymbol{\Sigma}^{i\ell} = \begin{cases} \sqrt{2\varepsilon}\ 1_{\{i=\ell\}}, & \text{if } \ell = 1, 2, \ldots, \mathrm{d}, \\ \hat{\sigma}_{(\varepsilon)}^{ik}, & \text{if } \ell = \mathrm{d} + k,\ k = 1, \ldots, \mathrm{d}_0, \\ \sigma_{(\varepsilon)}^{il}, & \text{if } \ell = \mathrm{d} + \mathrm{d}_0 + l,\ l = 1, \ldots, \mathrm{d}_1. \end{cases}$$

Consider the diffusion process ${}_{\varepsilon}\mathcal{X}$ defined by

$$
{}_{\varepsilon}\mathcal{X}^i(t) = x_0^i + \int\limits_{[0,t]} \mathcal{B}^i_{(\varepsilon)}\big(s, {}_{\varepsilon}\mathcal{X}(s)\big)\, ds + \int\limits_{[0,t]} {}_{\varepsilon}\boldsymbol{\Sigma}^{i\ell}\big(s, {}_{\varepsilon}\mathcal{X}(s)\big)\, d\tilde{\mathcal{W}}^{\ell}(s), \tag{5.4.20}
$$

$$
i = 1, 2, \ldots, \mathrm{d},\ t \in [0, T].
$$

By Lemma 1.2, $\mathcal{B}^i_{(\varepsilon)}$ and $\boldsymbol{\Sigma}_{(\varepsilon)}$, as functions of x, belong to $\mathbb{C}_b^{\infty}(\mathbb{R}^{\mathrm{d}})$, for all $i, l, \varepsilon, s, \omega$ and are uniformly bounded with respect to ε, s, ω together with their first-order derivatives. Therefore, by Theorem 1.11, system (5.4.20) has a unique solution.

Define

$$
{}_{\varepsilon}a^{ij} := \frac{1}{2}\, {}_{\varepsilon}\boldsymbol{\Sigma}^{i\ell}\, {}_{\varepsilon}\boldsymbol{\Sigma}^{j\ell}
$$

and consider the problem

$$
\begin{aligned}
du^{\varepsilon}(t,x) =& \Big(\big({}_{\varepsilon}a^{ij} u^{\varepsilon}\big)_{ij} - \big(b^i_{(\varepsilon)} u^{\varepsilon}\big)_i + c_{(\varepsilon)} u^{\varepsilon} \Big)(t,x)\, dt \\
&+ \Big(- \big(\sigma^{il}_{(\varepsilon)} u^{\varepsilon}\big)_i + h^l_{(\varepsilon)} u^{\varepsilon} \Big)(t,x)\, dw^l(t), \\
&(t, x, \omega) \in (0, T] \times \mathbb{R}^{\mathrm{d}} \times \Omega;
\end{aligned} \tag{5.4.21}
$$

$$
u(0, x) = \varphi(x),\ (x, \omega) \in \mathbb{R}^{\mathrm{d}} \times \Omega. \tag{5.4.22}
$$

The coefficients in (5.4.21) are infinitely differentiable with respect to x and uniformly in (t, x, ω) bounded, together with their derivatives. Further, similar to Sect. 4.2.6, we use $(\xi_{(\varepsilon)})^2 \le \xi^2_{(\varepsilon)}$ to conclude that Eq. (5.4.21) is super-parabolic: for all t, x, ω,

$$
2\, {}_{\varepsilon}a^{ij}(t,x,\omega) y^i y^j \; - \; \sum_{l=1}^{\mathrm{d}_1} \Big| \sigma^{il}_{(\varepsilon)}(t,x,\omega) y^i \Big|^2 \ge \varepsilon |y|^2,
$$

$$
y \in \mathbb{R}^{\mathrm{d}},\ \varepsilon\ \in \mathbb{R}_+.
$$

Thus problem (5.4.21), (5.4.22) satisfies the condition of Lemma 5.3 and consequently

$$
\mathbb{E}\Big[\psi\big({}_{\varepsilon}\mathcal{X}(t)\big)\, {}_{\varepsilon}\rho(t) \big| \mathfrak{F}_t \Big] = \big(\psi, u^{\varepsilon}(t)\big)_0,\ t \in [0, T],\ (\mathbb{P}\text{-a.s.}) \tag{5.4.23}
$$

where

$$
{}_{\varepsilon}\rho(t) := \exp\Bigg(\int\limits_{[0,t]} \Big(c_{(\varepsilon)} - \frac{1}{2} h^l_{(\varepsilon)} h^l_{(\varepsilon)} \Big)\big(s, {}_{\varepsilon}\mathcal{X}(s)\big)\, ds + \int\limits_{[0,t]} h^l_{(\varepsilon)}\big(s, {}_{\varepsilon}\mathcal{X}(s)\big)\, dw^l(s) \Bigg).
$$

We now pass to the limit as $\varepsilon \to 0$. The objective is to show that

$$\mathbb{E}\Big[\psi\big({}_\varepsilon\mathcal{X}(t)\big){}_\varepsilon\rho(t)\big|\mathfrak{F}_t\Big] \to \mathbb{E}\Big[\psi\big(\mathcal{X}(t)\big)\rho(t)\big|\mathfrak{F}_t\Big]$$

and

$$\big(\psi, u^\varepsilon(t)\big)_0 \to \big(\psi, u(t)\big)_0,$$

possibly along a subsequence ε_n. To simplify the notation, we will not use special symbols for subsequences and keep ε throughout; the convergence will be indicated by $\to$, without additional comments.

To specify the mode of convergence, and to handle the conditional expectation, we take an $\mathfrak{F}_t$-measurable bounded random variable ζ so that (5.4.23) becomes

$$\mathbb{E}\Big(\zeta\,\psi\big({}_\varepsilon\mathcal{X}(t)\big){}_\varepsilon\rho(t)\Big) = \mathbb{E}\Big(\zeta\,\big(\psi, u^\varepsilon(t)\big)_0\Big). \tag{5.4.24}$$

Then, to finish the proof, it remains to show that

$$\mathbb{E}\Big(\zeta\,\big(\psi, u^\varepsilon(t)\big)_0\Big) \to \mathbb{E}\Big(\zeta\,(\psi, u(t))_0\Big) \tag{5.4.25}$$

and

$$\mathbb{E}\Big(\zeta\,\psi\big({}_\varepsilon\mathcal{X}(t)\big){}_\varepsilon\rho(t)\Big) \to \mathbb{E}\Big(\zeta\,\psi\big(\mathcal{X}(t)\big)\rho(t)\Big). \tag{5.4.26}$$

From Lemma 1.2 it follows that, for all t, x, ω, i, j, l, and ℓ, ${}_\varepsilon a^{ij} \to a^{ij}$, ${}_\varepsilon\boldsymbol{\Sigma}^{i\ell} \to \boldsymbol{\Sigma}^{i\ell}$, $b^i_{(\varepsilon)} \to b^i$, $c_{(\varepsilon)} \to c$, $h^l_{(\varepsilon)} \to h^l$, and $\mathcal{B}^i_{(\varepsilon)} \to \mathcal{B}^i$. Next, by Corollary 4.6, we have

$$\lim_{\varepsilon\to 0} \mathbb{E} \sup_{t\in[0,T]} \big\|u^\varepsilon(t) - u(t)\big\|_0^2 = 0, \tag{5.4.27}$$

where u is the generalized solution of problem (5.4.1), (5.4.2). Then (5.4.27) implies (5.4.25).

Next, similar to (5.4.27),

$$\mathbb{E} \sup_{t\in[0,T]} \big|{}_\varepsilon\mathcal{X}(t) - \mathcal{X}(t)\big|^2 \to 0;$$

see [35, Section 1.2.7] for details. Then (5.4.26) follows, concluding the proof of Theorem 5.6. □

5.4.7. In this paragraph, we prove Corollary 5.6. Since the result is about a particular version u of the generalized solution of (5.4.1), (5.4.2), we will construct this version starting with a generalized solution v.

First, by the usual density/continuity arguments, we argue that the equality

$$\tilde{\Phi}_t[\psi] = \big(v(t), \psi\big)_0 \tag{5.4.28}$$

and the inequalities $\tilde{\Phi}_t[1] < \infty$, $\big(v(t), \psi\big)_0 \geq 0$, if $\psi \geq 0$, hold on the same set $\Omega' \subset \Omega$ with $\mathbb{P}(\Omega') = 1$ for all $t \in [0, T]$ and all $\psi \in \mathbb{C}_0^\infty(\mathbb{R}^d)$. Taking a sequence $\{\psi_n,\ n \geq 1\}$ such that $0 \leq \psi_n \leq 1$ and $\psi_n(x) \to 1$ for every x, we conclude by the dominated convergence theorem that

$$\tilde{\Phi}_t[1] = \big(v(t), 1\big)_0 < \infty, \tag{5.4.29}$$

that is, $v(t, \omega) \in \mathbb{L}_1$ for all $(t, \omega) \in [0, T] \times \Omega'$.

Next, define

$$u(t, \omega) := \begin{cases} v(t, \omega), & \text{if } \omega \in \Omega', \\ 0, & \text{if } \omega \notin \Omega'. \end{cases}$$

By construction, u is a version of v and thus a generalized solution of (5.4.1), (5.4.2), and also, for all $(t, \omega) \in [0, T] \times \Omega$, u belongs to the cone of non-negative functions in $\mathbb{L}_1$. By the Pettis theorem, u is a $\mathscr{P}(\mathfrak{F})$-measurable mapping from $[0, T] \times \Omega$ to $\mathbb{L}_1$; recall that the topological dual of $\mathbb{L}_1$ is isometrically isomorphic to $\mathbb{L}_\infty$. Also, equality (5.4.3) implies $\mathbb{E}\big(u(t), 1\big)_0 = \mathbb{E}\rho(t) \leq e^{KT}$; therefore,

$$u \in \mathbb{L}_1\big([0, T]; \mathscr{P}(\mathfrak{F}); \mathbb{L}_1\big). \tag{5.4.30}$$

It remains to show that the mapping $t \mapsto u(t)$ is weakly continuous in $\mathbb{L}_1$, that is, for every $\psi \in \mathbb{L}_\infty$, the real-valued function $t \mapsto \big(u(t), \psi\big)_0$ is continuous on $[0, T]$, $\mathbb{P}$- a.s.

For $R > 0$ and $\psi \in \mathbb{L}_\infty$, define

$$\psi_R(x) := \begin{cases} \psi(x), & \text{if } |x| \leq R \\ 0, & \text{if } |x| > R. \end{cases}$$

Then

$$\big(u(t), \psi\big)_0 = \big(u(t), \psi_R\big)_0 + \int\limits_{|x|>R} u(t, x)\psi(x)dx.$$

The first term on the right is continuous in t by Theorem 4.5, because $\psi_R \in \mathbb{L}_2$; the second term can be made arbitrarily small because of (5.4.30). □

Remark 5.9 If the matrix $(2a^{ij} - \sigma^{il}\sigma^{jl})$ is uniformly positive definite, that is, equations (4.1.1), (4.4.3), and (5.4.1) are super-parabolic, then the corresponding AOC formulas (5.2.2), (5.2.3), and (5.4.3) hold under less restrictive regularity assumptions on the coefficients, free terms, and the initial conditions. Working out the details could be an interesting exercise for the reader.

□

Chapter 6
Filtering, Interpolation and Extrapolation of Diffusion Processes

6.1 Introduction

6.1.1. Recall that the filtering problem for diffusion processes first appeared in Sect. 1.2.2, where we discussed the motivation and general setting. Accordingly, we now go directly to the mathematical formulation of the problem.

Warning 6.1 *Warning* 4.1 *(regarding various notations, including the summation convention and subscripts for partial derivatives) is to be in force throughout this chapter.* □

Fix a non-random number $T \in \mathbb{R}_+$ and a stochastic basis $\mathbb{F} := (\Omega, \mathscr{F}, \{\mathscr{F}_t\}_{t\in[0,T]}, \mathbb{P})$ with the usual assumptions. Let $\mathcal{W} = \mathcal{W}(t)$ be a standard Wiener process defined on $\mathbb{F}$ and taking values in $\mathbb{R}^{\mathrm{d}+\mathrm{d}_1}$.

Consider an $\mathbb{R}^{\mathrm{d}+\mathrm{d}_1}$-dimensional diffusion process $Z = Z(t)$, with

$$Z^i(t) = z_0^i + \int_{[0,t]} {}_Z b^i\big(s, Z(s)\big)\, ds + \int_{[0,t]} {}_Z\boldsymbol{\Sigma}^{i\ell}\big(s, Z(s)\big)\, d\mathcal{W}^\ell(s), \tag{6.1.1}$$

$$i, \ell = 1, 2, \ldots, \mathrm{d} + \mathrm{d}_1,\ t \in [0, T].$$

To ensure existence and uniqueness of the solution of (6.1.1), we assume throughout this chapter that $z_0 := (z_0^1, \ldots, z_0^{\mathrm{d}+\mathrm{d}_1})$ is an $\mathscr{F}_0$-measurable random vector and the functions ${}_Z b$ and ${}_Z\boldsymbol{\Sigma}$ satisfy the conditions of Theorem 1.11.

Suppose that only some of the components of the process Z are observable. In what follows, we denote by Y the observable components of Z and by X, the un-observable components. With no loss of generality, and thinking of vectors as columns, we assume that

$$X(t) = \big(Z^1(t), \ldots, Z^{\mathrm{d}}(t)\big)^*.$$

B. L. Rozovsky, S. V. Lototsky, *Stochastic Evolution Systems*, Probability Theory and Stochastic Modelling 89, https://doi.org/10.1007/978-3-319-94893-5_6

Next, we introduce the corresponding σ-algebras:

- $\mathscr{X}_t^s$, the σ-algebra generated by $X(\tau)$, $\tau \in [s,t]$, and completed with respect to $\mathbb{P}$;
- $\mathscr{Y}_t^s$, the σ-algebra generated by $Y(\tau)$, $\tau \in [s,t]$, and completed with respect to $\mathbb{P}$;
- $\mathscr{Z}_t^s$, the σ-algebra generated by $Z(\tau)$, $\tau \in [s,t]$, and completed with respect to $\mathbb{P}$.

Note that $\mathscr{Z}_t^s = \mathscr{X}_t^s \vee \mathscr{Y}_t^s$.

We now describe the problem of filtering and the related problems of interpolation, and extrapolation. Take a function $f: \mathbb{R}^{\mathrm{d}} \to \mathbb{R}^1$ such that $\mathbb{E}|f(X(t))|^2 < \infty$, $t \in [0,T]$, and the numbers T_0, T_1, T_2 from the interval $[0,T]$ such that $T_0 \le T_1 \wedge T_2$.

General properties of the conditional expectation imply that, in the mean square sense, $\mathbb{E}\big[f\big(X(T_2)\big)|\mathscr{Y}_{T_1}^{T_0}\big]$ is the best estimator of $f\big(X(T_2)\big)$ given $Y(t)$, $t \in [T_0, T_1]$.

Calculation of $\mathbb{E}\big[f\big(X(T_2)\big)|\mathscr{Y}_{T_1}^{T_0}\big]$ is called

- **filtering** if $T_2 = T_1$,
- **interpolation** (or **smoothing**) if $T_2 < T_1$,
- **extrapolation** (or **prediction**) if $T_2 > T_1$.

Accordingly, the corresponding conditional distribution $\mathbb{P}\big[X(T_2) \in \cdot|\mathscr{Y}_{T_1}^{T_0}\big]$ is called the filtering, interpolation or extrapolation measure. Analysis of these measures is the main goal of this chapter. In particular, we will show that, under mild spatial regularity of the coefficients in (6.1.1), the measures are absolutely continuous with respect to the Lebesgue measure on $\mathbb{R}^{\mathrm{d}}$, and derive and investigate the equations satisfied by the densities.

6.1.2. The objective of this paragraph is to reduce Eq. (6.1.1) to a canonical form that is the most convenient for the study of the filtering problem. Along the way, we will introduce additional notation and assumptions to be used throughout the rest of the chapter.

Following our partition of the process Z into unobservable component X and observable component Y,

$$Z = \begin{pmatrix} X \\ Y \end{pmatrix},$$

we similarly partition the Brownian motion $\mathcal{W}$ and the functions ${}_{\mathrm{Z}}b$ and ${}_{\mathrm{Z}}\boldsymbol{\Sigma}$:

$$\mathcal{W} = (\bar{w}^1, \ldots, \bar{w}^{\mathrm{d}}, w^1, \ldots, w^{\mathrm{d}_1})^* := (\overline{W}^*, W^*),\ {}_{\mathrm{Z}}b = \begin{pmatrix} {}_{\mathrm{X}}b \\ {}_{\mathrm{Y}}b \end{pmatrix},\ {}_{\mathrm{Z}}\boldsymbol{\Sigma} = \begin{pmatrix} {}_{\mathrm{X}}\boldsymbol{\Sigma} \\ {}_{\mathrm{Y}}\boldsymbol{\Sigma} \end{pmatrix}.$$

Then, given a point $z \in \mathbb{R}^{\mathrm{d}+\mathrm{d}_1}$, we write $dz := dz^1 \cdots dz^{\mathrm{d}+\mathrm{d}_1}$ and $f_\ell(z) := \frac{\partial}{\partial z^\ell} f(z)$, while keeping in mind that $z^\ell = x^i$ for $\ell = i = 1, \ldots, \mathrm{d}$ and $z^\ell = y^l$ for $\ell = \mathrm{d} + l$, $l = 1, \ldots, \mathrm{d}_1$.

To proceed, we need to assume that the matrix

$$ {}_0\hat{\sigma} := \left({}_{\mathrm{Y}}\boldsymbol{\Sigma}\, {}_{\mathrm{Y}}\boldsymbol{\Sigma}^* \right)^{1/2} $$

is invertible. Then we define the matrices

$$ \hat{\sigma} = {}_{\mathrm{X}}\boldsymbol{\Sigma}\, {}_{\mathrm{Y}}\boldsymbol{\Sigma}^*\, {}_0\hat{\sigma}^{-1}, $$

$$ {}_1\hat{\sigma} = \left({}_{\mathrm{X}}\boldsymbol{\Sigma}\, {}_{\mathrm{X}}\boldsymbol{\Sigma}^* - \hat{\sigma}\hat{\sigma}^* \right)^{1/2}, \quad \widehat{\boldsymbol{\Sigma}} = \begin{pmatrix} {}_1\hat{\sigma} & \hat{\sigma} \\ 0 & {}_0\hat{\sigma} \end{pmatrix}, $$

where 0 is the zero $\mathrm{d}_1 \times \mathrm{d}$-matrix, A^* is the transpose of the matrix A, and, for a symmetric non-negative definite matrix B, the matrix $B^{1/2}$ denotes the symmetric non-negative square root of B. Direct computations show that the matrix ${}_{\mathrm{X}}\boldsymbol{\Sigma}\, {}_{\mathrm{X}}\boldsymbol{\Sigma}^* - \hat{\sigma}\hat{\sigma}^*$ is indeed symmetric and non-negative definite; cf. [93, Lemma 13.2].

By construction,

$$ {}_{\mathrm{Z}}\boldsymbol{\Sigma}\, {}_{\mathrm{Z}}\boldsymbol{\Sigma}^* = \widehat{\boldsymbol{\Sigma}}\, \widehat{\boldsymbol{\Sigma}}^*, $$

and therefore there exists a standard $\mathbb{R}^{\mathrm{d}+\mathrm{d}_1}$-dimensional Wiener process $\widehat{\mathcal{W}} = \widehat{\mathcal{W}}(t)$ on $\mathbb{F}$ such that, for every t,

$$ \int_{[0,t]} {}_{\mathrm{Z}}\boldsymbol{\Sigma}\big(s, Z(s)\big)\, d\mathcal{W}(s) = \int_{[0,t]} \widehat{\boldsymbol{\Sigma}}\big(s, Z(s)\big) d\widehat{\mathcal{W}}(s) \quad (\mathbb{P}\text{- a.s.}); $$

cf. [93, Lemma 10.4]. As a result, with no loss of generality, we will assume from now on that

$$ {}_{\mathrm{Z}}\boldsymbol{\Sigma} = \begin{pmatrix} {}_1\sigma & \sigma \\ 0 & {}_0\sigma \end{pmatrix}, \tag{6.1.2} $$

with symmetric non-negative definite matrices ${}_1\sigma$, ${}_0\sigma$, so that Eq. (6.1.1) takes the canonical form

$$ X(t) = x_0 + \int_{[0,t]} {}_{\mathrm{X}}b\big(s, Z(s)\big)\, ds + \int_{[0,t]} {}_1\sigma\big(s, Z(s)\big)\, d\overline{W}(s) + \int_{[0,t]} \sigma\big(s, Z(s)\big)\, dW(s), $$

$$ Y(t) = y_0 + \int_{[0,t]} {}_{\mathrm{Y}}b\big(s, Z(s)\big)\, ds + \int_{[0,t]} {}_0\sigma\big(s, Z(s)\big)\, dW(s). $$

Recall that our objective is to compute the conditional distribution of $X(T_2)$ given the observations $Y(s)$, $s \in [0, T_1]$. If the matrix ${}_0\sigma$ is singular, then the noise in the observation process will be vanishing at some points, making it possible to obtain additional information about X from the equation satisfied by Y; singularity of ${}_0\sigma$ can also complicate the reduction of the original equation (6.1.1) to the canonical form. Similarly, if the function ${}_0\sigma$ depends on X, then additional information about X will be contained in the quadratic variation process of Y. As a result, to ensure that the conditional distribution indeed contains all the available information about the X component, we make the following key assumptions about the matrix ${}_0\sigma$:

$$
{}_0\sigma(t,x,y) = {}_0\sigma(t,y), \quad {}_0\sigma^{ik}(t,y){}_0\sigma^{jk}(t,y)\xi^i\xi^j \geq \delta\,|\xi|^2
$$

for some $\delta > 0$ and all $t \in [0, T]$, $x, \xi \in \mathbb{R}^{\mathrm{d}}$, and $y \in \mathbb{R}^{\mathrm{d}_1}$.

Finally, we use the matrix ${}_Z\boldsymbol{\Sigma}$ from (6.1.2) to define three more matrices:

$$
A := \frac{1}{2}\,{}_Z\boldsymbol{\Sigma}\,{}_Z\boldsymbol{\Sigma}^*, \quad a := \frac{1}{2}\left({}_1\sigma\,{}_1\sigma^* + \sigma\sigma^*\right), \quad D := \begin{pmatrix} \sigma \\ {}_0\sigma \end{pmatrix}
$$

and a vector

$$
h := {}_0\sigma^{-1}\,{}_Yb.
$$

Here is another key assumption to hold throughout the rest of the chapter:

(H) *The functions* $h^l = h^l(t,z,\omega)$, $l = 1, 2, \ldots, \mathrm{d}_1$, *are uniformly bounded in* $(t,z,\omega) \in [0,T] \times \mathbb{R}^{d+\mathrm{d}_1} \times \Omega$.

6.2 The Bayes Formula and the Conditional Markov Property

6.2.1. This section investigates the structure of the conditional expectation

$$
\mathbb{E}\big[f\big(X(T_2)\big)\big|\mathscr{Y}^0_{T_1} \vee \mathfrak{U}\big],
$$

where $\mathfrak{U}$ is a (possibly trivial) sub-σ-algebra of $\mathscr{X}^0_{T_0}$. Of special interest is the case when Z is a Markov process and $\mathfrak{U} = \mathscr{X}^{T_0}_{T_0}$.

Using a suitable change of measure and a version of the Bayes formula, we connect $\mathbb{E}\big[f\big(X(T_2)\big)\big|\mathscr{Y}^0_{T_1} \vee \mathfrak{U}\big]$ with the averaging over the characteristics formulas from the previous chapter. In the subsequent sections, we use this connection to derive equations describing the dynamics of the filtering, interpolation and extrapolation densities, as well as the backward filtering equations.

6.2.2. For $0 \le s \le t \le T$, define

$$\rho(t,s) = \exp\left\{ \int_{[s,t]} h^l\big(\tau, Z(\tau)\big)\, dw^l(\tau) + \frac{1}{2} \int_{[s,t]} |h|^2(\tau, Z(\tau))\, d\tau \right\}.$$

Assumption (H) implies $\mathbb{P}\big(0 < \rho(T,0) < \infty\big) = 1$ and $\mathbb{E}\big(1/\rho(T,0)\big) = 1$. Therefore it is possible to introduce a new probability measure $\widetilde{\mathbb{P}}$ on $(\Omega, \mathscr{F})$ by

$$\widetilde{\mathbb{P}}(\Gamma) = \int_{\Gamma} \frac{d\mathbb{P}(\omega)}{\rho(T,0)}, \quad \Gamma \in \mathscr{F},$$

so that the measures $\mathbb{P}$ and $\widetilde{\mathbb{P}}$ are mutually absolutely continuous and $d\mathbb{P}/d\widetilde{\mathbb{P}} = \rho(T,0)$. Accordingly, we use $\widetilde{\mathbb{E}}$ to denote the expectation with respect to $\widetilde{\mathbb{P}}$.

By the Girsanov theorem, there exists an $\mathbb{R}^{d_1}$-valued standard Wiener process $\widetilde{W} = \widetilde{W}(t)$ on the stochastic basis $\widetilde{\mathbb{F}} := (\Omega, \mathscr{F}, \{\mathscr{F}_t\}_{t\in[0,T]}, \widetilde{\mathbb{P}})$ such that the process $Z = (X,Y)$, considered on $\widetilde{\mathbb{F}}$, satisfies

$$X(t) = x_0 + \int_{[0,t]} ({}_Xb - \sigma h)\big(s, Z(s)\big)\, ds + \int_{[0,t]} {}_1\sigma\big(s, Z(s)\big)\, d\overline{W}(s) \qquad (6.2.1)$$

$$+ \int_{[0,t]} \sigma\big(s, Z(s)\big)\, d\widetilde{W}(s),$$

$$Y(t) = y_0 + \int_{[0,t]} {}_0\sigma\big(s, Y(s)\big)\, d\widetilde{W}(s), \ t \in [0,T]. \qquad (6.2.2)$$

Note that the stochastic basis $\widetilde{\mathbb{F}}$ satisfies the usual assumptions and, because

$$d\widetilde{W} = dW + h dt,$$

we have

$$\rho(t,s) = \exp\left\{ \int_{[s,t]} h^l\big(\tau, Z(\tau)\big)\, d\widetilde{w}^l(\tau) - \frac{1}{2} \int_{[s,t]} \big|h\big(\tau, Z(\tau)\big)\big|^2\, d\tau \right\}. \qquad (6.2.3)$$

Theorem 6.1 *Assume that $\rho(T_0, 0)$ is $\mathfrak{U}$-measurable. If ψ is a $\mathscr{Z}_T^{T_0}$-measurable and $\mathbb{P}$-integrable random variable, then*

$$\mathbb{E}[\psi|\mathscr{Y}_{T_1}^0 \vee \mathfrak{U}] = \frac{\widetilde{\mathbb{E}}[\psi\rho(T_1 \vee T_2, T_0)|\mathscr{Y}_{T_1}^0 \vee \mathfrak{U}]}{\widetilde{\mathbb{E}}[\rho(T_1 \vee T_2, T_0)|\mathscr{Y}_{T_1}^0 \vee \mathfrak{U}]} \quad (\mathbb{P}\text{- a.s.}). \tag{6.2.4}$$

□

The proof relies on the following version of the **Bayes formula**; cf. [94, §24.4].

Lemma 6.1 *Let Q and $\widetilde{Q}$ be mutually absolutely continuous probability measures on $(\Omega, \mathscr{F})$. If ξ is a Q-integrable random variable and $\mathscr{G}$ is a sub-σ-algebra of $\mathscr{F}$, then*

$$\mathbb{E}_Q[\xi|\mathscr{G}] = \frac{\mathbb{E}_{\widetilde{Q}}[\xi dQ/d\widetilde{Q}|\mathscr{G}]}{\mathbb{E}_{\widetilde{Q}}[dQ/d\widetilde{Q}|\mathscr{G}]}, \tag{6.2.5}$$

where $\mathbb{E}_Q[\cdot|\mathscr{G}]$ (resp. $\mathbb{E}_{\widetilde{Q}}[\cdot|\mathscr{G}]$) is the conditional, relative to $\mathscr{G}$, expectation with respect to the measure Q (resp. $\widetilde{Q}$). Equality (6.2.5) *holds on the set having Q- and $\widetilde{Q}$-measure one.* □

Given the importance of the result, here is an outline of the proof. To make the formulas shorter, write $U = \mathbb{E}_Q[\xi|\mathscr{G}] \cdot \mathbb{E}_{\widetilde{Q}}[dQ/d\widetilde{Q}|\mathscr{G}]$, $V = \mathbb{E}_{\widetilde{Q}}[\xi dQ/d\widetilde{Q}|\mathscr{G}]$, and let η be a bounded $\mathscr{G}$-measurable random variable. Then (6.2.5) is equivalent to

$$\mathbb{E}_{\widetilde{Q}}(\eta U) = \mathbb{E}_{\widetilde{Q}}(\eta V) \equiv \mathbb{E}_{\widetilde{Q}}(\eta\xi dQ/d\widetilde{Q}) \equiv \mathbb{E}_Q(\eta\xi).$$

On the other hand,

$$\begin{aligned}\mathbb{E}_{\widetilde{Q}}(\eta U) &\equiv \mathbb{E}_{\widetilde{Q}}\Big(\eta\mathbb{E}_{\widetilde{Q}}[dQ/d\widetilde{Q}|\mathscr{G}]\mathbb{E}_Q[\xi|\mathscr{G}]\Big) = \mathbb{E}_{\widetilde{Q}}\Big(\mathbb{E}_{\widetilde{Q}}\big(\eta dQ/d\widetilde{Q}\mathbb{E}_Q[\xi|\mathscr{G}]|\mathscr{G}\big)\Big)\\ &= \mathbb{E}_{\widetilde{Q}}\big(\eta dQ/d\widetilde{Q}\mathbb{E}_Q[\xi|\mathscr{G}]\big) = \mathbb{E}_Q\big(\eta\mathbb{E}_Q[\xi|\mathscr{G}]\big) = \mathbb{E}_Q(\eta\xi).\end{aligned}$$

Proof of Theorem 1 We apply Lemma 6.1 with $Q = \mathbb{P}$, $\widetilde{Q} = \widetilde{\mathbb{P}}$, $\xi = \psi$, and $\mathscr{G} = \mathscr{Y}_{T_1}^0 \vee \mathfrak{U}$. Then

$$\mathbb{E}[\psi|\mathscr{Y}_{T_1}^0 \vee \mathfrak{U}] = \frac{\widetilde{\mathbb{E}}[\psi\rho(T, 0)|\mathscr{Y}_{T_1}^0 \vee \mathfrak{U}]}{\widetilde{\mathbb{E}}[\rho(T, 0)|\mathscr{Y}_{T_1}^0 \vee \mathfrak{U}]}. \tag{6.2.6}$$

Note that $\rho(T, 0) = \rho(T_0, 0)\rho(T_1 \vee T_2, T_0)\rho(T,\ T_1 \vee T_2)$. By assumption, $\rho(T_0, 0)$ is $\mathfrak{U}$-measurable; by (6.2.3), $\widetilde{\mathbb{E}}[\rho(T, T_1 \vee T_2)|\mathscr{F}_{T_1 \vee T_2}] = 1$. Then (6.2.4) follows from (6.2.6). □

6.2.3. While the solution Z of (6.1.1) is usually called a diffusion process, it is not, in general, a Markov process unless all the coefficients in (6.1.1) are non-random. Theorem 6.1 does not require Z to be Markov, but essentially relies on the structure of Z as a diffusion process, and more precisely, on Eq. (6.2.2) satisfied by the Y component. Further analysis of (6.2.4) requires additional information about the process Z. A satisfactory solution of the filtering problem, including the implementation of the program outlined at the beginning of the chapter, is possible if Z is both a diffusion process and a Markov process; this is easily achieved by assuming that the coefficients in (6.1.1) are non-random, that is, do not depend on ω.

Warning 6.2 *Throughout the rest of the chapter, we assume that the coefficients of system* (6.1.1) *are non-random.* □

Denote by $\tilde{\mathscr{F}}_s^t$ the σ-algebra generated by the random variables $\widetilde{w}(\tau_1) - \widetilde{w}(\tau_2)$, $\tau_1, \tau_2 \in [s, t]$, and completed with respect to the measure $\mathbb{P}$.

Theorem 6.2 *Let ψ be a $\mathscr{Z}_{T_1 \vee T_2}$-measurable, $\mathbb{P}$-integrable random variable. Then, with probability one,*

$$\mathbb{E}[\psi|\mathscr{Y}_{T_1}^0 \vee \mathscr{X}_{T_0}^0] = \mathbb{E}[\psi|\mathscr{Y}_{T_1}^0 \vee \mathscr{X}_{T_0}^{T_0}] = \frac{\widetilde{\mathbb{E}}[\psi\rho(T_1 \vee T_2, T_0)|\tilde{\mathscr{F}}_{T_1}^{T_0} \vee \mathscr{Z}_{T_0}^{T_0}]}{\widetilde{\mathbb{E}}[\rho(T_1 \vee T_2, T_0)|\tilde{\mathscr{F}}_{T_1}^{T_0} \vee \mathscr{Z}_{T_0}^{T_0}]}. \quad \square$$

The proof relies on the following two auxiliary statement, each of independent interest.

Lemma 6.2 *If the coefficients in* (6.1.1) *are non-random, then the σ-algebras $\mathscr{Y}_{T_1}^{T_0}$ and $\tilde{\mathscr{F}}_{T_1}^{T_0} \vee \mathscr{Y}_{T_0}^{T_0}$ coincide.* □

Proof By uniqueness of solution of (6.2.2) (see Theorem 1.11),

$$\mathbb{P}\Big(Y(t) = Y\big(t, T_0, Y(T_0)\big),\ t \in [T_0, T]\Big) = 1,$$

where

$$Y\big(t, T_0, Y(T_0)\big) = Y(T_0) + \int\limits_{[T_0,t]} {}_0\sigma\Big(s, Y\big(s, T_0, Y(T_0)\big)\Big)\, d\widetilde{W}(s),\ t \in [T_0, T]. \tag{6.2.7}$$

The construction of Y by successive approximations (cf. [65, Theorem 2.5.7]) implies that, for $t \in [T_0, T_1]$, the random variable $Y\big(t, T_0, Y(T_0)\big)$ is $\tilde{\mathscr{F}}_{T_1}^{T_0} \vee \mathscr{Y}_{T_0}^{T_0}$-measurable, and consequently $\mathscr{Y}_{T_1}^{T_0} \subseteq \tilde{\mathscr{F}}_{T_1}^{T_0} \vee \mathscr{Y}_{T_1}^{T_0}$.

On the other hand, the matrix ${}_0\sigma$ is invertible, so that (6.2.2) implies

$$\widetilde{W}(t) - \widetilde{W}(T_0) = \int_{[T_0,t]} {}_0\sigma^{-1}\big(s, Y(s)\big)\, dY(s),\ t \in [T_0, T].$$

In particular, if $t \in [T_0, T_1]$, then $\widetilde{W}(t) - \widetilde{W}(T_0)$ is $\mathscr{Y}_{T_0}^{T_1}$-measurable, from which the reverse inclusion $\widetilde{\mathscr{F}}_{T_1}^{T_0} \vee \mathscr{Y}_{T_0}^{T_0} \subseteq \mathscr{Y}_{T_1}^{T_0}$ follows. □

Proposition 6.1 *If ξ is a $\mathscr{Z}_T^{T_0}$-measurable $\mathbb{P}$-integrable, real-valued random variable, then, with probability one,*

$$\mathbb{E}[\xi \,|\, \mathscr{Y}_{T_0}^0 \vee \mathscr{X}_{T_0}^0] = \mathbb{E}[\xi \,|\, \mathscr{Y}_{T_1}^{T_0} \vee \mathscr{X}_{T_0}^{T_0}] = \mathbb{E}[\xi \,|\, \mathscr{Y}_{T_1}^0 \vee \mathscr{X}_{T_0}^{T_0}]. \tag{6.2.8}$$

□

Proof In what follows, we verify the first equality in (6.2.8); then the second equality follows directly from the first because $\mathscr{Y}_{T_1}^{T_0} \vee \mathscr{X}_{T_0}^{T_0} \subseteq \mathscr{Y}_{T_1}^0 \vee \mathscr{X}_{T_0}^{T_0}$.

The Markov property of Z means

$$\mathbb{E}[\psi \,|\, \mathscr{Z}_{T_0}^0] = \mathbb{E}[\psi \,|\, \mathscr{Z}_{T_0}^{T_0}] \tag{6.2.9}$$

for every $\mathscr{Z}_T^{T_0}$-measurable and $\mathbb{P}$-integrable random variable ψ.

Let ζ and η be bounded random variables measurable such that ζ is $\mathscr{Z}_{T_0}^0$-measurable and η is $\mathscr{Y}_{T_1}^0$-measurable. It follows from (6.2.9) and general properties of the conditional expectation that

$$\begin{aligned}
\mathbb{E}[\zeta\eta\mathbb{E}[\xi \,|\, \mathscr{Y}_{T_1}^0 \vee \mathscr{X}_{T_0}^0] &= \mathbb{E}(\zeta\eta\xi) = \mathbb{E}\big(\zeta\mathbb{E}[\eta\xi \,|\, \mathscr{Z}_{T_0}^0]\big) \\
&= \mathbb{E}\big(\zeta\mathbb{E}[\eta\xi \,|\, \mathscr{Z}_{T_0}^{T_0}]\big) = \mathbb{E}\Big(\zeta\mathbb{E}\big[\eta\mathbb{E}[\xi \,|\, \mathscr{Y}_{T_1}^{T_0} \vee \mathscr{Z}_{T_0}^{T_0}] \,|\, \mathscr{Z}_{T_0}^{T_0}\big]\Big) \\
&= \mathbb{E}\big(\zeta\eta\mathbb{E}[\xi \,|\, \mathscr{Y}_{T_1}^{T_0} \vee \mathscr{X}_{T_0}^{T_0}]\big),
\end{aligned}$$

completing the proof of the first equality in (6.2.8). □

Remark 6.1 It is natural to call (6.2.8) the **conditional Markov property**. A version of this property exists for all multi-dimensional Markov processes, in both discrete and continuous time and with both discrete and continuous state space. □

We can now prove Theorem 6.1. The first equality in the statement of the theorem is the same as the first equality in Proposition 6.1.

On the other hand, using Theorem 6.1 with $\mathfrak{U} = \mathscr{Z}^0_{T_0}$,

$$\mathbb{E}[\psi|\mathscr{Y}^0_{T_1} \vee \mathscr{X}^0_{T_0}] = \mathbb{E}[\psi|\mathscr{Y}^0_{T_1} \vee \mathscr{Z}^0_{T_0}] = \frac{\widetilde{\mathbb{E}}[\psi\rho(T_1 \vee T_2, T_0)|\mathscr{Y}^0_{T_1} \vee \mathscr{Z}^0_{T_0}]}{\widetilde{\mathbb{E}}[\rho(T_1 \vee T_2, T_0)|\mathscr{Y}^0_{T_1} \vee \mathscr{Z}^0_{T_0}]}$$
$$= \frac{\widetilde{\mathbb{E}}[\psi\rho(T_1 \vee T_2, T_0)|\mathscr{Y}^0_{T_1} \vee \mathscr{X}^0_{T_0}]}{\widetilde{\mathbb{E}}[\rho(T_1 \vee T_2, T_0)|\mathscr{Y}^0_{T_1} \vee \mathscr{X}^0_{T_0}]}. \tag{6.2.10}$$

Next, equality (6.2.2) implies that Z is a Markov process with respect to the measure $\widetilde{\mathbb{P}}$. Accordingly, to establish the second equality in the statement of Theorem 6.2, we first apply Proposition 6.1 to (6.2.10) and replace $\mathscr{Y}^0_{T_1} \vee \mathscr{X}^0_{T_0}$ with $\mathscr{Y}^{T_0}_{T_1} \vee \mathscr{X}^{T_0}_{T_0}$. Then we use Lemma 6.2 to conclude that

$$\mathscr{Y}^{T_0}_{T_1} \vee \mathscr{X}^{T_0}_{T_0} = \tilde{\mathscr{F}}^{T_0}_{T_1} \vee \mathscr{Z}^{T_0}_{T_0}.$$ □

Let us point out again that our analysis of the filtering and related problems relies on two assumptions, namely, that Z satisfies (6.1.1) and that Z is a Markov process. In particular, both assumptions are explicitly used in the proof of Lemma 6.2. Still, there are related results that require only one of the properties of Z. For example, Theorem 6.1 only requires a particular form of the Y component of Z, whereas Proposition 6.1 only relies on the Markov property of Z.

6.3 The Forward Filtering Equation

6.3.1. In this section we show that, under some regularity assumptions, the conditional distribution $\mathbb{P}[X(t) \in \cdot\,|\mathscr{Y}^0_t]$ is absolutely continuous with respect to the Lebesgue measure on $\mathbb{R}^{\mathrm{d}}$, derive the equation satisfied by the corresponding filtering density $\pi = \pi(t,x)$, and investigate this equation.

We keep the notation and assumptions introduced in Sects. 6.1.1, 6.1.2, 6.2.2 (in particular, cf. Warning 6.2), and make the following additional assumptions:

(i) The regular conditional distribution $P_0(X(0) \in \cdot)$ of $X(0)$ given $\mathscr{Y}^0_0$ is absolutely continuous with respect to the Lebesgue measure on $\mathbb{R}^{\mathrm{d}}$ and the density $\pi_0 := dP_0/dx$ satisfies $\pi_0 \in \mathbb{L}_2(\Omega, \mathscr{Y}^0_0, \mathbb{P}; \mathbb{H}^1)$;
(ii) For every $(t,y) \in [0,T] \times \mathbb{R}^{\mathrm{d}_1}$, all the components of the matrix functions $a = a(t,x,y)$ and $\sigma = \sigma(t,x,y)$ are three times differentiable in x, whereas the components of the vector functions ${}_{\mathrm{x}}b = {}_{\mathrm{x}}b(t,x,y)$ and $h = h(t,x,y)$ are twice differentiable in x. All these functions and their derivatives are bounded uniformly in $(t,x,y) \in [0,T] \times \mathbb{R}^{\mathrm{d}} \times \mathbb{R}^{\mathrm{d}_1}$.

Define the operators $\mathcal{L}^* = \mathcal{L}^*(t, x, Y(t))$ and $\mathcal{M}^{l,*} = \mathcal{M}^{l,*}(t, x, Y(t))$ by

$$\mathcal{L}^* f(t,x) := \left(a^{ij}(t,x,Y(t)) f(t,x)\right)_{ij} - \left({}_x b^i(t,x,Y(t)) f(t,x)\right)_i,$$
$$\mathcal{M}^{l,*} f(t,x) := -\left(\sigma^{il}(t,x,Y(t)) f(t,x)\right)_i + h^l(t,x,Y(t)) f(t,x);$$

as usual, $f_i = \partial f/\partial x^i$, $i = 1, \ldots, \mathrm{d}$.

6.3.2. In this paragraph, the underlying stochastic basis will be

$$\widetilde{\mathbb{F}}_{\mathscr{Y}} := \left(\Omega, \mathscr{F}, \{\mathscr{Y}_t^0\}_{t\in[0,T]}, \widetilde{\mathbb{P}}\right)$$

with the σ-algebra of predictable sets $\mathscr{P}(\mathscr{Y})$. Note that $\widetilde{\mathbb{F}}_{\mathscr{Y}}$ satisfies the usual assumptions and, by Lemma 6.2, $\mathscr{Y}_t^0 = \tilde{\mathscr{F}}_t^0 \vee \mathscr{Y}_0^0$.

Consider the following Cauchy problem

$$du(t,x) = \mathcal{L}^* u(t,x)\, dt + \mathcal{M}^{l,*} u(t,x)\, d\widetilde{w}^l(t), \tag{6.3.1}$$
$$(t,x,w) \in [0,T] \times \mathbb{R}^{\mathrm{d}} \times \Omega,$$
$$u(0,x) = \pi_0(x),\ (x,\omega) \in \mathbb{R}^{\mathrm{d}} \times \Omega. \tag{6.3.2}$$

By construction, all coefficients in Eq. (6.3.1) are $\mathscr{P}(\mathscr{Y})$-measurable for every $x \in \mathbb{R}^{\mathrm{d}}$ and $\widetilde{W}$ is a standard Wiener process on $\widetilde{\mathbb{F}}_{\mathscr{Y}}$. Therefore, by Theorem 4.5 and Remark 5.6, problem (6.2.2), (6.3.2) has a unique generalized solution

$$u \in \mathbb{L}_2\left([0,T] \times \Omega, \mathscr{P}(\mathscr{Y}); \mathbb{H}^1\right) \bigcap \mathbb{L}_2\left(\Omega; \mathbb{C}[0,T]; \mathbb{L}_2\right).$$

Furthermore, problem (6.3.1), (6.3.2) satisfies the conditions of Theorem 5.6 on the stochastic basis $\widetilde{\mathbb{F}}_{\mathscr{Y}}$, with $\mathfrak{F}_t = \mathscr{Y}_t^0$ so that, by Corollary 5.6, the solution is measure-valued. Thus the following result holds.

Theorem 6.3 *Let $u = u(t,x)$ be the measure-valued generalized solution of problem* (6.3.1), (6.3.2). *Then, for every $f \in \mathbb{L}_\infty$ and $t \in [0,T]$,*

$$\widetilde{\mathbb{E}}[f(X(t))\rho(t,0)|\mathscr{Y}_t^0] = \int_{\mathbb{R}^{\mathrm{d}}} f(x) u(t,x)\, dx \ \ (\widetilde{\mathbb{P}}\text{- a.s.}). \tag{6.3.3}$$

From this and Theorem 6.1 we immediately get

Corollary 6.1 *If $u = u(t,x)$ is the measure-valued generalized solution of problem (6.3.1), (6.3.2), then, for every $f \in \mathbb{L}_\infty$ and every $t \in [0,T]$,*

$$\mathbb{E}[f(X(t))|\mathscr{Y}_t^0] = \frac{\int_{\mathbb{R}^d} f(x)u(t,x)\,dx}{\int_{\mathbb{R}^d} u(t,x)\,dx} \quad (\mathbb{P}\text{- a.s.}). \tag{6.3.4}$$

6.3.3. The objective of this paragraph is to establish the existence of the filtering density. We continue to work on the stochastic basis $\widetilde{\mathbb{F}}_{\mathscr{Y}}$.

For a bounded measurable function $g = g(t,z)$ define

$$\pi_s[g] := \mathbb{E}\big[g\big(s, Z(s)\big)|\mathscr{Y}_s^0\big].$$

Proposition 6.2 *For every $t \in [0,T]$, the process $V = V(t)$ defined by*

$$V(t) = \exp\left\{ \int_{[0,t]} \pi_s[h^l]\,d\widetilde{w}^l(s) - \frac{1}{2}\int_{[0,t]} \sum_{l=1}^{d_1} |\pi_s[h^l]|^2 ds \right\}$$

is a modification of $\widetilde{\mathbb{E}}[\rho(t,0)|\mathscr{Y}_t^0] \equiv \int_{\mathbb{R}^d} u(t,x)dx$. □

Proof By Theorem 5.7, with $f(t,x) \equiv 1$, the conditional expectations

$$\widetilde{\mathbb{E}}[\rho(t,0)|\mathscr{Y}_t^0], \text{ and } \widetilde{\mathbb{E}}\big[h^l\big(t, Z(t)\big)\rho(t,0)|\mathscr{Y}_t^0\big]$$

have $\mathscr{P}(\mathscr{Y})$-measurable versions satisfying $\mathbb{P}$-a.s.

$$\widetilde{\mathbb{E}}[\rho(t,0)|\mathscr{Y}_t^0] = 1 + \int_{[0,t]} \widetilde{\mathbb{E}}[h^l(s,Z(s))\rho(s,0)|\mathscr{Y}_s^0]\,d\widetilde{w}^l(s), \quad t \in [0,T];$$

in what follows, we always consider these versions without introducing new notation. Combining this equality with Corollary 6.1 leads to

$$\begin{aligned}\widetilde{\mathbb{E}}[\rho(t,0)|\mathscr{Y}_t^0] &= 1 + \int_{[0,t]} \frac{\widetilde{\mathbb{E}}\big[h^l\big(s,Z(s)\big)\rho(s,0)|\mathscr{Y}_s^0\big]}{\widetilde{\mathbb{E}}\big[\rho(s,0)|\mathscr{Y}_s^0\big]}\,\widetilde{\mathbb{E}}\big[\rho(s,0)|\mathscr{Y}_s^0\big]\,d\widetilde{w}^l(s)\\ &= 1 + \int_{[0,t]} \pi_s[h^l]\,\widetilde{\mathbb{E}}[\rho(s,0)|\mathscr{Y}_s^0]\,d\widetilde{w}^l(s),\end{aligned}$$

which is an equation for $\widetilde{\mathbb{E}}[\rho(t,0)|\mathscr{Y}_t^0]$. By the Itô formula, the process $V = V(t)$ is the unique solution of this equation. □

Denote by $\mathscr{M}(\mathbb{R}^d)$ the space of probability measures on $\mathbb{R}^d$.

Theorem 6.4 *There exists a function* $P^t_{\mathscr{Y}} : \Omega \to \mathscr{M}(\mathbb{R}^d)$*, called the* ***filtering measure****, with the following properties.*

(a) *For every* $t \in [0, T]$, $P^t_{\mathscr{Y}}(\cdot, \omega)$ *is regular conditional probability of* $X(t)$ *relative to* $\mathscr{Y}^0_t$.

(b) *For all* $(t, \omega) \in [0, t] \times \Omega'$ *with* $\mathbb{P}(\Omega') = 1$*, the measure* $P^t_{\mathscr{Y}}(\cdot, \omega)$ *is absolutely continuous with respect to the Lebesgue measure on* $\mathbb{R}^d$ *and the Radon–Nikodym derivative* $\pi = \pi(t, x, \omega)$*, known as the* ***filtering density****, satisfies*

$$\pi(t, x, \omega) = \frac{u(t, x, \omega)}{\int\limits_{\mathbb{R}^d} u(t, x, \omega) dx};$$

where u *is the generalized solution of problem* (6.3.1), (6.3.2) *and the equality is in* $\mathbb{L}_1$*, for each* (t, ω).

(c) *For every* $\Gamma \in \mathscr{B}(\mathbb{R}^d)$,

$$(t, \omega) \mapsto P^t_{\mathscr{Y}}(\Gamma, \omega)$$

is a $\mathscr{P}(\mathscr{Y})$*-measurable, continuous stochastic process.* □

Proof Let u be the measure-valued solution of (6.3.1), (6.3.2). By Theorem 6.3, Proposition 6.2, and Corollary 5.6, there exists a set $\Omega' \subset \Omega$ with $\widetilde{\mathbb{P}}(\Omega') = 1$, such that, for $t \in [0, T]$ and $\omega \in \Omega'$, $\int\limits_{\mathbb{R}^d} u(t, x, \omega) dx > 0$. Then, for all $t \in [0, T]$ and $\omega \in \Omega$, the function

$$P^t_{\mathscr{Y}}(\Gamma, \omega) := \begin{cases} \int_\Gamma \pi(t, x, \omega)\, dx, & \omega \in \Omega', \\ 1, & \omega \notin \Omega', \end{cases}$$

is a probability measure on $\mathscr{B}(\mathbb{R}^d)$ and is the regular conditional probability of $X(t)$ relative to $\mathscr{Y}^0_t$, establishing parts (a) and (b) of the theorem. Note that $\widetilde{\mathbb{P}}(\Omega') = 1$ implies $\mathbb{P}(\Omega') = 1$.

Part (c) follows from weak continuity and predictability of u as an $\mathbb{L}_1$-valued process; cf. Corollary 5.6. □

Problem (6.3.1), (6.3.2) is called the (direct or forward) **linear filtering equation**. The $\mathscr{P}(\mathscr{Y})$-measurable solution $u \in \mathbb{L}_1([0, T] \times \Omega; \mathbb{C}_w\mathbb{L}_1)$ of this problem, which belongs to the cone of non-negative functions in $\mathbb{L}_1$ (see Corollary 5.6) is called the **unnormalized filtering density**. The function π defined in part (ii) of Theorem 6.2 is called the (normalized) **filtering density**.

The unnormalized filtering density, being a generalized solution of problem (6.3.1), (6.3.2), has a variety of analytical properties described in Theorems 4.5, 4.7 and Corollaries 4.4, 4.5, 4.8, and 5.6. In particular, from Corollary 4.8 it follows that, under some additional assumptions on the smoothness of the coefficients and the initial data of problems (6.3.1), (6.3.2), the function u is a classical solution.

Remark 6.2 In applications, it is more natural to consider the linear filtering equation in a different-looking but mathematically equivalent form

$$du(t,x) = \mathcal{L}^* u(t,x)\,dt + \hat{\mathcal{M}}^{l,*} u(t,x) dY^l(t),\ (t,x,\omega) \in (0,T] \times \mathbb{R}^{\mathrm{d}} \times \Omega, \tag{6.3.1'}$$

$$u(0,x) = \pi_0(x),\ (x,\omega) \in \mathbb{R}^{\mathrm{d}} \times \Omega, \tag{6.3.2'}$$

where, for $l = 1, \ldots, \mathrm{d}_1$,

$$\hat{\mathcal{M}}^{l,*} v(t,x) = {}_0\sigma^{-jl}\big(t, Y(t)\big)\Big(-\big(\sigma^{ij}\big(t, xY(t)\big)v(t,x)\big)_i + h^j\big(t,x,Y(t)\big)v(t,x)\Big)$$

and ${}_0\sigma^{-jl}$ is the corresponding element of the matrix ${}_0\sigma^{-1}$.

Given a real-valued $\mathscr{Y}^0_t$-adapted process $f = f(t)$, equality (6.2.2) suggests that the stochastic integral of f with respect to Y can be defined by

$$\int_{[0,t]} f(s)\,dY^l(s) = \int_{[0,t]} f(s)\,{}_0\sigma^{lj}\big(s, Y(s)\big)\,d\widetilde{w}^j(s).$$

If the process f is continuous and $\{0 = t_0^n < t_1^n < \ldots < t_{k(n)}^n = t\}$ is a sequence of nested partitions of the interval $[0,t]$ with diameter tending to zero, then, by direct computation,

$$\int_{[0,t]} f(s) dY^l(s) = \lim_{n\to\infty} \sum_{i=0}^{k(n)-1} f(t_i^n)\Big(Y^l(t_{i+1}^n) - Y^l(t_i^n)\Big)$$

in probability.

Together with the last equality, the linear filtering equation in the form (6.3.1′), (6.3.2′) can be used for actual computations, because the observed process is Y rather than $\widetilde{W}$. On the other hand, the linear filtering equation in the form (6.3.1), (6.3.2) is more convenient from an analytical point of view. Accordingly, we will study the filtering equation in the form (6.3.1), (6.3.2). □

6.3.4. In this paragraph we derive an equation for the filtering density $\pi = \pi(t,x)$ on the original stochastic basis $\mathbb{F}$ and establish the main analytical properties of π.

Recall the notation

$$\pi_t[g] = \mathbb{E}\big[g\big(t, Z(t)\big)|\mathscr{Y}_t^0\big] \equiv \int_{\mathbb{R}^d} \pi(t,x) g\big(t,x,Y(t)\big)\,dx.$$

We also define the processes

$$\breve{w}^l(t) = \widetilde{w}^l(t) - \int_{[0,t]} \pi_s[h^l]\,ds,$$

$$V(t) = \exp\left\{ \int_{[0,t]} \pi_s[h^l]\,d\widetilde{w}^l(s) - \frac{1}{2}\int_{[0,t]} \sum_{l=1}^{d_1} \big|\pi_s[h^l]\big|^2 ds \right\},$$

and recall that, by Proposition 6.1, $V(t) = \int_{\mathbb{R}^d} u(t,x)dx$ ($\mathbb{P}$-a.s.).

Note that

$$\breve{w}^l(t) = w^l(t) + \int_{[0,t]} \Big(h^l\big(s, Z(s)\big) - \pi_s[h^l]\Big)\,ds,$$

and $W = (w^1, \ldots, w^{d_1})$ is a standard Wiener process on $\mathbb{F}$. Direct computations (cf. [93, Theorem 7.17]) show that each $\breve{w}^l$ is a continuous square-integrable martingale on $\mathbb{F}$ and

$$\langle \breve{w}^k, \breve{w}^l \rangle_t = \begin{cases} t, & k = l, \\ 0, & k \neq l. \end{cases}$$

A version of the Lévy characterization theorem then implies that $\breve{W} = (\breve{w}^1, \ldots, \breve{w}^{d_1})$ is a standard Wiener process on $\mathbb{F}$. The difference between the processes W and $\breve{W}$ is that $\breve{W}$ is the Brownian motion relative to a smaller family of sigma-algebras $\{\mathscr{Y}_t^0\}$. Using (6.2.2),

$$\breve{W}(t) = \int_{[0,t]} {}_0\sigma^{-1}\big(s, Y(s)\big)\,dY(s) - \int_{[0,t]} \pi_s[h]\,ds.$$

Accordingly, $\breve{W}$ is sometimes called the **innovation process**.

Theorem 6.5 *The filtering density has the following properties:*

(i) *it is an element of the space*

$$\mathbb{L}_2^{\omega}\big([0,T]\times\Omega, \mathscr{P}(\mathscr{Y}); \mathbb{H}^1\big) \bigcap \mathbb{L}_2\big(\Omega, \mathscr{P}(\mathscr{Y}); \mathbb{C}([0,T]; \mathbb{L}_2)\big),$$

it is $\mathscr{P}(\mathscr{Y})$-measurable and weakly continuous as an $\mathbb{L}_1$-valued process, and, for all $t \in [0, T] \times \Omega$ and $\eta \in \mathbb{C}_0^{\infty}(\mathbb{R}^{\mathrm{d}})$, satisfies the equality

$$\pi_t[\eta] = \pi_0[\eta] + \int_{[0,t]} \pi_s[\mathcal{L}\eta]\, ds + \int_{[0,t]} \Big(\pi_s[\mathcal{M}^l \eta] - \pi_s[h^l]\pi_s[\eta]\Big)\, d\breve{w}^l(s) \tag{6.3.5}$$

on the same ω-set of $\mathbb{P}$-probability 1.

(ii) *If π and $\bar{\pi}$ are two elements of*

$$\mathbb{L}_2^{\omega}([0,T], \mathscr{P}(\mathscr{Y}); \mathbb{H}^1) \bigcap \mathbb{L}_2^{\omega}([0,T], \mathscr{P}(\mathscr{Y}); \mathbb{L}_1) \bigcap \mathbb{L}_2(\Omega, \mathscr{P}(\mathscr{Y}); \mathbb{C}([0,T]; \mathbb{L}_2))$$

and both satisfy Eq. (6.3.5) *for every $t \in [0, T]$ and $\eta \in \mathbb{C}_0^{\infty}(\mathbb{R}^{\mathrm{d}})$, then*

$$\mathbb{P}\left(\sup_{t\in[0,T]} \|\pi - \bar{\pi}\|_0^2(t) > 0 \right) = 0. \qquad \square$$

Proof To establish (6.3.5), apply the Itô formula to the process

$$t \mapsto \frac{\int_{\mathbb{R}^{\mathrm{d}}} u(t,x)\eta(x)\, dx}{V(t)}$$

and keep in mind that u satisfies (6.3.1), (6.3.2) and $\pi(t,x) = u(t,x)/V(t)$. The computations are straightforward but somewhat long. Here is an outline. With

$$M(t) := \int_{\mathbb{R}^{\mathrm{d}}} u(t,x)\eta(x)\, dx, \quad N(t) := 1/V(t),$$

we have $d(MN) = N dM + M dN + (dM)(dN)$. By (6.3.1), (6.3.2),

$$N dM = \pi_t[\mathcal{L}\eta]dt + \pi_t[\mathcal{M}^l\eta]d\widetilde{w}^l,$$

and, by Proposition 6.1,

$$dN = \big(-\pi_t[h^l]d\widetilde{w}^l + \pi_t[h^l]\pi_t[h^l]dt\big)/V,$$

so that

$$M dN = -\pi_t[\eta]\pi_t[h^l]d\widetilde{w}^l + \pi_t[\eta]\pi_t[h^l]\pi_t[h^l]dt$$

and

$$dMdN = -\pi_t[\mathcal{M}^l\eta]\pi_t[h^l]dt.$$

The rest of part (i) follows by Theorems 6.3 and 6.4.

Let us now prove uniqueness of solution of Eq. (6.3.5), that is, part (ii) of the theorem. Similar to the process V, define

$$\bar{V}(t) = \exp\left\{ \int\limits_{[0,t]} \bar{\pi}_s[h^l]\, d\widetilde{w}^l(s) - \frac{1}{2}\int\limits_{[0,t]} \sum_{l=1}^{d_1} \left|\bar{\pi}_s[h^l]\right|^2 ds \right\},$$

and then set $u(t,x) := \pi(t,x)V(t)$, $\bar{u}(t,x) := \bar{\pi}(t,x)\bar{V}(t)$.

The Itô formula applied to u and $\bar{u}$ shows that each of these functions is a generalized solution of the linear filtering equation (6.3.1), (6.3.2), whose solution is unique by Theorem 4.5. In particular, keeping in mind that the measures $\mathbb{P}$ and $\widetilde{\mathbb{P}}$ are mutually absolutely continuous,

$$\mathbb{P}\left(\sup_{t\in[0,T]} \|u-\bar{u}\|_0^2(t) > 0\right) = \widetilde{\mathbb{P}}\left(\sup_{t\in[0,T]} \|u-\bar{u}\|_0^2(t) > 0\right) = 0.$$

Because by Proposition 6.1

$$V(t) = \int\limits_{\mathbb{R}^d} u(t,x)dx,\ \bar{V}(t) = \int\limits_{\mathbb{R}^d} \bar{u}(t,x)dx,$$

it follows that

$$\mathbb{P}\left(\sup_{t\in[0,T]} |V(t)-\bar{V}(t)| > 0\right) = 0.$$

Consequently,

$$\mathbb{P}\left(\sup_{t\in[0,T]} \|\pi-\bar{\pi}\|_0^2(t) > 0\right) = \mathbb{P}\left(\sup_{t\in[0,T]} \|(u/V)-(\bar{u}/\bar{V})\|_0^2(t) > 0\right) = 0,$$

completing the proof. □

While equality (6.3.5) is a non-linear equation, it is possible to treat $t \mapsto \pi_t[h^l]$ as a known process (e.g. by first finding the function u from (6.3.1), (6.3.2) and then computing $\pi_t[h^l]$ using (6.3.4)) and re-write (6.3.5) as a homogeneous linear equation for the filtering density $\pi = \pi(t, x)$ on the stochastic basis $\widetilde{\mathbb{F}}_{\mathscr{Y}}$:

$$d\pi(t,x) = \Big(\mathcal{L}^*\pi(t,x) - \pi_t[h^l]\mathcal{M}^{l,*}\pi(t,x) + \pi_t[h^l]\pi_t[h^l]\,\pi(t,x)\Big)\,dt \tag{6.3.6}$$

$$+ \Big(\mathcal{M}^{l,*}\pi(t,x) - \pi_t[h^l]\pi(t,x)\Big)d\widetilde{w}^l(t),\ (t,x,\omega) \in (0,T] \times \mathbb{R}^{\mathrm{d}} \times \Omega;$$

$$\pi(0,x) = \pi_0(x), \quad (x,\omega) \in \mathbb{R}^{\mathrm{d}} \times \Omega. \tag{6.3.7}$$

Then one can use the results of Chaps. 4 and 5 to establish various regularity properties of $\pi = \pi(t,x)$ depending on regularity of the coefficients of the original diffusion model (6.1.1). Indeed, problem (6.3.6), (6.3.7) is of the type (4.1.1), (4.2.2) (see Remark 5.6) and satisfies the conditions of Theorem 4.5 and Corollary 4.4, leading to the following

Corollary 6.2 *In addition to conditions* (i) *and* (ii) *on page 221, assume that the coefficients in* (6.3.6) *have the following regularity with respect to the variable* x, *for all* $(t,y) \in [0,T] \times \mathbb{R}^{\mathrm{d}_1}$*:*

- *the functions* $\sigma^{il} = \sigma^{il}(t,x,y)$ *have* $m+2$ *derivatives,*
- *the functions* $a^{ij} = a^{ij}(t,x,y)$ *and* $h^l = h^l(t,x,y)$ *have* $m+1$ *derivatives,*
- *the functions* ${}_{\mathrm{X}}b^i = {}_{\mathrm{X}}b^i(t,x,y)$ *have* m *derivatives,*

where $i, j = 1, 2, \ldots, \mathrm{d}$, $l = 1, 2, \ldots, \mathrm{d}_1$, *and* m *is a positive integer, and all these functions and their derivatives are uniformly bounded by the number* K. *Assume also that* $\pi_0 \in \mathbb{L}_{p'}(\Omega; \mathbb{W}^m_{p'})$ *for* $p' = 2$ *and* $p' = p \in (2,\infty)$.

Then the function $\pi = \pi(t,x)$ *has the following properties:*

1. $\pi \in \mathbb{L}_2\big(\Omega; \mathbb{C}([0,T]; \mathbb{H}^{m-1})\big) \cap \mathbb{L}_2\big([0,T] \times \Omega; \mathbb{C}_w\mathbb{H}^m\big) \cap \mathbb{L}_p\big([0,T] \times \Omega; \mathbb{C}_w\mathbb{W}^m_p\big)$;
2. $\widetilde{\mathbb{E}} \sup_{t\in[0,T]} \|\pi(t)\|^{p'}_{m,p'} \leq N\widetilde{\mathbb{E}}\|\pi_0\|^{p'}_{m,p'}$, $p' = 2$ and $p' = p$, *where* N *depends only on* p', d, d_1, m, T, *and* K. □

6.3.5. The following result is derived from Theorem 6.5 and Corollary 6.2 using the same arguments as in the proofs of Theorem 4.3 and Corollary 4.3.

Corollary 6.3 *Suppose that the assumptions of Corollary 6.2 are satisfied and there exists a non-negative integer* n *such that* $(m-n)p > \mathrm{d}$.

Then the filtering density $\pi = \pi(t,x)$ *has a version, also denoted by* π, *such that*

(a) *For every* $x \in \mathbb{R}^{\mathrm{d}}$, $t \mapsto \pi(t,x,\omega)$ *is a predictable real-valued stochastic process.*
(b) *For all* ω, $\pi \in \mathbb{C}^{0,n}_b([0,T] \times \mathbb{R}^{\mathrm{d}})$.
(c) π *possesses all the properties from the previous paragraph.*
(d) $\mathbb{E}\sup_{t\in[0,T]} \|\pi(t)\|^p_{\mathbb{C}^n_b(\mathbb{R}^{\mathrm{d}})} < \infty$.

(e) *If $n \geq 2$, then π is a classical solution of (6.3.5), that is, for every $x \in \mathbb{R}^{\mathrm{d}}$ and for all $(t,\omega) \in [0,T] \times \Omega'$ with $\mathbb{P}(\Omega') = 1$, the following equality holds*

$$\pi(t,x) = \pi_0(x) + \int_{[0,t]} \mathcal{L}_0^* \pi(s,x)\,ds + \int_{[0,t]} \Big(\mathcal{M}^{l,*}\pi(s,x) - \pi_s[h^l]\pi(s,x)\Big)\,d\check{w}^l(s),$$

where $d\check{w}^l(s) = d\widetilde{w}^l(s) \ - \pi_s[h^l]ds$.

(f) *If both π and $\bar{\pi}$ satisfy properties (a), (b), and (e) and belong to $\mathbb{L}_2^{\omega}([0,T];\mathbb{H}^1) \cap \mathbb{L}_2^{\omega}([0,T];\mathbb{L}_1)$, then*

$$\mathbb{P}\left(\sup_{(t,x)\in[0,T]\times\mathbb{R}^{\mathrm{d}}} |\pi(t,x,\omega) - \bar{\pi}(t,x,\omega)| > 0\right) = 0. \qquad \square$$

6.3.6. In this paragraph it is supposed that the assumptions of Corollary 6.3 are satisfied for $n \geq 2$. Using the results of Sect. 5.3, we will derive an alternative representation of the unnormalized filtering density, one that involves random, as opposed to stochastic, parabolic equations.

Consider the system of Itô's equations,

$$\eta^i(t,x) = x^i + \int_{[0,t]} \sigma^{il}\big(s,\eta(s,x),Y(s)\big)\,d\widetilde{w}^l(s),\ t \in [0,T],\ i = 1,2,\ldots,\mathrm{d}.$$

Corollary 5.3 implies that the process $\eta(t,x) := \big(\eta^1(t,x),\ldots,\eta^{\mathrm{d}}(t,x)\big)$ is a flow of $\mathbb{C}^{0,2}$-diffeomorphisms of $\mathbb{R}^{\mathrm{d}}$ onto $\mathbb{R}^{\mathrm{d}}$. Denote by $\left|\frac{\partial\eta(t,x)}{\partial x}\right|$ the Jacobian of the transformation $x \mapsto \eta(t,x)$.

Define

$$\begin{aligned}
{}_1\psi(t,x) :=& \exp\Bigg\{-\int_{[0,t]} h^l\big(s,\eta(s,x),Y(s)\big)\,d\widetilde{w}^l(s)\\
&+\frac{1}{2}\int_{[0,t]} (h^l h^l)\big(s,\eta(s,x),Y(s)\big)\,ds\Bigg\},\\
{}_1\widetilde{\psi}(t,x) :=& \exp\Bigg\{\int_{[0,t]} h^l\big(s,x,Y(s)\big)\,d\widetilde{w}^l(s) - \frac{1}{2}\int_{[0,t]} (h^l h^l)\big(s,x,Y(s)\big)\,ds\Bigg\},
\end{aligned}$$

$$\begin{aligned}
{}_1\mathcal{L}u :=& a^{ij}\big(t,x,Y(t)\big)u_{ij} + {}_1b^i\big(t,x,Y(t)\big)u_i + {}_1c\big(t,x,Y(t)\big)u,\\
{}_1\mathcal{M}^l u :=& -\sigma^{ij}\big(t,x,Y(t)\big)u_i + {}_1h^l\big(t,x,Y(t)\big)u,
\end{aligned}$$

where ${}_1b^i := 2a^{ij}_j - {}_{\mathrm{X}}b^i$, ${}_1c := a^{ij}_{ij} - {}_{\mathrm{X}}b^i_i$, ${}_1h^l := h^l - \sigma^{il}_i$.

Let ${}_1\widetilde{\mathcal{L}}$ be the differential operator constructed using the operators ${}_1\mathcal{L}$, ${}_1\mathcal{M}^l$ and the functions ${}_1\psi$, ${}_1\widetilde{\psi}$ in the same way as the operator $\widetilde{L}$ is constructed in Sect. 5.3.3 using the operators L, B^l and the functions ψ, $\widetilde{\psi}$. Denote by v the classical solution of the problem

$$\frac{\partial v(t,x)}{\partial t} = {}_1\widetilde{\mathcal{L}}v(t,x),$$
$$v(0,x) = \pi_0(x).$$

Theorem 5.5, Theorem 6.3, and Corollary 6.3 imply

$$u(t,x) = {}_1\widetilde{\psi}(t,x)v\big(t,\eta^{-1}(t,x)\big),$$

where η^{-1} is the inverse of η: $\eta^{-1}\big(t,\eta(t,x)\big) = x$. This leads to the following version of Corollary 6.1.

Corollary 6.4 *For every $f \in \mathbb{L}_\infty$, there exists a set $\Omega' \in \mathcal{F}$ with $\mathbb{P}(\Omega') = 1$ such that the equality*

$$\mathbb{E}\big[f\big(X(t)\big)|\mathcal{Y}_t^0\big] = \frac{\displaystyle\int_{\mathbb{R}^d} f\big(\eta(t,x)\big)\,{}_1\widetilde{\psi}\big(t,\eta(t,x)\big)v(t,x)\left|\frac{\partial\eta(t,x)}{\partial x}\right|dx}{\displaystyle\int_{\mathbb{R}^d} {}_1\widetilde{\psi}\big(t,\eta(t,x)\big)v(t,x)\left|\frac{\partial\eta(t,x)}{\partial x}\right|dx}$$

holds for all $t \in [0,T]$ and all $\omega \in \Omega'$. □

It is possible to show that, unlike the function u, the function v depends continuously on the trajectory $Y = Y(s), s \le t$: small, in a certain sense, perturbations of Y lead to small changes of v. As a result, while not especially convenient for actual computations, the above result can be useful for certain theoretical investigations of the filtering problem.

Remark 6.3 Looking back at Itô's equation for the process $Z = Z(t)$, one can see that the analysis of the filtering problem does not change if the coefficients in the equation depend, in a measurable way, on the whole past of Y, that is, each coefficient at time t depends on $Y(s)$, $s \le t$. Of course, extra conditions are necessary to ensure the existence of a solution of the resulting equation, cf. [93, Theorem 4.6]. □

6.4 The Backward Filtering Equation, Interpolation, and Extrapolation

6.4.1. In this section we derive the backward filtering equation describing the dynamics of $\mathbb{E}[f(\mathbb{X}(T_1))|\mathscr{Y}_{T_1}^s \vee \mathscr{X}_s^s]$ with respect to s for a fixed $T_1 \in (s, T]$. We also establish absolute continuity of the interpolation and the extrapolation measures with respect to the Lebesgue measure and investigate the structure of the corresponding densities.

The notation introduced previously in this chapter will remain in this section. In particular, the reader is encouraged to review the definitions of the matrices ${}_{\mathrm{Z}}\boldsymbol{\Sigma}, \sigma, {}_0\sigma, {}_1\sigma, A, a$, and D, vectors ${}_{\mathrm{X}}b$ and h, and the Brownian motions $\overline{W}$ and $\widetilde{W}$, as well as two different equations satisfied by the process Z, on $\mathbb{F}$ and $\widetilde{\mathbb{F}}$.

Warning 6.3 *As in most of this chapter, we assume that* ${}_{\mathrm{Z}}\boldsymbol{\Sigma}, {}_{\mathrm{X}}b$, *and* h *are functions of* (t, z) *only and do not depend on* ω. *In addition, for every* $t \in [0, T]$, $i, j = 1, 2, \ldots, \mathrm{d}+\mathrm{d}_1$, *and* $l = 1, 2, \ldots, \mathrm{d}_1$, *the functions* $A^{ij}(t, z)$, $D^{il}(t, z)$, *and* $h^l(t, z)$ *are twice differentiable in* z, *the functions* ${}_{\mathrm{X}}b^i(t, z)$ *are once differentiable in* z, *and all the derivatives are uniformly bounded.* □

The key background result used in this section is Theorem 5.1, which makes the following observation useful:

$$2A \equiv {}_{\mathrm{Z}}\boldsymbol{\Sigma}\,{}_{\mathrm{Z}}\boldsymbol{\Sigma}^* = DD^* + \hat{D}\hat{D}^*, \text{ with } \hat{D} = \begin{pmatrix} {}_1\sigma \\ 0 \end{pmatrix}.$$

We also define

$${}_{\mathrm{Z}}\tilde{b} = \begin{pmatrix} {}_{\mathrm{X}}b \\ 0 \end{pmatrix}.$$

6.4.2. The objective of this paragraph is to derive the equation satisfied by $\mathbb{E}[f(X(T_1))|\mathscr{Y}_{T_1}^s \vee \mathscr{X}_s^s]$ as a function of s.

For $\varphi \in \mathrm{C}_b^1(\mathbb{R}^{\mathrm{d}+\mathrm{d}_1})$ and $r < -(\mathrm{d} + \mathrm{d}_1)/2$, denote by $v_\varphi = v_\varphi(s, z, \omega)$ the continuous in (s, z) version of the r-generalized solution of the equation

$$\begin{aligned} -dv(s, z, \omega) &= \big(A^{ij}(s, z)v_{ij}(s, z, \omega) + {}_{\mathrm{Z}}\tilde{b}^i(s, z)v_i(s, z, \omega)\big)\, ds \\ &\quad + \big(D^{il}(s, z)v_i(s, z, \omega) + h^l(s, z)v(s, z, \omega)\big) * d\widetilde{w}^l(s, \omega), \qquad (6.4.1) \\ &\quad (s, z, \omega) \in [0, T_1) \times \mathbb{R}^{\mathrm{d}+\mathrm{d}_1} \times \Omega, \end{aligned}$$

with the terminal condition

$$v(T_1, z) = \varphi(z), \; (z, \omega) \in \mathbb{R}^{d+\mathrm{d}_1} \times \Omega. \qquad (6.4.2)$$

To avoid confusion with the partial derivative $\partial v/\partial x^1$, the function v_φ corresponding to $\varphi(x) \equiv 1$ will be denoted by $v_{[1]}$. As usual, we will typically omit the argument ω from the corresponding notation.

Existence and uniqueness of v_φ follow from Theorem 4.7 and Corollary 4.8 (see also Warning 4.5), combined with the inclusion $\mathbb{C}_b^1(\mathbb{R}^{\mathrm{d}+\mathrm{d}_1}) \subset \mathbb{W}_p^1(r, \mathbb{R}^{d+\mathrm{d}_1})$ for every $p \geq 2$ and in particular for $p > \mathrm{d} + \mathrm{d}_1$ (cf. Lemma 3.6).

We will now generalize Theorem 5.1 to allow a random initial condition in the corresponding diffusion process.

Theorem 6.6 *For every* $f \in \mathbb{C}_b^1(\mathbb{R}^{\mathrm{d}+\mathrm{d}_1})$ *and* $T_1 \in [T_0, T]$,

$$\mathbb{E}\big[f\big(Z(T_1)\big)\big|\mathscr{Y}_{T_1}^0 \vee \mathscr{X}_{T_0}^0\big] = \frac{v_f\big(T_0, Z(T_0)\big)}{v_{[1]}\big(T_0, Z(T_0)\big)} \quad (\mathbb{P}\text{- a.s.}). \tag{6.4.3}$$

□

Proof In addition to the process $Z = Z(t)$ we consider, on the stochastic basis $\widetilde{\mathbb{F}}$, the process $\mathcal{Z} = \mathcal{Z}(t, s, z)$ defined for fixed $s \in [0, T)$ and $z \in \mathbb{R}^{\mathrm{d}+\mathrm{d}_1}$ by

$$\mathcal{Z}(t, s, z) = z + \int_{[s,t]} {}_{Z}\mathcal{B}\big(\tau, \mathcal{Z}(\tau, s, z)\big)\, d\tau + \int_{[s,t]} {}_{Z}\boldsymbol{\Sigma}\big(\tau, \mathcal{Z}(\tau, s, z)\big)\, d\widetilde{\mathcal{W}}(\tau),$$

$t \in [s, T]$, where

$${}_{Z}\mathcal{B} = \begin{pmatrix} {}_{X}b - \sigma h \\ 0 \end{pmatrix}, \quad {}_{Z}\boldsymbol{\Sigma} = \begin{pmatrix} {}_{1}\sigma & \sigma \\ 0 & {}_{0}\sigma \end{pmatrix}, \quad \widetilde{\mathcal{W}} = \begin{pmatrix} \overline{W} \\ \widetilde{W} \end{pmatrix}.$$

If

$$\rho(t, s, z) := \exp\left\{ \int_{[s,t]} h^l\big(\tau, \mathcal{Z}(\tau, s, z)\big)\, d\widetilde{w}^l(\tau) - \frac{1}{2} \int_{[s,t]} |h|^2\big(\tau, \mathcal{Z}(\tau, s, z)\big)\, d\tau \right\},$$

then, by Theorem 5.1 with $\mathcal{Z}$ instead of $\mathcal{X}$,

$$v_\varphi(s, z) = \widetilde{\mathbb{E}}\big[\varphi\big(\mathcal{Z}(T_1, s, z)\big)\rho(T_1, s, z)\big|\mathscr{F}_{T_1}^s\big]. \tag{6.4.4}$$

We now make the following observations:

1. By Theorem 6.4 with $T_1 = T_2$,

$$\mathbb{E}\big[f\big(Z(T_1)\big)\big|\mathscr{Y}_{T_1}^0 \vee \mathscr{X}_{T_0}^0\big] = \frac{\widetilde{\mathbb{E}}\big[f\big(Z(T_1)\big)\rho(T_1, T_0)\big|\tilde{\mathscr{F}}_{T_1}^{T_0} \vee \mathscr{Z}_{T_0}^{T_0}\big]}{\widetilde{\mathbb{E}}\big[\rho(T_1, T_0)\big|\tilde{\mathscr{F}}_{T_1}^{T_0} \vee \mathscr{Z}_{T_0}^{T_0}\big]};$$

recall that $\rho(t, s)$ is defined in (6.2.4).

2. By uniqueness, $\mathcal{Z}\big(t, T_0, Z(T_0)\big) = Z(t)$, $t \in [T_0, T]$, and so $\rho(T_1, T_0) = \rho\big(T_1, T_0, Z(T_0)\big)$.

To complete the proof of the theorem, it remains to show that it is indeed possible to use Theorem 5.1 when the diffusion process starts from a random point:

$$\widetilde{\mathbb{E}}[\varphi(\mathcal{Z}(T_1, T_0, Z(T_0)))\rho(T_1, T_0, Z(T_0))|\tilde{\mathscr{F}}_{T_1}^{T_0} \vee \mathscr{Z}_{T_0}^{T_0}] = v_\varphi(T_0, Z(T_0)). \tag{6.4.5}$$

The proof of (6.4.5) uses (6.4.4) and a piece-wise constant approximation of $Z(T_0)$.

For a non-negative integer n and a real number a, define

$$x_n(a) := 2^{-n}[2^n a],$$

where $[a]$ is the integer part of a, and then introduce the following objects:

- the vector $x_n\big(Z(T_0)\big)$ with components $x_n\big(Z^1(T_0)\big), \ldots, x_n\big(Z^{\mathrm{d}+\mathrm{d}_1}(T_0)\big)$;
- the countable set $\Gamma_n \subset \mathbb{R}^{\mathrm{d}+\mathrm{d}_1}$, which is the collection of possible values of $x_n\big(Z(T_0, \omega)\big)$ for $\omega \in \Omega$;
- the process $Z^n = Z^n(t)$, with $Z^n(t) = \mathcal{Z}\big(t, x_n\big(Z(T_0)\big)\big)$;
- the random variable $\rho^n(T_1, T_0)$, constructed using (6.3.3), but with process Z^n instead of Z.

By uniqueness,

$$Z^n(t) = \sum_{z \in \Gamma^n} 1_{\{x_n(Z(T_0))=z\}} Z(t, z, T_0). \tag{6.4.6}$$

Continuity of Z with respect to the initial condition implies the existence of a sequence $\{n'\}$ such that

$$\lim_{n' \to \infty} \sup_{s \in [T_0, T]} \left| Z^{n'}(s) - Z(s) \right| = 0 \qquad (\mathbb{P}\text{- a.s.}),$$

and then, possibly along a further subsequence $\{k\}$,

$$\lim_{k \to \infty} f\big(Z^k(T_1)\big)\rho^k(T_1, T_0) = \varphi\big(Z(T_1)\big)\rho(T_1, T_0) \quad (\mathbb{P}\text{- a.s.}).$$

Let η be a bounded $\tilde{\mathscr{F}}_{T_1}^{T_0} \vee \mathscr{Z}_{T_0}^{T_0}$-measurable random variable. The definition of ρ^k and boundedness of φ and η imply

$$\sup_k \widetilde{\mathbb{E}} \big|\varphi\big(Z^k(T_1)\big)\rho^k(T_1, T_0)\eta\big|^2 < \infty,$$

meaning that the collection $\{\varphi(Z^k(T_1))\rho^k(T_1, T_0)\eta,\ k \geq 1\}$ is uniformly $\widetilde{\mathbb{P}}$-integrable and hence

$$\lim_{k\to\infty} \widetilde{\mathbb{E}}\Big(\varphi(Z^k(T_1))\rho^k(T_1, T_0)\eta\Big) = \widetilde{\mathbb{E}}\Big(\varphi(Z(T_1))\rho(T_1, T_0)\eta\Big). \tag{6.4.7}$$

Combining (6.4.6) and (6.4.7) yields

$$\begin{aligned}
&\widetilde{\mathbb{E}}\Big(\varphi(Z(T_1))\rho(T_1, T_0)\eta\Big)\\
&= \lim_{k\to\infty} \widetilde{\mathbb{E}}\left(\sum_{z\in\Gamma_k} 1_{\{x_k(Z(T_0))=z\}}\varphi\big(\mathcal{Z}(T_1, T_0, z)\big)\rho(T_1, T_0, z)\eta\right)\\
&= \lim_{k\to\infty} \widetilde{\mathbb{E}}\left(\sum_{z\in\Gamma_k} 1_{\{x_k(Z(T_0))=z\}}\widetilde{\mathbb{E}}\Big[\varphi\big(\mathcal{Z}(T_1, T_0, z)\big)\rho(T_1, T_0, z)\Big|\mathcal{F}^{T_0}_{T_1}\vee\mathcal{Z}^{T_0}_{T_0}\Big]\eta\right).
\end{aligned}$$

On the other hand, (6.4.4) means that, for each k,

$$\begin{aligned}
&v_\varphi\big(T_0, Z^k(T_0)\big)\eta\\
&= \sum_{z\in\Gamma_k} 1_{\{x_k(Z(T_0))=z\}}\widetilde{\mathbb{E}}\Big[\varphi\big(\mathcal{Z}(T_1, T_0, z)\big)\rho(T_1, T_0, z)\Big|\mathcal{F}^{T_0}_{T_1}\vee\mathcal{Z}^{T_0}_{T_0}\Big]\eta,
\end{aligned} \tag{6.4.8}$$

whereas continuity of the function $z \mapsto v_\varphi(T_0, z)$ implies

$$\lim_{k\to\infty} v_\varphi\big(T_0, Z^k(T_0)\big)\eta = v_f\big(T_0, Z(T_0)\big)\eta.$$

Now (6.4.5) will follow after passing to the limit as $k \to \infty$ in (6.4.8). To pass to this limit, it is enough to show uniform $\widetilde{\mathbb{P}}$-integrability of the family $\{v_\varphi(T_0, Z^k(T_0))\eta,\ k \geq 1\}$, a sufficient condition for which is

$$\sup_k \widetilde{\mathbb{E}}|v_\varphi\big(T_0, Z^k(T_0)\big)\eta|^2 < \infty. \tag{6.4.9}$$

By (6.4.8),

$$\begin{aligned}
&\widetilde{\mathbb{E}}\left|v_\varphi(T_0, Z^k(T_0))\eta\right|^2\\
&= \widetilde{\mathbb{E}}\sum_{z\in\Gamma_k} 1_{\{x_k(Z(T_0))=z\}}\left(\widetilde{\mathbb{E}}\Big[\varphi\big(\mathcal{Z}(T_1, T_0, z)\big)\rho(T_1, T_0, z)\Big|\mathcal{F}^{T_0}_{T_1}\vee\mathcal{Z}^{T_0}_{T_0}\Big]\eta\right)^2.
\end{aligned}$$

By the Cauchy–Schwarz inequality,

$$\Big(\widetilde{\mathbb{E}}\Big[\varphi\big(\mathcal{Z}(T_1,T_0,z)\big)\rho(T_1,T_0,z)\Big|\tilde{\mathscr{F}}^{T_0}_{T_1}\vee\mathscr{Z}^{T_0}_{T_0}\Big]\eta\Big)^2$$
$$\leq\widetilde{\mathbb{E}}\Big[\Big(\varphi\big(\mathcal{Z}(T_1,T_0,z)\big)\rho(T_1,T_0,z)\Big)^2\Big|\tilde{\mathscr{F}}^{T_0}_{T_1}\vee\mathscr{Z}^{T_0}_{T_0}\Big]\eta^2,$$

and then

$$\widetilde{\mathbb{E}}\big|v_\varphi\big(T_0,Z^k(T_0)\big)\eta\big|^2$$
$$\leq\widetilde{\mathbb{E}}\sum_{z\in\Gamma_k}1_{\{x_k(Z(T_0))=z\}}\widetilde{\mathbb{E}}\Big[\Big(\varphi\big(Z(T_1,T_0,z)\big)\rho(T_1,T_0,z)\eta\Big)^2\Big|\tilde{\mathscr{F}}^{T_0}_{T_1}\vee\mathscr{Z}^{T_0}_{T_0}\Big]$$
$$=\widetilde{\mathbb{E}}\Big(\varphi\big(Z^k(T_1)\big)\rho^k(T_1,T_0)\eta\Big)^2,$$

from which (6.4.9) follows, completing the proof of the theorem. The reader who tried to follow the exceptional sets in each of the $\mathbb{P}$-a.s. (or $\widetilde{\mathbb{P}}$-a.s.) equalities will notice that there were only countably many of those sets, so that (6.4.3) indeed holds. □

Remark 6.4 From (6.4.5) and Lemmas 6.1 and 6.2 we obtain that, under the assumptions of the theorem, the following equality holds $\mathbb{P}$- a.s.:

$$\widetilde{\mathbb{E}}\big[\varphi\big(Z(T_1)\big)\rho(T_1,T_0)\big|\mathscr{Y}^0_{T_1}\vee\mathscr{X}^0_{T_0}\big]=\widetilde{\mathbb{E}}\big[\varphi\big(Z(T_1)\big)\rho(T_1,T_0)\big|\mathscr{Y}^0_{T_1}\vee\mathscr{X}^{T_0}_{T_0}\big]$$
$$=v_\varphi\big(T_0,Z(T_0)\big).$$

Accordingly, by analogy with (6.3.1), we will call (6.4.1) the **backward filtering equation**. □

If the functions ${}_{\mathrm{X}}b$, h, and ${}_{\mathrm{Z}}\boldsymbol{\Sigma}$ do not depend on y, then $A^{ij}=a^{ij}$ and $D^{il}=\sigma^{il}$ for $i,j=1,2,\ldots,\mathrm{d}$, $l=1,2,\ldots,\mathrm{d}_1$, so that (6.4.1), turns into

$$\begin{aligned}-dv(s,x)&=\big(a^{ij}(s,x)v_{ij}(s,x)+{}_{\mathrm{X}}b(s,x)v_i(s,x)\big)\,ds\\&\quad+\big(\sigma^{il}(s,x)v_i(s,x)+h^l(s,x)v(s,x)\big)*d\widetilde{w}^l(s).\end{aligned}\tag{6.4.10}$$

Given a function $\varphi\in\mathbb{C}^1_b(\mathbb{R}^{\mathrm{d}})$, we denote by $v^X_\varphi=v^X_\varphi(s,x)$ the continuous in (s,x) version of the solution of (6.4.10) satisfying the terminal condition $v^X_\varphi(T_1,x)=\varphi(x)$. As before, we write $v^X_{[1]}$ when $\varphi\equiv1$.

Because now the function $\rho=\rho(t,s,z)$ does not depend on y either, equality (6.4.4) becomes

$$v^X_\varphi(s,x)=\widetilde{\mathbb{E}}\big[\psi\big(X(T_1,s,x)\big)\rho(T_1,s,x)\big|\tilde{\mathscr{F}}^s_{T_1}\big].$$

Corollary 6.5 *Let the assumptions of Theorem 6.1 hold and assume that the functions* ${}_{\mathrm{X}}b$, h, *and* ${}_{\mathrm{Z}}\boldsymbol{\Sigma}$ *do not depend on* y. *Then* $X = X(t)$ *is a Markov process on both* $\mathbb{F}$ *and* $\widetilde{\mathbb{F}}$, *and, for every* $f \in \mathbb{C}_b^1(\mathbb{R}^{\mathrm{d}})$,

$$\mathbb{E}\big[f\big(X(T_1)\big)\big|\mathscr{Y}_{T_1}^0 \vee \mathscr{X}_{T_0}^0\big] = \frac{v_f^X\big(T_0, X(T_0)\big)}{v_{[1]}^X\big(T_0, X(T_0)\big)} \quad (\mathbb{P}\text{-a.s.}). \qquad \square$$

Proof The Markov property of X follows from the equations satisfied by X:

$$dX(t) = {}_{\mathrm{X}}b\big(t, X(t)\big)\,dt + {}_{1}\sigma\big(t, X(t)\big)d\overline{W}(t) + \sigma\big(t, X(t)\big)dW(t)$$

on $\mathbb{F}$ and

$$\begin{aligned} dX(t) = \Big({}_{\mathrm{X}}b\big(t, X(t)\big) - \sigma\big(t, X(t)\big)h\big(t, X(t)\big)\Big)\,dt \\ + {}_{1}\sigma\big(t, X(t)\big)d\overline{W}(t) + \sigma\big(t, X(t)\big)\,d\widetilde{W}(t) \end{aligned}$$

on $\widetilde{\mathbb{F}}$. The rest follows from (6.4.3) and uniqueness of an r-generalized solution for (6.4.1), (6.4.2). $\square$

6.4.3. In what follows, we keep the assumptions and notation first introduced earlier in this chapter. In particular we denote the unnormalized filtering density by $u = u(t, x)$ and recall that $\mathscr{M}(\mathbb{R}^{\mathrm{d}})$ stands for the collection of all probability measures on $\big(\mathbb{R}^{\mathrm{d}}, \mathscr{B}(\mathbb{R}^{\mathrm{d}})\big)$.

In connection with the function $v_{[1]}$, that is, the continuous version of the r-generalized solution of the backward filtering equation (6.4.1) with terminal condition $v_{[1]}(T_1, x) \equiv 1$, we also note that (6.4.3) implies $v_{[1]}(s, x) > 0$ for all s, x on the same set of probability one.

Theorem 6.7 *There exists a function* $P_{\mathscr{Y}}^{T_1,T_0} : \Omega \to \mathscr{M}(\mathbb{R}^{\mathrm{d}})$, *called* ***the interpolation measure***, *with the following properties:*

(a) $P_{\mathscr{Y}}^{T_1,T_0}$ *is the regular conditional probability distribution of* $X(T_0)$ *with respect to* $\mathscr{Y}_{T_1}^0$.

(b) $P_{\mathscr{Y}}^{T_1,T_0}$ *is* $\mathbb{P}$-a.s. *absolutely continuous with respect to the Lebesgue measure on* $\mathbb{R}^{\mathrm{d}}$, *with the Radon–Nikodym derivative*

$$\pi^{T_1,T_0}(x) := \frac{v_{[1]}\big(T_0, x, Y(T_0)\big)u(T_0, x)}{\displaystyle\int_{\mathbb{R}^{\mathrm{d}}} v_{[1]}\big(T_0, x, Y(T_0)\big)u(T_0, x)\,dx},$$

called ***the interpolation density***. $\square$

Proof Compared to Corollary 6.5, we now need to compute $\mathbb{E}\big[f\big(X(T_0)\big)\big|\mathscr{Y}_{T_1}^0\big]$. The plan is to compute $\mathbb{E}\big[f\big(X(T_0)\big)\big|\mathscr{Y}_{T_0}^0\big]$ using the forward filtering equation, and then combine the result with Corollary 6.5.

The first step is to extend Theorem 6.3 to random functions f.

Lemma 6.3 *Let $\psi : \mathbb{R}^{\mathrm{d}} \times \Omega \to \mathbb{R}^1$ be a $\mathscr{B}(\mathbb{R}^{\mathrm{d}}) \otimes \mathscr{Y}_{T_1}^0$-measurable function with $\mathbb{E}\big|\psi\big(X(T_0)\big)\big| < \infty$.*

Then

$$\widetilde{\mathbb{E}}\big[\psi\big(X(T_0)\big)\rho(T_0,0)\big|\mathscr{Y}_{T_1}^0\big] = \int_{\mathbb{R}^{\mathrm{d}}} \psi(x)u(T_0,x)\,dx. \tag{6.4.11}$$

Proof The argument is very similar to the derivation of (6.4.3). In particular, we use the functions $x_n(a) = 2^{-n}[2^n a]$, the vectors $x_n\big(X(T_0)\big) \in \mathbb{R}^{\mathrm{d}}$ with components $\Big(x_n\big(X^1(T_0)\big), \dots, x_n\big(X^{\mathrm{d}}(T_0)\big)\Big)$, and the countable sets $\Gamma_n \subset \mathbb{R}^{\mathrm{d}}$ of the values of $x_n\big(X(T_0,\omega)\big)$, $\omega \in \Omega$.

With no loss of generality, assume that ψ is bounded and continuous as a function of x, and $\psi \geq 0$.

Due to the independence of $\mathscr{Y}_{T_0}^0$ and $\tilde{\mathscr{F}}_{T_1}^{T_0}$, we obtain by Theorem 6.3 that, for $\mathbb{P}$-a.a. ω,

$$\begin{aligned}
&\widetilde{\mathbb{E}}\big[\psi\big(X(T_0)\big)\rho(T_0,0)\big|\mathscr{Y}_{T_1}^0\big] \\
&= \lim_{n\to\infty} \widetilde{\mathbb{E}}\big[\psi\big(x_n(X(T_0))\big)\rho(T_0,0)\big|\mathscr{Y}_{T_1}^0\big] \\
&= \lim_{n\to\infty} \widetilde{\mathbb{E}}\left[\sum_{x\in\Gamma_n} \psi(x)1_{\{x_n(X(T_0))=x\}}\rho(T_0,0)\big|\mathscr{Y}_T^0 \vee \tilde{\mathscr{F}}_{T_1}^{T_0}\right] \\
&= \lim_{n\to\infty} \sum_{x\in\Gamma_n} \psi(x)\widetilde{\mathbb{E}}\left[1_{\{x_n(X(T_0))=x\}}\rho(T_0,0)\big|\mathscr{Y}_{T_0}^0\right] \\
&= \lim_{n\to\infty} \sum_{x\in\Gamma_n} \psi(x)\int_{\mathbb{R}^{\mathrm{d}}} 1_{\{x_n(x)=x\}}u(T_0,x)\,dx \\
&= \lim_{n\to\infty} \int_{\mathbb{R}^{\mathrm{d}}} \psi\big(x_n(x)\big)u(T_0,x)\,dx = \int_{\mathbb{R}^{\mathrm{d}}} \psi(x)u(T_0,x)\,dx,
\end{aligned}$$

completing the proof of the lemma. □

Coming back to the proof of the theorem, note that, by Theorem 6.1, for every $f \in \mathbb{L}_\infty$ the following equality holds $\mathbb{P}$-a.s.:

$$\mathbb{E}\big[f\big(X(T_0)\big)\big|\mathscr{Y}_{T_1}^0\big] = \frac{\widetilde{\mathbb{E}}\big[f\big(X(T_0)\big)\rho(T_1,0)\big|\mathscr{Y}_{T_1}^0\big]}{\widetilde{\mathbb{E}}\big[\rho(T_1,0)\big|\mathscr{Y}_{T_1}^0\big]}. \tag{6.4.12}$$

We now use (6.2.4) (to split ρ into a product) and general properties of the conditional expectation to write

$$\begin{aligned} I &:= \widetilde{\mathbb{E}}\big[f\big(X(T_0)\big)\rho(T_1,0)\big|\mathscr{Y}^0_{T_1}\big] \\ &= \widetilde{\mathbb{E}}\Big[f\big(X(T_0)\big)\rho(T_0,0)\widetilde{\mathbb{E}}\big[\rho(T_1,T_0)\big|\mathscr{Y}^0_{T_1}\vee\mathscr{X}^0_{T_0}\big]\Big|\mathscr{Y}^0_{T_1}\Big], \end{aligned}$$

and then combine the result with equality (6.4.11) and Remark 6.4:

$$\begin{aligned} I &= \widetilde{\mathbb{E}}\big[f\big(X(T_0)\big)\rho(T_0,0)v_{[1]}\big(T_0,Z(T_0)\big)\big|\mathscr{Y}^0_{T_1}\big] \\ &= \int_{\mathbb{R}^d} f(x)v_{[1]}\big(T_0,x,Y(T_0)\big)u(T_0,x)\,dx. \end{aligned} \tag{6.4.13}$$

Next, applying (6.4.13) with $f\equiv 1$,

$$\widetilde{\mathbb{E}}[\rho(T_1,0)|\mathscr{Y}^0_{T_1}] = \int_{\mathbb{R}^d} v_{[1]}(T_0,x,Y(T_0)u(T_0,x)\,dx. \tag{6.4.14}$$

Collecting (6.4.12)–(6.4.14), we complete the proof of the theorem by observing that

$$P^{T_1,T_0}_{\mathscr{Y}}(\Gamma) = \frac{\int_{\Gamma} v_{[1]}\big(T_0,x,Y(T_0)\big)u(T_0,x)\,dx}{\int_{\mathbb{R}^d} v_{[1]}(T_0,x,Y(T_0))u(T_0,x)\,dx}.$$

The reader is welcome to track down all the exceptional sets. □

Remark 6.5 By Theorem 6.3,

$$\widetilde{\mathbb{E}}[\rho(T_1,0)|\mathscr{Y}^0_{T_1}] = \int_{\mathbb{R}^d} u(T_1,x)\,dx,$$

which, together with (6.4.14), leads to an alternative representation of the interpolation density

$$\pi^{T_1,T_0}(x) = \frac{v_{[1]}\big(T_0,x,Y(T_0)\big)u(T_0,x)}{\int_{\mathbb{R}^d} u(T_1,x)\,dx}.$$

Note that, not surprisingly, $\pi^{T_0,T_0}(x) = \pi(T_0,x)$. □

6.4.4.

Remark 6.6 The interpolation problem can be reduced to the filtering problem by introducing an additional component in the unobservable process X. Specifically, we define

$$\hat{X}(t) = 1_{\{[0,T_0]\}}(t)X(t) + 1_{\{[T_0,T]\}}(t)X(T_0),$$
$$\check{X}(t) = \begin{pmatrix} X(t) \\ \hat{X}(t) \end{pmatrix}, \ \check{Z}(t) = \begin{pmatrix} \check{X}(t) \\ Y(t) \end{pmatrix}.$$

Then, for $f \in \mathbb{L}_\infty$,

$$\mathbb{E}\big[f\big(X(T_0)\big)\big|\mathscr{Y}^0_{T_1}\big] = \mathbb{E}\big[f\big(\hat{X}(T_1)\big)\big|\mathscr{Y}^0_{T_1}\big].$$

6.4.5. Recall that the extrapolation problem involves estimation of the process X over a period when the observations are not available. The natural way to proceed is to carry out filtering while the observations are available, and then continue with the unconditional distribution of the process X.

More precisely, the Markov property of the process Z means that, for every $f \in \mathbb{L}_\infty$,

$$\begin{aligned}\mathbb{E}\big[f\big(X(T_1)\big)\big|\mathscr{Y}^0_{T_0}\big] &\equiv \mathbb{E}\Big[\mathbb{E}\big[f\big(X(T_1)\big)\big|\mathscr{Z}^0_{T_0}\big]\Big|\mathscr{Y}^0_{T_0}\Big] \\ &= \mathbb{E}\Big[\mathbb{E}\big[f\big(X(T_1)\big)\big|\mathscr{Z}^{T_0}_{T_0}\big]\Big|\mathscr{Y}^0_{T_0}\Big].\end{aligned}$$

If $f \in \mathbb{C}(\mathbb{R}^{\mathrm{d}})$, then, by combining Theorem 5.1 with (6.4.3),

$$\mathbb{E}\big[f\big(X(T_1)\big)\big|\mathscr{Z}^{T_0}_{T_0}\big] = v\big(T_0, Z(T_0)\big),$$

where $v = v(t,z)$ is a continuous in (t,z) version of the r-generalized solution of the backward Kolmogorov equation

$$-\frac{\partial v(t,z)}{\partial t} = A^{ij}(t,z)v_{ij}(t,z) + {}_{\mathrm{X}}b^i(t,z)v_i(t,z),$$
$$(t,z) \in [0,T_1) \times \mathbb{R}^{\mathrm{d}+\mathrm{d}_1},$$
$$v(T_1,z) = f(x), \quad z = (x,y) \in \mathbb{R}^{\mathrm{d}+\mathrm{d}_1}.$$

In other words,

$$\mathbb{E}\big[f\big(X(T_1)\big)\big|\mathscr{Y}^0_{T_0}\big] = \int_{\mathbb{R}^{\mathrm{d}}} v\big(T_0, x, Y(T_0)\big)\pi(T_0,x)\,dx,$$

where $\pi(T_0,x)$ is the filtering density at time T_0.

Theorem 6.8 *Assume that process Z has transition density $p = p(t, z'; s, z)$, that is, for every $\Gamma \in \mathscr{B}(\mathbb{R}^{d+d_1})$,*

$$\mathbb{P}\big(Z(t) \in \Gamma | Z(s) = z\big) = \int_{\Gamma} p(t, z'; s, z)\, dz'.$$

Then there exists a function $P_{\mathscr{Y}}^{T_0,T_1} : \Omega \to \mathscr{M}(\mathbb{R}^d)$, called the ***extrapolation measure****, with the following properties:*

(a) *$P_{\mathscr{Y}}^{T_0,T_1}$ is the regular conditional probability distribution of $X(T_1)$ with respect to $\mathscr{Y}_{T_0}^0$.*
(b) *$P_{\mathscr{Y}}^{T_0,T_1}$ is $\mathbb{P}$-a.s. absolutely continuous with respect to the Lebesgue measure on $\mathbb{R}^d$, with the Radon–Nikodym derivative*

$$\pi^{T_0,T_1}(x) = \int_{\mathbb{R}^{d_1}} \int_{\mathbb{R}^d} p\big(T_1, x, y; T_0, x', Y(T_0)\big)\pi(T_0, x')\, dx'dy,$$

called the ***extrapolation density****.* □

Proof By definition, the transition density p is the fundamental solution of the Kolmogorov equation (both forward and backward), so that

$$v(T_0, z) = \int_{\mathbb{R}^{d+d_1}} p(T_1, z'; T_0, z) f(x')\, dz'.$$

As a result,

$$\mathbb{E}\big[f\big(X(T_1)\big)|\mathscr{Y}_{T_0}^0\big] = \int_{\mathbb{R}^d} \int_{\mathbb{R}^{d_1}} \int_{\mathbb{R}^d} p\big(T_1, x', y'; T_0, x, Y(T_0)\big)\pi(T_0, x) f(x')\, dx\, dx'\, dy'$$
$$= \int_{\mathbb{R}^d} \pi^{T_0,T_1}(x') f(x')\, dx',$$

completing the proof. Note that $\pi^{T_0,T_0}(x) = \pi(T_0, x)$. □

6.4.6. On the one hand, the smoothness assumptions about the coefficients and the initial and terminal values in the forward and backward filtering equations can be relaxed if the matrix ${}_1\sigma{}_1\sigma^*$ is uniformly non-singular so that the corresponding equations become super-parabolic; cf. [70].

On the other hand, the assumption of uniform non-singularity of the matrix ${}_1\sigma{}_1\sigma^*$ is rather restrictive and, in particular, excludes non-random processes X as well as the possibility to connect the interpolation and filtering problems; cf. Remark 6.6.

Chapter 7
Hypoellipticity of Itô's Second Order Parabolic Equations

7.1 Introduction

7.1.1. Smoothness of solutions of deterministic parabolic equations increases as the smoothness assumptions on their coefficients increase. This is a typical feature of parabolic equations. Moreover, under wide assumptions, the smoothness of solutions for $t > 0$ depends only on the smoothness of coefficients and does not depend on the smoothness of the initial functions. This is important, for example, in the study of the fundamental solution of a parabolic equation, since we can consider this solution as a solution of the corresponding Cauchy problem where the initial function is the Dirac delta function. Hypoellipticity is a particular case of the growth of smoothness property mentioned above.

Definition 7.1 A parabolic equation possesses the hypoellipticity property if every generalized solution of the equation has a modification that is infinitely differentiable with respect to the space variable.

As is well known (e.g. [88]) a non-degenerate deterministic parabolic equation with smooth coefficients possesses this property; it also has been known for a while that the non-degeneracy condition is not necessary to have hypoellipticity. Hörmander [49] established a general necessary and sufficient condition for hypoellipticity of second-order deterministic parabolic equations.

While solutions of deterministic equations can also be infinitely differentiable in time, we cannot expect this to happen for stochastic equations because of the presence of Brownian motion. This is why, for stochastic equations, hypoellipticity is used only in connection with the space variable.

7.1.2. In Chaps. 3 and 4 we studied regularity of the solution of Itô's second-order parabolic and super-parabolic equations in the Sobolev spaces $\mathbb{H}^\gamma$. In this chapter we study regularity of the solution in the traditional spaces of continuously

B. L. Rozovsky, S. V. Lototsky, *Stochastic Evolution Systems*, Probability Theory and Stochastic Modelling 89, https://doi.org/10.1007/978-3-319-94893-5_7

differentiable functions. Moreover, to establish hypoellipticity, we must enlarge the class of admissible initial conditions to include finite measures on $\mathbb{R}^d$.

We derive a generalization of Hörmander's condition for hypoellipticity of Itô's second-order parabolic equation and prove that Itô's second-order super-parabolic equation satisfies this condition.

Because of the well-known connection between second-order parabolic equations and diffusion processes (see e.g. [35]), hypoellipticity of a second-order deterministic parabolic equation implies the existence of an infinitely smooth transition density for the corresponding diffusion process. In Sect. 7.2 we prove a similar result in the case of Itô's second-order parabolic equation.

In Sect. 7.3 we apply this result to establish hypoellipticity of the filtering equation and prove that, under a Hörmander-type condition, there exists a smooth conditional transition density for the corresponding diffusion process.

7.2 Measure-Valued Solution and Hypoellipticity Under a Generalized Hörmander Condition

7.2.1. As in Sect. 5.4, we consider the equation

$$du(t,x,\omega) = L^*u(t,x,\omega)\,dt + B^{l,*}u(t,x,\omega)\,dw^l(t), \tag{7.2.1}$$

where $0 < t \le T$, $x \in \mathbb{R}^d$, $\omega \in \Omega$, and

$$\begin{aligned} L^*u(t,x,\omega) &:= \big(a^{ij}(t,x,\omega)u(t,x,\omega)\big)_{ij} - \big(b^i(t,x,\omega)u(t,x,\omega)\big)_i \\ &\quad + c(t,x,\omega)u(t,x,\omega), \\ B^{l,*}u(t,x,\omega) &:= \big(\sigma^{il}(t,x,\omega)u(t,x,\omega)\big)_i + h^l(t,x,\omega)u(t,x,\omega), \end{aligned}$$

and also

$$\begin{aligned} Lu(t,x,\omega) &:= a^{ij}(t,x,\omega)u_{ij}(t,x,\omega) + b^i(t,x,\omega)u_i(t,x,\omega) \\ &\quad + c(t,x,\omega)u(t,x,\omega), \\ B^l u(t,x,\omega) &:= \sigma^{il}(t,x,\omega)u_i(t,x,\omega) + h^l(t,x,\omega)u(t,x,\omega). \end{aligned}$$

Warning 7.1 *Throughout this chapter, unless otherwise stated, the main assumptions and notation in Sects. 5.1.2, 5.4.1 and 5.4.2 will be in force, although eventually much higher regularity of the coefficients in space will be necessary.*

We fix a σ-algebra $\mathfrak{F}_0 \subset \mathscr{F}$ completed with respect to $\mathbb{P}$, but, contrary to Sects. 5.4.1, 5.4.2 we no longer assume that the regular conditional probability $P_{\mathfrak{F}_0}(\cdot)$ has a density with respect to the Lebesgue measure. □

7.2.2. Denote by $\mathcal{M}_c(\mathbb{R}^d)$ the collection of countably additive finite measures on $\mathbb{R}^d$. For $\mu \in \mathcal{M}_c(\mathbb{R}^d)$ and $f \in \mathbb{L}_\infty$,

$$\mu[f] := \int_{\mathbb{R}^d} f(x)\mu(dx).$$

Definition 7.2 A family of measures $\mu_t(\omega, \cdot)$, $(t, \omega) \in [0, T] \times \Omega$, is called a measure-valued solution of Eq. (7.2.1) with initial condition

$$\mu_0(\omega, \cdot) = P_{\mathfrak{F}_0}(\cdot) \tag{7.2.2}$$

if

(i) for every $(t, \omega) \in (0, T] \times \Omega$, $\mu_t(\omega, \cdot) \in \mathcal{M}_c(\mathbb{R}^d)$, and

$$\int_{[0,T]} \mathbb{E}\Big(\mu_t(\mathbb{R}^d)\Big)^2 < \infty;$$

(ii) for every $f \in \mathbb{L}_\infty$, the process $t \mapsto \mu_t[f]$ is $\mathcal{F}_t$-adapted and has a continuous in t version;

(iii) for every $g \in \mathbb{C}_b^2(\mathbb{R}^d)$, there exists a set $\tilde{\Omega} \subset \Omega$ with $\mathbb{P}(\tilde{\Omega}) = 1$ such that, for all $t \in [0, T]$ and $\omega \in \tilde{\Omega}$,

$$\mu_t[g] = P_{\mathfrak{F}_0}[g] + \int_{[0,t]} \mu_s[Lg](s)\, ds + \int_{[0,t]} \mu_s[B^l g](s)\, dw^l(s). \tag{7.2.3}$$

□

Note that the integrals on the right-hand side of (7.2.3) are well defined because the functions Lg and $B^l g$ are bounded and continuous in x and predictable for every x, so that the processes $t \mapsto \mu_t[Lg](t)$ and $t \mapsto \mu_t[B^l g](t)$ have predictable versions.

If it exists, the density $\pi = \pi(t, x, \omega)$ of μ_t with respect to the Lebesgue measure satisfies

$$\big(\pi(t), g\big)_0 = P_{\mathfrak{F}_0}[g] + \int_{[0,t]} \big(\pi, Lg\big)_0(s)\, ds + \int_{[0,t]} \big(\pi, B^l g\big)_0(s)\, dw^l(s),$$

which, with $(\cdot, \cdot)_0$ denoting the inner product in $\mathbb{L}_2(\mathbb{R}^d)$, is the same as (7.2.3).

As a result, the hypoellipticity of Eq. (7.2.1) is equivalent to the existence and infinite differentiability with respect to x of the density of the measure-valued solution.

7.2.3. In this paragraph we prove the existence of the measure-valued solution. In subsequent paragraphs, we will show that, for $t > 0$ and under a Hörmander-type condition, this solution has a density with respect to the Lebesgue measure and the density is infinitely differentiable with respect to x.

Let $\mathcal{X} = \mathcal{X}(t)$ be a diffusion process which is a solution of the following system of Itô's equations

$$d\mathcal{X}(t) = \mathcal{B}\big(t, \mathcal{X}(t)\big)\, dt + \boldsymbol{\Sigma}\big(t, \mathcal{X}(t)\big)\, d\mathcal{W}(t), \quad 0 < t \le T, \tag{7.2.4}$$

$$\mathcal{X}(0) = x_0, \tag{7.2.5}$$

where $\mathcal{W}$ is a standard Wiener process with values in $\mathbb{R}^{\mathrm{d}_0+\mathrm{d}_1}$, x_0 is an $\mathscr{F}_0$-measurable random vector in $\mathbb{R}^{\mathrm{d}}$,

$$\mathcal{B}^i = b^i - \sigma^{il} h^l, \; i = 1, 2, \dots, \mathrm{d}, \tag{7.2.6}$$

and $\boldsymbol{\Sigma}$ is the matrix with d rows and $\mathrm{d}_0 + \mathrm{d}_1$ columns, obtained by combining σ and $\hat{\sigma}$:

$$\boldsymbol{\Sigma}^{i\ell} = \begin{cases} \hat{\sigma}^{il}, & \text{if } \ell = l = 1, \dots, \mathrm{d}_0, \\ \sigma^{il}, & \text{if } \ell = \mathrm{d}_0 + l = \mathrm{d}_0 + 1, \dots, \mathrm{d}_0 + \mathrm{d}_1. \end{cases}$$

Recall that we assume representation (5.1.1) to hold for the matrix a, that is, $2a = \sigma\sigma^* + \hat{\sigma}\hat{\sigma}^*$. As in Chap. 5, the Brownian motion $\mathcal{W}$ is the column vector with components

$$(\hat{w}^1, \dots, \hat{w}^{\mathrm{d}_0}, w^1, \dots, w^{\mathrm{d}_1}) := (\hat{W}, W),$$

and $P_{\mathfrak{F}_0}(\cdot)$ is the regular conditional distribution of x_0 relative to $\mathfrak{F}_0$, where $\mathfrak{F}_t = \mathscr{F}_0 \vee \mathscr{F}_t^0$ and $\mathscr{F}_t^0$ is the σ-algebra generated by $w(s)$ for $s \in [0, t]$ and completed with respect to $\mathbb{P}$.

Define

$$\rho(t) = \exp\Bigg\{ \int\limits_{[0,t]} c\big(s, \mathcal{X}(s)\big)\, ds + \int\limits_{[0,t]} h^l\big(s, \mathcal{X}(s)\big)\, dw^l(s) - \frac{1}{2} \int\limits_{[0,t]} h^l h^l\big(s, \mathcal{X}(s)\big)\, ds \Bigg\}. \tag{7.2.7}$$

Theorem 7.1 *The problem* (7.2.1), (7.2.2) *has a unique measure-valued solution* μ_t, *and, for every* $\psi \in \mathbb{L}_\infty$ *and* $t \in [0, T]$, *the equality*

$$\mu_t[\psi] = \mathbb{E}\Big[\psi\big(\mathcal{X}(t)\big)\rho(t)\big|\mathfrak{F}_t\Big] \tag{7.2.8}$$

holds with probability one.

Remark 7.1 The only difference between this theorem and Theorem 5.6 is the assumption about the initial condition: in Theorem 5.6 it is assumed that $P_{\mathfrak{F}_0}$ has a density $\varphi \in \mathbb{L}_2(\Omega, \mathfrak{F}_0; \mathbb{H}^1)$, whereas this theorem drops this assumption. As as a result, formula (7.2.8) is a bona fide generalization of the AOC formula (5.4.3). □

Proof We only prove the existence of μ_t. The intuition behind the uniqueness is that every measure-valued solution has to satisfy (7.2.8); a complete argument is beyond the scope of this work, partly because the subsequent additional assumptions about Eq. (7.2.1) will ensure that, for every $t > 0$, the measure-valued solution has a density with respect to the Lebesgue measure and, because this density must be a generalized solution of (5.4.1), uniqueness will follow from Theorem 4.5.

To construct μ_t, we start with a family of (random) continuous linear functionals $\tilde{\Phi} = \tilde{\Phi}_t$, $t \geq 0$, on $\mathbb{C}^0_b(\mathbb{R}^d)$ such that, for every $\psi \in \mathbb{C}^0_b(\mathbb{R}^d)$, the process $t \mapsto \tilde{\Phi}_t[\psi]$ is a continuous in t version of $\mathbb{E}\Big[\psi\big(\mathcal{X}(t)\big)\rho(t)\big|\mathfrak{F}_t\Big]$; the existence of $\tilde{\Phi}$ follows from Theorem 5.8. Then there exists a set $\Omega' \subseteq \Omega$ with $\mathbb{P}(\Omega') = 1$ such that, for all $\omega \in \Omega'$ and all $\psi \in \mathbb{C}^0_b(\mathbb{R}^d)$ and $t \geq 0$, $\tilde{\Phi}_t[\psi] \geq 0$.

Next, with the set Ω' in mind, let K_i be a sequence of compact sets in $\mathbb{R}^d$ such that $K_i \subseteq K_{i+1}$ and $\bigcup_{i=1}^{\infty} K_i = \mathbb{R}^d$. Consider the restriction of $\tilde{\Phi}_t[\cdot](\omega)$ to $\mathbb{C}^0(K_i)$. By the Riesz representation theorem [26, Theorem IV. 6.3], for every $(t, \omega) \in (0, T] \times \Omega$, there exists a countably additive measure $\mu^i_t(\omega, dx)$ on K_i such that, for every $\psi \in \mathbb{C}^0(K_i)$,

$$\tilde{\Phi}_t[\psi](\omega) = \int_{K_i} \psi(x)\, \mu^i_t(\omega, dx). \tag{7.2.9}$$

By construction, the measure $\mu^i_t(\omega, dx)$ is non-negative for $\omega \in \Omega'$.

For $\omega \in \Omega'$ and $\Gamma \in \mathscr{B}(\mathbb{R}^d)$ we define

$$\tilde{\mu}_t(\omega, \Gamma) := \lim_{i \to \infty} \mu^i_t(\omega, \Gamma \cap K_i).$$

This definition implies the existence of a set Ω'' with $\mathbb{P}(\Omega'') = 1$ such that, for all $\omega \in \Omega''$ and $t > 0$,

$$\tilde{\mu}_t(\omega, \mathbb{R}^d) = \tilde{\Phi}_t[1] = \mathbb{E}[\rho(t)|\mathfrak{F}_t] < \infty.$$

Now let $\Omega''' = \Omega' \cap \Omega''$ and define $\mu_t(\omega, \cdot)$ by

$$\mu_t(\omega, \Gamma) = \begin{cases} \tilde{\mu}_t(\omega, \Gamma) & \text{if } \omega \in \Omega''', \\ 0 & \text{if } \omega \notin \Omega'''. \end{cases}$$

By construction, the measure $\mu_t(\omega, \cdot)$ is positive and finite for every t, ω, and $\mu_t(\omega, \Gamma)$ is $\mathfrak{F}_t$-measurable for every $\Gamma \in \mathscr{B}(\mathbb{R}^d)$ and $t \in [0, T]$. Also, because the function c and h are bounded,

$$\int_{[0,T]} \mathbb{E}\Big(\mu_t(\mathbb{R}^d)\Big)^2\,dt = \int_{[0,T]} \mathbb{E}\big|\tilde{\Phi}_t[1]\big|^2\,dt \le \int_{[0,T]} \mathbb{E}\rho^2(t)\,dt < \infty.$$

To consider μ_t as a possible measure-valued solution, it remains to verify that

- μ_t is countably additive;
- μ_t does not depend on the choice of the sequence of compacts K_i used in the definition.

To check the countable additivity of μ_t, take a sequence $\{\Gamma_n\}$ of disjoint sets from $\mathscr{B}(\mathbb{R}^d)$. By the monotone convergence theorem, for all t, ω,

$$\begin{aligned}\mu_t\big(\omega, \bigcup_{n=1}^{\infty}\Gamma_n\big) &= \lim_{i\to\infty}\mu_t^i\big(\omega, \bigcup_{n=1}^{\infty}\Gamma_n\cap K_i\big) = \lim_{i\to\infty}\sum_{n=1}^{\infty}\mu_t^i(\omega, \Gamma_n\cap K_i)\\ &= \sum_{n=1}^{\infty}\mu_t(\omega, \Gamma_n).\end{aligned}$$

To check that μ_t does not depend on the choice of the sequence $\{K_i\}$, take another sequence $\{K_i'\}$ of expanding compacts such that $\bigcup_{i=1}^{\infty}K_i' = \mathbb{R}^d$. By the Riesz representation theorem, there exists a sequence of countably additive measures $\nu_t^i(\omega, \cdot)$ on K_i' such that

$$\tilde{\Phi}_t[\psi](\omega) = \int_{K_i'}\psi(x)\,\nu_t^i(\omega, dx),\ \ \psi \in \mathbb{C}^0(K_i').$$

As a result, for every $\Gamma \in \mathscr{B}(\mathbb{R}^d)$ and $n \in \mathbb{N}$,

$$\nu_t^n(\Gamma\cap K_n\cap K_n') = \mu_t^n(\Gamma\cap K_n\cap K_n'). \tag{7.2.10}$$

Now define

$$\tilde{\nu}_t(\omega, \Gamma) := \lim_{i\to\infty}\nu_t^i(\omega, \Gamma\cap K_i).$$

From (7.2.10) and countable additivity of both measures it follows that, for all $\Gamma \in \mathscr{B}(\mathbb{R}^d)$ and $\mathbb{P}$- a.a. ω,

$$\tilde{\mu}_t(\omega, \Gamma) = \tilde{\nu}_t(\omega, \Gamma).$$

To show that μ_t is indeed a measure-valued solution of (7.2.1), it remains to verify that

- For every $f \in \mathbb{L}_\infty$, the process $t \mapsto \mu_t[f]$ is $\mathcal{F}_t$-adapted and has a version that is continuous in t;
- Equality (7.2.3) holds.

If f is continuous function with compact support, then (7.2.9) implies

$$\mathbb{E}\Big[f(\mathcal{X}(t))\rho(t)\Big|\,\mathfrak{F}_t\Big] = \int_{\mathbb{R}^d} f(x)\,\mu(\omega, dx), \quad (\mathbb{P}\text{-a.s.}) \tag{7.2.11}$$

that is, the process $t \mapsto \mu_t[f]$ is $\mathcal{F}_t$-adapted and has a version that is continuous in t. The general case is handled via a limiting procedure; cf. Sect. 5.4.7.

Equality (7.2.3) follows from (7.2.9) and Theorem 5.7.

Theorem 7.1 is proved. □

7.2.4. In this paragraph we make additional assumptions about the coefficients of Eq. (7.2.1) and formulate the main result of the section.

Consider the functions

$${}_{\mathrm{Y}}\mathcal{B} \in \mathbb{C}_b^\infty\big(\mathbb{R}^{d_1}; \mathbb{R}^{d_1}\big) \text{and } {}_{\mathrm{Y}}\boldsymbol{\Sigma} \in \mathbb{C}_b^\infty\big(\mathbb{R}^{d_1}; \mathbb{R}^{d_1 \times d_1}\big)$$

and define the diffusion process $\mathcal{Y} = \mathcal{Y}(t)$ by

$$\mathcal{Y}(t) = y_0 + \int_{[0,t]} {}_{\mathrm{Y}}\mathcal{B}\big(\mathcal{Y}(s)\big)\,ds + \int_{[0,t]} {}_{\mathrm{Y}}\boldsymbol{\Sigma}\big(\mathscr{Y}(s)\big)\,dW(s), \tag{7.2.12}$$

$t \in [0, T]$, $y_0 \in \mathbb{R}^{d_1}$.

Warning 7.2 *Throughout the rest of the section we make the following assumption:* (D) *Every coefficient $F = F(t, x, \omega)$ of Eq. (7.2.1) has the form*

$$F = \tilde{F}\big(x, \mathcal{Y}(t, \omega)\big)$$

for some $\tilde{F} \in \mathbb{C}_b^\infty(\mathbb{R}^{d+d_1})$.

In other words, all coefficients of (7.2.1) are smooth functions of x, and all dependence on t and ω comes through a smooth dependence on the process $\mathcal{Y}$.

Similar to Chap. 6, introduce the $\mathbb{R}^{d+d_1}$-valued process

$$\mathcal{Z}(t) = \begin{pmatrix} \mathcal{X}(t) \\ \mathcal{Y}(t) \end{pmatrix}.$$

Thinking of $\mathcal{X}$, $\mathcal{Y}$, $\mathcal{W}$, $\mathcal{B}$, and ${}_{\mathrm{Y}}\mathcal{B}$ as column vectors, we write the equation for $\mathcal{Z}$ as

$$\mathcal{Z}(t) = z_0 + \int_{[0,t]} V^0\big(\mathcal{Z}(s)\big)\,ds + \int_{[0,t]} \hat{V}\big(\mathcal{Z}(s)\big)\,d\mathcal{W}(s), \tag{7.2.13}$$

where

$$V^0(z) = \begin{pmatrix} \mathcal{B}(x,y) \\ {}_{\mathrm{Y}}\mathcal{B}(y) \end{pmatrix} \in \mathbb{R}^{\mathrm{d}+\mathrm{d}_1},\quad \hat{V}(z) = \begin{pmatrix} \hat{\sigma}(x,y) & \sigma(x,y) \\ 0 & {}_{\mathrm{Y}}\boldsymbol{\Sigma}(y) \end{pmatrix} \in \mathbb{R}^{(\mathrm{d}+\mathrm{d}_1)\times(\mathrm{d}_0+\mathrm{d}_1)}.$$

Next, we need some constructions from differential geometry.

Recall that a smooth vector field X in $\mathbb{R}^r$ is a function from $\mathbb{C}^\infty(\mathbb{R}^r;\mathbb{R}^r)$. Each $X \in \mathbb{C}^\infty(\mathbb{R}^r;\mathbb{R}^r)$ is a column vector and defines a differential operator ∂_X on $\mathbb{C}^\infty(\mathbb{R}^r;\mathbb{R}^1)$ by

$$f(x) \mapsto \partial_X f(x) = \big(X(x), \nabla f(x)\big),$$

where ∇f is the gradient of f. In coordinate form,

$$\partial_X f(x) = X^i(x) f_i(x) \equiv \sum_{i=1}^r X^i(x)\frac{\partial f(x)}{\partial x^i}.$$

For two smooth vector fields X, Y, the vector field $(X\cdot\nabla)Y$ is defined by the equality

$$\big((X\cdot\nabla)Y, \boldsymbol{n}\big) = \big(X, \nabla(Y,\boldsymbol{n})\big)$$

for every fixed unit vector $\boldsymbol{n} \in \mathbb{R}^{\mathrm{d}}$. Equivalent interpretations of $(X\cdot\nabla)Y$ are as follows:

1. In the coordinate form,

$$\big((X\cdot\nabla)Y\big)^j(x) = \partial_X Y^j(x) \equiv \sum_{i=1}^r X^i(x)\frac{\partial Y^j(x)}{\partial x^i}.$$

2. If $\boldsymbol{D}Y$ is the derivative of Y [the matrix with row number j equal to the gradient of the jth component of Y], then $(X\cdot\nabla)Y = \big(\boldsymbol{D}Y\big)X$, as a matrix-vector product.
3. If $R = R(t)$ is a function in $\mathbb{C}^1(\mathbb{R};\mathbb{R}^{\mathrm{d}})$, then

$$\frac{d}{dt}Y\big(R(t)\big) = \Big(\big(R'(t)\cdot\nabla\big)Y\Big)\big(R(t)\big).$$

Next, for smooth vector fields X, Y, their **Lie bracket** $[X, Y]$ is the vector field

$$[X,Y] = (X\cdot\nabla)Y - (Y\cdot\nabla)X.$$

In the coordinate form,

$$[X,Y]^j = X^i Y^j_i - Y^i X^j_i.$$

In terms of the corresponding differential operators,

$$\partial_{[X,Y]} = \partial_X \partial_Y - \partial_Y \partial_X.$$

Indeed, by the product rule,

$$\partial_X \partial_Y f = X^i \big(\partial_Y f\big)_i = X^i \big(Y^j f_j\big)_i = X^i Y^j_i f_j + X^i Y^j f_{ij}.$$

Direct computations confirm the following properties of the operation $[\cdot,\cdot]$: $\mathbb{C}^\infty(\mathbb{R}^r;\mathbb{R}^r)\times\mathbb{C}^\infty(\mathbb{R}^r;\mathbb{R}^r)\to\mathbb{C}^\infty(\mathbb{R}^r;\mathbb{R}^r)$

1. Anti-symmetry: $[Y,X] = -[X,Y]$;
2. Bi-linearity: $[aX+bY,Z] = a[X,Z]+b[Y,Z],\ [X,aY+bZ] = a[X,Y]+b[X,Z],\ a,b\in\mathbb{R}^1$;
3. Jacobi identity: $[[X,Y],Z]+[[Y,Z],X]+[[Z,X],Y]=0$;
4. Product rule: $\partial_{[X,fY]} = \big(\partial_X f\big)\partial_Y + f\partial_{[X,Y]},\ f\in\mathbb{C}^\infty(\mathbb{R}^{\mathrm{d}};\mathbb{R}^1)$.

Let $X^0, X^1,\ldots,X^m$ be smooth vector fields in $\mathbb{R}^r$. For $k=0,1,2,3,\ldots$ and $x\in\mathbb{R}^{\mathrm{d}}$, define the collections $\mathscr{V}_k(x)$ of vectors by

$$\mathscr{V}_0(x) = \big\{X^1(x), X^2(x),\ldots,X^m(x)\big\},$$
$$\mathscr{V}_k(x) = \big\{[Y,X^j](x),\ Y\in\mathscr{V}_{k-1},\ j=0,\ldots,m\big\},\ k\geq 1.$$

Note that X^0 is not included in $\mathscr{V}_0$.

Definition 7.3 The vector fields $X^0, X^1,\ldots,X^m$ satisfy the **parabolic Hörmander condition** in $\mathbb{R}^{\mathrm{d}}$ if, for every $x\in\mathbb{R}^r$, the set $\bigcup_{k\geq 0}\mathscr{V}_k(x)$ contains d linearly independent vectors.

The original result of Hörmander can now be stated in probabilistic terms for the process $\mathcal{Z}$ from (7.2.13): with $\hat{V}^l,\ l=1,\ldots,\mathrm{d}_0+\mathrm{d}_1$, denoting the columns of the matrix $\hat{V}(z)$ and

$$\hat{V}^0(z) = V^0(z) - \frac{1}{2}(\hat{V}^l\cdot\nabla)\hat{V}^l(z),$$

if the vector fields $\hat{V}^0,\hat{V}^1,\ldots,\hat{V}^{\mathrm{d}_0+\mathrm{d}_1}$ satisfy the parabolic Hörmander condition in $\mathbb{R}^{\mathrm{d}+\mathrm{d}_1}$, then, for every $t>0$, the random variable $\mathcal{Z}(t)$ has a density with respect to the Lebesgue measure in $\mathbb{R}^{\mathrm{d}+\mathrm{d}_1}$ and the density belongs to $\mathbb{C}^\infty_b(\mathbb{R}^{\mathrm{d}+\mathrm{d}_1})$.

Equivalently, the deterministic equation

$$\frac{\partial v(t,z)}{\partial t} = \Big(\partial_{\hat{V}^0} + \frac{1}{2}\partial_{\hat{V}^l}\partial_{\hat{V}^l}\Big)v(t,z)$$

is hypoelliptic in the sense of Definition 7.1. Note that, by the Itô formula, the operator

$$f \mapsto \Big(\partial_{\hat{V}^0} + \frac{1}{2}\partial_{\hat{V}^l}\partial_{\hat{V}^l}\Big)f$$

is the generator of the process $\mathcal{Z} = \mathcal{Z}(t)$.

Recall that, in this chapter, we investigate the stochastic PDE (7.2.1), and the solution of this equation is connected with the process $\mathcal{X}$ from (7.2.4). To state the appropriate version of the Hörmander condition, we therefore need to identify suitable vector fields in $\mathbb{R}^{\mathrm{d}}$; we also should expect these vector fields to be random and time-dependent through the dependence on the process $\mathcal{Y}$. Accordingly, our next step is to construct these vectors and to state the corresponding condition.

Define the vector fields $\hat{X}^0, \hat{X}^1, \ldots, \hat{X}^{\mathrm{d}_0+\mathrm{d}_1}$ as follows:

$$\hat{X}^\ell(t,x,\omega) \text{ is column number } \ell \text{ of the matrix } \boldsymbol{\Sigma}\big(x,\mathcal{Y}(t)\big),\ \ell = 1,\ldots,\mathrm{d}_0+\mathrm{d}_1;$$

$$\hat{X}^0(t,x,\omega) = \mathcal{B}\big(x,\mathcal{Y}(t)\big) - \frac{1}{2}\sum_{\ell=1}^{\mathrm{d}_0+\mathrm{d}_1}\sum_{i=1}^{\mathrm{d}} \boldsymbol{\Sigma}^{i\ell}\big(x,\mathcal{Y}(t)\big)\frac{\partial \hat{X}^\ell(t,x,\omega)}{\partial x^i}.$$

Next, for $k = 0,1,2,3,\ldots,\ t\in(0,T],\ \omega\in\Omega$, and $x\in\mathbb{R}^{\mathrm{d}}$, define the collections $\hat{\mathscr{V}}_k(t,x,\omega)$ of vectors by

$$\hat{\mathscr{V}}_0(t,x,\omega) = \big\{\hat{X}^1(t,x,\omega), \hat{X}^2(t,x,\omega), \ldots, \hat{X}^{\mathrm{d}_0}(t,x,\omega)\big\},$$
$$\hat{\mathscr{V}}_k(t,x,\omega) = \big\{[Y,\hat{X}^j](t,x,\omega),\ Y\in\hat{\mathscr{V}}_{k-1},\ j=0,\ldots,\mathrm{d}_0+\mathrm{d}_1\big\},\ k\geq 1.$$

Note that $\hat{X}^0$ and $\hat{X}^{\mathrm{d}_0+1},\ldots,\hat{X}^{\mathrm{d}_0+\mathrm{d}_1}$ are not included in $\hat{\mathscr{V}}_0$.

We now state the **generalized Hörmander condition.**

(H_1) *For every* $t\in(0,T],\ x\in\mathbb{R}^{\mathrm{d}}$, *and* $\omega\in\Omega$, *the set* $\bigcup_{k\geq 0}\hat{\mathscr{V}}_k(t,x,\omega)$ *contains* d *linearly independent vectors.*

Note that

1. Condition (H_1) holds if the rank of the matrix $\hat{\sigma} = \hat{\sigma}\big(x,\mathcal{Y}(t)\big)$ is equal to d for all t and x.
2. If $B^l \equiv 0$, ${}_{\mathrm{Y}}\boldsymbol{\Sigma} \equiv 0$, and ${}_{\mathrm{Y}}\mathcal{B} \equiv 0$, then Eq. (7.2.1) is the forward Kolmogorov equation for the process $\mathcal{X}$, and (H_1) is the (original) parabolic Hörmander condition for the (non-random) vector fields $\hat{X}^0, \hat{X}^1, \ldots, \hat{X}^{\mathrm{d}_0}$ in $\mathbb{R}^{\mathrm{d}}$.

3. If moreover $d_0 = d = 1$, then (7.2.1) becomes

$$du(t,x) = \frac{\partial^2\big(\sigma^2(x)u(t,x)\big)}{\partial x^2} - \frac{\partial\big(b(x)u(t,x)\big)}{\partial x}$$

and the parabolic Hörmander condition reduces to the following: for every $x \in \mathbb{R}^1$, there exists an $n \geq 1$ such that

$$|\sigma(x)|^2 + \left| b(x)\frac{d^n\sigma(x)}{dx^n}\right|^2 > 0.$$

Here is the main result of this section.

Theorem 7.2 *Under condition* (H_1)*, every measure-valued solution of* (7.2.1), (7.2.2) *has a version that, for all* $t \in (0,T)$*, is absolutely continuous with respect to the Lebesgue measure on* $\mathbb{R}^d$*, and the density belongs to* $\mathbb{C}^{0,\infty}\big((0,T)\times\mathbb{R}^d\big)$. □

In other words, under the generalized Hörmander condition, the parabolic Itô equation (7.2.1) is **hypoelliptic**.

The proof relies on the following result. Recall that, for a smooth function $f = f(x),\ x \in \mathbb{R}^d$, and a multi-index $\alpha = (\alpha_1, \ldots, \alpha_d)$ with $\alpha_i \in \{0, 1, 2, \ldots\}$,

$$f_\alpha := \frac{\partial^{\alpha_1+\ldots+\alpha_d} f}{\partial (x^1)^{\alpha_1}\ldots(x^d)^{\alpha_d}}.$$

Proposition 7.1 *Let* μ *be a measure from* $\mathcal{M}_c(\mathbb{R}^d)$ *with the following property: for every multi-index* α*, there exists a number* N_α *such that, for all* $f \in \mathbb{C}_b^\infty(\mathbb{R}^d)$,

$$\big|\mu[f_\alpha]\big| \leq N_\alpha \sup_{x\in\mathbb{R}^d} |f(x)|. \tag{7.2.14}$$

Then the measure μ *is absolutely continuous with respect to the Lebesgue measure on* $\mathbb{R}^d$ *and the corresponding density belongs to* $\mathbb{L}_2(\mathbb{R}^d) \cap \mathbb{C}_b^\infty(\mathbb{R}^d)$. □

Proof Consider the Fourier transform

$$\hat{\mu}(\lambda) = \int_{\mathbb{R}^d} e^{\mathrm{i}(\lambda,x)}\, \mu(dx)$$

of μ; $\lambda \in \mathbb{R}^d$, $\mathrm{i} = \sqrt{-1}$.

By (7.2.14), using $(1-\boldsymbol{\Delta})^k f$ with $f(x) = \cos(\lambda, x)$ and $f(x) = \sin(\lambda, x)$, it follows that, for every $k \in \mathbb{N}$, there exists a number $N(k)$ such that, for all $\lambda \in \mathbb{R}^d$,

$$\big|\hat{\mu}(\lambda)\big| \leq \frac{N(k)}{(1+|\lambda|^2)^k}. \tag{7.2.15}$$

Then, by the inversion formula for the Fourier transform, the function

$$v(x) = (2\pi)^{-\mathrm{d}} \int_{\mathbb{R}^{\mathrm{d}}} e^{-\mathrm{i}(\lambda,x)} \hat{\mu}(\lambda)\, d\lambda$$

is well defined, belongs to $\mathbb{C}_b^{0,\infty}\big((0,T)\times\mathbb{R}^{\mathrm{d}}\big)$, and is the density of μ with respect to the Lebesgue measure on $\mathbb{R}^{\mathrm{d}}$. □

Remark 7.2 By Theorem 7.1, the measure-valued solution $\mu = \mu_t$ of (7.2.1) satisfies

$$\mu_t[f] = \mathbb{E}\big[f\big(\mathcal{X}(t)\big)\rho(t)\big|\mathfrak{F}_t\big].$$

Therefore, to establish (7.2.14) for μ_t, it suffices to verify that, for every $t > 0$ and every multi-index α, there exists an $\mathfrak{F}_t$-measurable random variable $h^{(\alpha)}(t)$ such that $\mathbb{E}|h^{(\alpha)}(t)|^2 < \infty$ and

$$\mathbb{E}\big[f_\alpha\big(\mathcal{X}(t)\big)\rho(t)\eta\big] = \mathbb{E}\big[f\big(\mathcal{X}(t)\big)\rho(t)h^{(\alpha)}(t)\eta\big] \tag{7.2.16}$$

for every bounded $\mathfrak{F}_t$-measurable random variable η and every $f \in \mathbb{C}_b^\infty(\mathbb{R}^{\mathrm{d}})$.

In fact, we will see that (7.2.16) holds with $h^{(\alpha)}(t)$ satisfying

$$\mathbb{E} \sup_{t\in[\varepsilon,T]} |h^{(\alpha)}(t)|^p < \infty,\ 0<\varepsilon<T,\ p\geq 2. \tag{7.2.17}$$

As a result, if $u = u(t,x)$ is the density of μ_t, then (7.2.14) and (7.2.16) imply that all partial derivatives u_α of u satisfy

$$\mathbb{E} \sup_{t\in[\varepsilon,T],\ x\in\mathbb{R}^{\mathrm{d}}} |u_\alpha(t,x)|^p < \infty, \qquad p \geq 2.$$ □

7.2.5. In this and the following paragraphs we establish several stochastic integration by parts formulas that will eventually lead us to (7.2.16).

Consider a diffusion process $\xi = \xi(t) \in \mathbb{R}^r$ defined on $(\Omega, \mathcal{F}, \{\mathfrak{F}_t\}, \mathbb{P})$ by the Itô equation

$$d\xi(t) = U^0\big(t,\xi(t)\big)\,dt + U^\ell\big(t,\xi(t)\big)\,d\mathcal{W}^\ell(t),\ \ 0<t\leq T, \tag{7.2.18}$$

$$\xi(0) = \xi_0 \in \mathbb{R}^r, \tag{7.2.19}$$

where $\mathcal{W}$ is a standard $\mathbb{R}^{\mathrm{d}_0+\mathrm{d}_1}$-valued Wiener process and ξ_0 is non-random.

Definition 7.4 The process defined by (7.2.18), (7.2.19) is called a **standard diffusion process** in $\mathbb{R}^r$ if the functions $U^k = U^k(t,x,\omega)$, $k = 0,\dots,\mathrm{d}_0+\mathrm{d}_1$ have the following properties:

(a) For every $x \in \mathbb{R}^{\mathrm{d}}$, $U^k(\cdot, x, \cdot)$ is $\mathscr{B}([0,T]\times\mathbb{R}^r)\otimes\mathscr{F}$-measurable and predictable relative to $\mathfrak{F}_t$;

(b) For every t, ω, $U^k(t, \cdot, \omega)$ belongs to $\mathbb{C}^\infty(\mathbb{R}^r; \mathbb{R}^r)$, with all the derivatives bounded;

(c) For every $p \geq 2$,

$$\sum_{k=0}^{\mathrm{d}_0+\mathrm{d}_1} \int_{[0,T]} \mathbb{E}|U^k(t,0)|^p \, dt < \infty. \qquad \square$$

Remark 7.3 Note that the coefficients in (7.2.18) are not necessarily bounded, but, because their derivatives are, the coefficients are at most of linear growth. As a result, by Theorem 1.11, every standard diffusion process ξ belongs to $\bigcap_{p\geq 2} \mathbb{L}_p\big(\Omega; \mathbb{C}([0,T]; \mathbb{R}^r)\big)$. In particular,

$$\mathbb{E} \sup_{0<t<T} |\xi(t)|^p < \infty$$

for all $p \geq 2$ and $T > 0$. $\square$

Warning 7.3 *Throughout this paragraph, it is assumed that ξ is a standard diffusion process.* $\square$

Let $Q = Q(t) \in \bigcap_{p\geq 0} \mathbb{L}_p\big([0,T] \times \Omega; \mathscr{P}(\mathfrak{F}); \mathbb{R}^{r\times \mathrm{d}_0}\big)$ be a matrix-valued stochastic process and let $\boldsymbol{\varepsilon} \in \mathbb{R}^r$ be non-random. Denote by $Q^k, k = 1, \ldots, \mathrm{d}_0$ the columns of the matrix Q; each Q^k is a vector in $\mathbb{R}^r$.

In addition to the process ξ, define the process $\xi^{\boldsymbol{\varepsilon}} = \xi^{\boldsymbol{\varepsilon}}(t)$ by

$$d\xi^{\boldsymbol{\varepsilon}}(t) = U^0\big(t, \xi^{\boldsymbol{\varepsilon}}(t)\big)\, dt + \sum_{k=1}^{\mathrm{d}_0} U^k\big(t, \xi^{\boldsymbol{\varepsilon}}\big)\, \big(d\hat{w}^k(t) + (Q^k(t), \boldsymbol{\varepsilon})\, dt\big) \qquad (7.2.20)$$

$$+ \sum_{j=1}^{\mathrm{d}_1} U^{\mathrm{d}_0+j}\big(t, \xi^{\boldsymbol{\varepsilon}}(t)\big)\, dw^j(t), \quad 0 < t \leq T,$$

$$\xi^{\boldsymbol{\varepsilon}}(0) = \xi_0. \qquad (7.2.21)$$

By construction, $\xi^{\boldsymbol{\varepsilon}}$ is also a standard diffusion process in $\mathbb{R}^r$; the arguments similar to [65, Ch.II, §8] confirm that the process $\xi^{\boldsymbol{\varepsilon}}$ has a version that is differentiable with respect to $\boldsymbol{\varepsilon}$ and the derivative matrix $\mathcal{D}\xi^{\boldsymbol{\varepsilon}}$, of size $r \times r$, satisfies

$$\begin{aligned}\mathcal{D}\xi^{\boldsymbol{\varepsilon}}(t) &= \int\limits_{[0,t]} \boldsymbol{D}U^0\big(s, \xi^{\boldsymbol{\varepsilon}}(s)\big)\mathcal{D}\xi^{\boldsymbol{\varepsilon}}(s)\,ds \\ &+ \sum_{k=1}^{\mathrm{d}_0} \int\limits_{[0,t]} \boldsymbol{D}U^k\big(s, \xi^{\boldsymbol{\varepsilon}}(s)\big)\mathcal{D}\xi^{\boldsymbol{\varepsilon}}(s)\big(d\hat{w}^k(s) + (Q^k, \boldsymbol{\varepsilon})\,ds\big) \\ &+ \sum_{k=1}^{\mathrm{d}_0} \int\limits_{[0,t]} U^k\big(s, \xi^{\boldsymbol{\varepsilon}}(s)\big)\big(Q^k(s)\big)^* ds \\ &+ \sum_{j=1}^{\mathrm{d}_1} \int\limits_{[0,t]} \boldsymbol{D}U^{\mathrm{d}_0+j}\big(s, \xi^{\boldsymbol{\varepsilon}}(s)\big)\mathcal{D}\xi^{\boldsymbol{\varepsilon}}(s)\,dw^j(s),\ t \in [0, T];\end{aligned} \tag{7.2.22}$$

here and below, $\boldsymbol{D}$ denotes the derivative with respect to the space variable and $\mathcal{D}$ denotes the derivative with respect to $\boldsymbol{\varepsilon}$.

If all components of the process Q are bounded, then Girsanov's theorem (Theorem 1.14) implies the equivalence of the measures P^0 and $P^{\boldsymbol{\varepsilon}}$ generated by the process ξ and $\xi^{\boldsymbol{\varepsilon}}$ in $\mathbb{C}([0, T]; \mathbb{R}^r)$. Define the random variable

$$\Phi(\boldsymbol{\varepsilon}) = \frac{dP^0}{dP^{\boldsymbol{\varepsilon}}}(\xi^{\boldsymbol{\varepsilon}});$$

note that, given Q and $\mathcal{W}$, $\Phi = \Phi(\boldsymbol{\varepsilon})$ is indeed a random function of $\boldsymbol{\varepsilon}$. Let

$$\varphi : \Big(\mathbb{C}([0, T]; \mathbb{R}^r) \times \Omega, \mathscr{B}\big(\mathbb{C}([0, T]; \mathbb{R}^r)\big) \otimes \mathscr{F}\Big) \to \mathbb{R}^1$$

be such that $\mathbb{E}|\varphi(\xi)| < \infty$.

Proposition 7.2 (Stochastic Integration by Parts Formula I) *Assume that*

(a) $Q \in \mathbb{L}_\infty([0, T]; \mathbb{R}^{r\times \mathrm{d}_0})$;
(b) *Both* $\Phi(\boldsymbol{\varepsilon})$ *and* $\varphi(\xi^{\boldsymbol{\varepsilon}})$ *are differentiable with respect to* $\boldsymbol{\varepsilon}$ *at a point* $\boldsymbol{\varepsilon}_0$;
(c) *The following uniform integrability condition holds:*

$$\sup_{|\boldsymbol{\varepsilon}|_r<\delta} \frac{\mathbb{E}\Big|\varphi(\xi^{\boldsymbol{\varepsilon}_0+\boldsymbol{\varepsilon}})\Phi(\boldsymbol{\varepsilon}_0 + \boldsymbol{\varepsilon}) - \varphi(\xi^{\boldsymbol{\varepsilon}_0})\Phi(\boldsymbol{\varepsilon}_0)\Big|^p}{|\boldsymbol{\varepsilon}|^p} < \infty,\ p > 1,\ \delta > 0. \tag{7.2.23}$$

Then

$$\mathbb{E}\Big(\Phi(\boldsymbol{\varepsilon}_0)\,\mathcal{D}\varphi(\xi^{\boldsymbol{\varepsilon}_0})\Big) = -\mathbb{E}\Big(\varphi(\xi^{\boldsymbol{\varepsilon}_0})\,\mathcal{D}\Phi(\boldsymbol{\varepsilon}_0)\Big). \tag{7.2.24}$$

□

Proof By the change of variables formula,

$$\mathbb{E}\varphi(\xi) = \mathbb{E}\Big(\varphi(\xi^{\varepsilon})\Phi(\varepsilon)\Big).$$

Then (7.2.24) follows after applying $\mathcal{D}$ to both sides of the last equality and using the product rule. Assumption (7.2.23) ensures the possibility to exchange the $\mathcal{D}$ and $\mathbb{E}$ operations; cf. [108, Ch.2, §2, Theorem 22]. □

The next step is to establish (7.2.24) for a special class of functionals φ without assuming boundedness of Q. To this end, we introduce a few more objects:

1. $\zeta = \zeta_t$, a bounded, $\mathfrak{F}_t$-adapted random process;
2. $\mathfrak{a} = \mathfrak{a}(t, x, \omega)$, a $\mathscr{B}([0, T] \times \mathbb{R}^r) \otimes \mathscr{F}$-measurable function with values in $\mathbb{R}^{\mathrm{d}_0}$ that is predictable (relative to the family $\{\mathfrak{F}_t\}$) for every x and belongs to $\mathbb{C}^1_b(\mathbb{R}^r)$ for every t, ω;
3. The processes $\mathcal{E}^{\varepsilon} = \mathcal{E}^{\varepsilon}(t) \in \mathbb{R}^1$, $\mathfrak{e}^{\varepsilon} = \mathfrak{e}^{\varepsilon}(t) \in \mathbb{R}^r$, and $H^{\varepsilon} = H^{\varepsilon}(t) \in \mathbb{R}^r$:

$$\mathcal{E}^{\varepsilon}(t) = \zeta_t \exp\left\{\int\limits_{[0,t]} \Big(\mathfrak{a}\big(s, \xi^{\varepsilon}(s)\big)\, dw^l(s)\Big) - \frac{1}{2}\int\limits_{[0,t]} \big|\mathfrak{a}\big(s, \xi^{\varepsilon}(s)\big)\big|\, ds\right\},$$

$$\mathfrak{e}^{\varepsilon}(t) = \int\limits_{[0,t]} \mathcal{D}\Big(\mathfrak{a}\big(s, \xi^{\varepsilon}(s)\big)\, dW(s)\Big) - \frac{1}{2}\int\limits_{[0,t]} \mathcal{D}\Big|\mathfrak{a}\big(s, \xi^{\varepsilon}(s)\big)\Big|\, ds,$$

$$H^{\varepsilon}(t) = \int\limits_{[0,t]} Q(s)\, d\hat{W}(s).$$

In particular,

$$\mathcal{D}\mathcal{E}^{\varepsilon}(t) = \mathcal{E}^{\varepsilon}(t)\mathfrak{e}^{\varepsilon}(t).$$

To simplify the notation, the superscript ε will be omitted when $\varepsilon = 0$: $\xi(t) = \xi^{\varepsilon}(t)\big|_{\varepsilon=0}$, $\mathcal{D}\xi(t) = \mathcal{D}\xi^{\varepsilon}(t)\big|_{\varepsilon=0}$, $\mathcal{E}(t) = \mathcal{E}^{\varepsilon}(t)\big|_{\varepsilon=0}$, etc. This convention is consistent with the original definitions of ξ and ξ^{ε}. Note that, by (7.2.22),

$$\begin{aligned}
\mathcal{D}\xi(t) = &\int\limits_{[0,t]} \boldsymbol{D}U^0\big(s, \xi(s)\big)\mathcal{D}\xi(s)\, ds \\
&+ \sum_{k=1}^{\mathrm{d}_0} \int\limits_{[0,t]} \boldsymbol{D}U^k\big(s, \xi(s)\big)\mathcal{D}\xi(s)\, d\hat{w}^k(s) + \int\limits_{[0,t]} \widetilde{U}\big(s, \xi(s)\big)Q^*(s)\, ds \\
&+ \sum_{j=1}^{\mathrm{d}_1} \int\limits_{[0,t]} \boldsymbol{D}U^{\mathrm{d}_0+j}\big(s, \xi(s)\big)\mathcal{D}\xi(s)\, dw^j(s),\ t \in [0, T];
\end{aligned}$$

where $\widetilde{U}$ is the matrix with columns $U^1, \dots, U^{d_0}$. If we consider ξ as a function of the initial condition $\xi(0) \in \mathbb{R}^r$, then (cf. Sect. 5.3.2) the derivative $\boldsymbol{D}\xi$ of ξ with respect to this initial condition satisfies

$$\boldsymbol{D}\xi(t) = \mathbb{I} + \int\limits_{[0,t]} \boldsymbol{D}U^0\big(s, \xi(s)\big)\boldsymbol{D}\xi(s)\, ds + \sum_{k=1}^{d_0} \int\limits_{[0,t]} \boldsymbol{D}U^k\big(s, \xi(s)\big)\boldsymbol{D}\xi(s)\, d\hat{w}^k(s)$$

$$+ \sum_{j=1}^{d_1} \int\limits_{[0,t]} \boldsymbol{D}U^{d_0+j}\big(s, \xi(s)\big)\boldsymbol{D}\xi(s)\, dw^j(s), \ t \in [0, T].$$

In other words, $\boldsymbol{D}\xi$ is the fundamental matrix for the equation satisfied by $\mathcal{D}\xi$ and

$$\mathcal{D}\xi(t) = \boldsymbol{D}\xi(t) \int\limits_{[0,t]} \Big(\boldsymbol{D}\xi(s)\Big)^{-1} \widetilde{U}\big(s, \xi(s)\big) Q^*(s)\, ds. \tag{7.2.25}$$

Lemma 7.1 (Stochastic Integration by Parts Formula II) *Let $f \in \mathbb{C}^1(\mathbb{R}^r; \mathbb{R}^1)$ and $|f(x)| + |\boldsymbol{D}f(x)| \le C(1 + |x|^p)$ for some $p > 0$ and $C < \infty$. Then*

$$\mathbb{E}\Big(\mathcal{E}(t)\boldsymbol{D}f\big(\xi(t)\big)\mathcal{D}\xi(t)\Big) = \mathbb{E}\Big(\mathcal{E}(t) f\big(\xi(t)\big)\big(\mathfrak{e}(t) - H(t)\big)\Big). \tag{7.2.26}$$

□

Proof Define

$$\varphi(\xi^{\boldsymbol{\varepsilon}}) = f\big(\xi^{\boldsymbol{\varepsilon}}(t)\big)\mathcal{E}^{\boldsymbol{\varepsilon}}(t). \tag{7.2.27}$$

The Girsanov theorem (Theorem 1.14) and equality (7.2.22) imply

$$\Phi(0) = 1, \ \mathcal{D}\Phi(0) = H(T).$$

Then, as long as Proposition 7.2 can be applied, direct computations show that (7.2.26) is a particular case of (7.2.24), when $\varphi(\xi^{\boldsymbol{\varepsilon}})$ is from (7.2.27) and $\boldsymbol{\varepsilon}_0 = 0$.

Accordingly, we prove (7.2.26) in two steps: first, we impose additional assumptions that allow a direct application of Proposition 7.2, and then we remove those assumptions.

Step 1. Assume that $Q \in \mathbb{L}_\infty([0, T]; \mathbb{R}^{d_0})$ and $f \in \mathbb{C}_b^1(\mathbb{R}^r; \mathbb{R}^1)$. Then, to apply Proposition 7.2, we only need to verify that φ from (7.2.27) satisfies (7.2.23) when $\boldsymbol{\varepsilon}_0 = 0$, which, in turn, will follow from

$$\sup_{|\boldsymbol{\varepsilon}| \le 1} \frac{\mathbb{E}\Big| f\big(\xi^{\boldsymbol{\varepsilon}}(t)\big)\Phi(\boldsymbol{\varepsilon})\mathcal{E}^{\boldsymbol{\varepsilon}}(t) - f\big(\xi(t)\big)\mathcal{E}(t)\Big|^p}{|\boldsymbol{\varepsilon}|^p} < \infty, \ \ p > 1. \tag{7.2.28}$$

By the triangle inequality, verification of (7.2.28) is reduced to the analysis of several expressions; one of them is

$$\mathfrak{T} := \sup_{|\boldsymbol{\varepsilon}|\le 1} \frac{\mathbb{E}\Big| f\big(\xi^{\boldsymbol{\varepsilon}}(t)\big)\Phi(\boldsymbol{\varepsilon})\big(\mathcal{E}^{\boldsymbol{\varepsilon}}(t) - \mathcal{E}(t)\big)\Big|^p}{|\boldsymbol{\varepsilon}|^p},$$

the others are similar and can be studied in the same way.

To study $\mathfrak{T}$, we use the fundamental theorem of calculus in the form

$$\mathcal{E}^{\boldsymbol{\varepsilon}}(t) - \mathcal{E}(t) = \int\limits_{[0,1]} \Big(\mathcal{D}\mathcal{E}^{\tau\boldsymbol{\varepsilon}}, \boldsymbol{\varepsilon}\Big)(t)\, d\tau$$

so that

$$\sup_{|\boldsymbol{\varepsilon}|\le 1} \frac{|\mathcal{E}^{\boldsymbol{\varepsilon}}(t) - \mathcal{E}(t)|}{|\boldsymbol{\varepsilon}|} \le \sup_{|\boldsymbol{\varepsilon}|\le 1} |\mathcal{D}\mathcal{E}^{\boldsymbol{\varepsilon}}(t)|.$$

By Theorem 1.11,

$$\sup_{|\boldsymbol{\varepsilon}|\le 1}\ \sup_{t\in[0,T]}\ \mathbb{E}\|\mathcal{D}\xi^{\boldsymbol{\varepsilon}}(t)\|^p < \infty,$$

for every $p > 1$ and every matrix norm $\|\cdot\|$; then, using the product rule and assumptions about $\mathfrak{a}$,

$$\sup_{|\boldsymbol{\varepsilon}|\le 1}\ \sup_{t\in[0,T]}\ \mathbb{E}|\mathcal{D}\mathcal{E}^{\boldsymbol{\varepsilon}}(t)|^p < \infty,$$

so that $\mathfrak{T} < \infty$ by Hölder's inequality and boundedness of f. This concludes the proof of (7.2.26) under the additional assumptions of boundedness of f and Q.
Step 2. To remove the assumptions $Q \in \mathbb{L}_\infty([0,T];\mathbb{R}^{r\times d_0})$ and $f \in \mathcal{C}_b^1(\mathbb{R}^d)$, we use a suitable limiting procedure. Note that, by Theorem 1.11, we have

$$\mathbb{E}\sup_{t\in[0,T]} |\xi(t)|^p < \infty,\ \ \mathbb{E}\sup_{t\in[0,t]} \|\mathcal{D}\xi\|^p < \infty,\ \ p > 1,$$

as long as $Q \in \mathbb{L}_p([0,T];\mathbb{R}^{r\times d_0})$ for every $p > 1$.

Let $\zeta(x)$ be the function defined in Sect. 1.5.11. For $R, N \in \mathbb{R}_+$, define $f^R(x) = f(x)\zeta(x/R)$, $Q_N^{il}(t) = (Q^{il}(t)\wedge N)\vee(-N)$, and the processes $\xi_N^{\boldsymbol{\varepsilon}}$, $\mathcal{D}\xi_N^{\boldsymbol{\varepsilon}}$, $\mathcal{E}_N$, etc. using Q_N instead of Q.

This construction implies

$$\mathbb{E}\sup_{t\in[0,T]} |\xi_N(t) - \xi(t)|^2 = 0,\ \ \mathbb{E}\sup_{t\in[0,T]} \|\mathcal{D}\xi_N(t) - \mathcal{D}\xi(t)\|^2,$$

and it remains to pass to the limit $\lim_{N\to\infty}\lim_{R\to\infty}$ in the equality

$$\mathbb{E}\Big(\mathcal{E}_N(t)\boldsymbol{D}f^R\big(\xi_N(t)\big)\mathcal{D}\xi_N(t)\Big) = \mathbb{E}\Big(\mathcal{E}_N(t)f^R\big(\xi_N(t)\big)\big(\mathfrak{e}_N(t) - H_N(t)\big)\Big).$$

This concludes the proof of Lemma 7.1. □

7.2.6. In this paragraph we derive the particular form of the integration by parts formula necessary for the proof of Theorem 7.5, and deduce from it an equality of the type of (7.2.16).

Denote by $\mathbb{C}^n_{pl}(\mathbb{R}^p, \mathbb{R}^q)$ the space of n times differentiable functions from $\mathbb{R}^p$ to $\mathbb{R}^q$ such that, for every $f \in \mathbb{C}^n_{pl}(\mathbb{R}^p; \mathbb{R}^q)$ and every multi-index α with $0 \le |\alpha| \le n$, there exist a positive number N and a positive integer k such that $|f_\alpha(x)| \le N(1 + |x|)^k)$.

Let $\mathbb{M}^r$ be the linear space of $r \times r$-dimensional matrices with norm

$$\|u\|^2 = \sum_{i,j=1}^{r} |u^{ij}|^2,$$

so that $\mathbb{M}^r$ can be identified with $\mathbb{R}^{r^2}$.

Consider two additional standard diffusion processes $\eta = \eta(t)$ and $\overline{\eta} = \overline{\eta}(t)$ with values in $\mathbb{R}^n$ and $\mathbb{R}^{\overline{n}}$ respectively. Define

$$\xi(t) = \big(\eta(t), \mathcal{D}\eta(t), \overline{\eta}(t)\big)$$

as a column vector, so that

$$\mathcal{D}\xi(t) = \big(\mathcal{D}\eta(t), \hat{\eta}(t)\big)$$

as the block matrix, and recall the corresponding functions $\mathfrak{a}$, $\mathcal{E}$, $\mathfrak{e}$, and H from the previous paragraph, with an additional assumption that the function $\mathfrak{a}$ satisfies

$$\mathfrak{a}\big(t, \xi(t)\big) = \mathfrak{a}\big(t, \eta(t)\big).$$

The following condition is the key to carrying out the proof of Theorem 7.5.

(L) *There exists a matrix*

$$Q \in \bigcap_{p>0} \mathbb{L}_p([0, T], \mathscr{P}(\mathfrak{F}); \mathbb{R}^{n\times \mathrm{d}_0})$$

such that, for every $t > 0$, the corresponding matrix $\mathcal{D}\eta(t)$ is invertible and

$$(\mathcal{D}\eta)^{-1}(t) \in \bigcap_{p>0} \mathbb{L}_p(\Omega; \mathbb{M}^n). \tag{7.2.29}$$

In fact, the main step in the proof of Theorem 7.5 is verifying condition (L) for the process $\mathcal{X}$.

Given a differentiable scalar field $\psi = \psi(x)$ and a vector v of the same size as x,

$$\partial_v \psi(x) := \psi_i(x) v^i \equiv \big(D\psi(x)\big)v.$$

Theorem 7.3 (Stochastic Integration by Parts Formula, III) *Assume that η satisfies condition* (L). *Then, for every $\psi \in \mathbb{C}^1_{pl}(\mathbb{R}^n; \mathbb{R}^1)$, $\varphi \in \mathbb{C}^\infty_{pl}(\mathbb{M}^n \times \mathbb{R}^{\overline{n}}; \mathbb{R}^1)$, and $v \in \mathbb{R}^n$, is possible to construct a standard diffusion process ${}_1\overline{\eta}$ with values in $\mathbb{R}^{\overline{n}_1}$ for suitable $\overline{n}_1$ and a function ${}_1\varphi \in \mathbb{C}^\infty_{pl}(\mathbb{M}^n \times \mathbb{R}^{\overline{n}_1}; \mathbb{R}^1)$ such that, for every $t \in [0, T]$,*

$$\begin{aligned} &\mathbb{E}\Big(\mathcal{E}(t)\partial_v\psi\big(\eta(t)\big)\varphi\big((\mathcal{D}\eta)^{-1}(t), \overline{\eta}(t)\big)\Big) \\ &\qquad = \mathbb{E}\Big(\mathcal{E}(t)\psi\big(\eta(t)\big)\,{}_1\varphi\big((\mathcal{D}\eta)^{-1}(t), {}_1\overline{\eta}(t)\big)\Big). \end{aligned} \tag{7.2.30}$$

□

Proof Consider the vector function

$$f\big(\xi(t)\big) = \psi\big(\eta(t)\big)\varphi\big((\mathcal{D}\eta)^{-1}(t), \overline{\eta}(t)\big)(\mathcal{D}\eta)^{-1}(t)v$$

and imagine that we can apply Lemma 7.1 to every component of this function. Then, keeping in mind that $\mathcal{E}$ and $\mathfrak{e}$ depend on ξ only through η, we would get (7.2.30) after adding all the components in the result and re-arranging the terms. In particular, ${}_1\overline{\eta}$ becomes the column vector consisting of all the components of $\overline{\eta}$, $\hat{\eta}$, and $\mathfrak{e} - H$; recall that $\hat{\eta}$ denotes all the components of $\mathcal{D}\xi$ other than $\mathcal{D}\eta$.

Because a direct application of Lemma 7.1 is not possible, we use a suitable limiting procedure. Let $S \in \mathbb{C}^\infty_b(\mathbb{M}^n; \mathbb{M}^n)$ be such that $S(y) = y^{-1}$ if $\|y^{-1}\| \le 1$ and $S(y) = 0$ if $\det(y) = 0$ or if $\|y^{-1}\| \ge 2$. For $R > 0$, define $S_R(y) = RS(Ry)$ so that

$$\lim_{R\to\infty} S_R(y) = \begin{cases} y^{-1}, & \text{if } \det(y) \ne 0; \\ 0, & \text{if } \det(y) = 0. \end{cases}$$

Next, define

$$\varphi_R(y, \overline{y}) = S_R(y)\varphi\big(S_R(y), \overline{y}\big), \quad \varphi_\infty(y, \overline{y}) = y^{-1}\varphi(y^{-1}, \overline{y}).$$

Then we apply Lemma 7.1 to each component of the function

$$f_R\big(\xi(t)\big) = \psi\big(\eta(t)\big)\varphi_R\big(\mathcal{D}\eta(t), \overline{\eta}(t)\big)v,$$

add all the components in the result and rearrange the terms. We interpret $\boldsymbol{D}\psi$ as a row vector, use Tr to denote the trace of a matrix, and write $\bar{H} := \mathfrak{e} - H$. The result is

$$\begin{aligned}\mathfrak{T} :=& \mathbb{E}\Big(\mathcal{E}(t)\boldsymbol{D}\psi\big(\eta(t)\big)\varphi_R\big(\mathcal{D}\eta(t), \bar{\eta}(t)\big)\mathcal{D}\eta(t)v\Big)\\ =& -\mathbb{E}\Big(\mathcal{E}(t)\psi\big(\eta(t)\big)\mathrm{Tr}\Big(\boldsymbol{D}\Big(\varphi_R\big(\mathcal{D}\eta(t), \bar{\eta}(t)\big)v\Big)\hat{\eta}(t)\Big)\Big)\\ &- \mathbb{E}\Big(\mathcal{E}(t)\psi\big(\eta(t)\big)v^*\varphi_R\big(\mathcal{D}\eta(t), \bar{\eta}(t)\big)\bar{H}(t)\Big) := -\mathfrak{T}_1 - \mathfrak{T}_2.\end{aligned}$$

We will now pass to the limit as $R \to +\infty$, first in $\mathfrak{T}$, and then in $\mathfrak{T}_1$ and $\mathfrak{T}_2$. Making use of the structure of the function S_R,

$$\begin{aligned}\mathfrak{T} =& \mathbb{E}\Big(\mathcal{E}(t)\boldsymbol{D}\psi\big(\eta(t)\big)\varphi_\infty\big(\mathcal{D}\eta(t), \bar{\eta}(t)\big)\mathcal{D}\eta(t)v\ 1_{\{\|(\mathcal{D}\eta)^{-1}(t)\| \le R\}}\Big)\\ &+ \mathbb{E}\Big(\mathcal{E}(t)\boldsymbol{D}\psi\big(\eta(t)\big)\varphi_R\big(\mathcal{D}\eta_g(t), \bar{\eta}(t)\big)\mathcal{D}\eta(t)v\ 1_{\{\|(\mathcal{D}\eta)^{-1}(t)\| > R\|\}}\Big) := U_1 + U_2.\end{aligned}$$

Condition (L), together with the Cauchy–Schwarz and Chebyshev inequalities, then implies

$$\lim_{R\to\infty} U_1 = \mathbb{E}\Big(\mathcal{E}(t)\partial_v\psi\big(\eta(t)\big)\varphi\big(\mathcal{D}\eta(t), \bar{\eta}(t)\big)\Big), \quad \lim_{R\to\infty} U_2 = 0.$$

Similarly,

$$\begin{aligned}\lim_{R\to\infty} \mathfrak{T}_1 &= \mathbb{E}\Big(\mathcal{E}(t)\psi\big(\eta(t)\big)\mathrm{Tr}\Big(\boldsymbol{D}\Big(\varphi_\infty\big(\mathcal{D}\eta(t), \bar{\eta}(t)\big)v\Big)\hat{\eta}(t)\Big)\Big),\\ \lim_{R\to\infty} \mathfrak{T}_2 &= \mathbb{E}\Big(\mathcal{E}(t)\psi\big(\eta(t)\big)\varphi\big(\mathcal{D}\eta(t), \bar{\eta}(t)\big)v^*(\mathcal{D}\eta)^{-1}(t)\bar{H}(t)\Big),\end{aligned}$$

completing the proof. □

Corollary 7.1 *Assume that $f \in \mathbb{C}^k_{pl}(\mathbb{R}^n; \mathbb{R}^1)$ and α is a multi-index with $|\alpha| = k$. Then there exist a standard diffusions process ${}_\alpha\bar{\eta}(t) \in \mathbb{R}^{\bar{n}_k}$ and a function ${}_\alpha\varphi \in \mathbb{C}^\infty_{pl}(\mathbb{M}^n \times \mathbb{R}^{\bar{n}_k}; \mathbb{R}^1)$ such that, for every $t \in [0, T]$,*

$$\mathbb{E}\Big(\mathcal{E}(t)f_\alpha\big(\eta(t)\big)\Big) = \mathbb{E}\Big(\mathcal{E}(t)f\big(\eta(t)\big)_\alpha\varphi\big((\mathcal{D}\eta)^{-1}(t), {}_\alpha\bar{\eta}(t)\big)\Big). \qquad (7.2.31)$$

□

Proof Writing

$$\alpha = \alpha_1 + \epsilon_i,\ |\epsilon_i| = 1,$$

we apply Theorem 7.3 to the functions $\psi = f_{\alpha_1}$ and $\varphi \equiv 1$, and take v a suitable unit vector. The result is

$$\mathbb{E}\Big(\mathcal{E}(t) f_\alpha\big(\eta(t)\big)\Big) = \mathbb{E}\Big(\mathcal{E}(t) f_{\alpha_1}\big(\eta(t)\big)_{\alpha_1}\varphi\big((\mathcal{D}\eta)^{-1}(t), {}_{\alpha_1}\overline{\eta}(t)\big)\Big). \tag{7.2.32}$$

A similar application of the theorem to the right-hand side of (7.2.32) yields

$$\mathbb{E}\Big(\mathcal{E}(t) f_\alpha\big(\eta(t)\big)\Big) = \mathbb{E}\Big(\mathcal{E}(t) f_{\alpha_2}\big(\eta(t)\big)_{\alpha_2}\varphi\big((\mathcal{D}\eta)^{-1}(t), {}_{\alpha_2}\overline{\eta}(t)\big)\Big),$$

where $|\alpha_2| = |\alpha| - 2$. After k steps like this, we obtain formula (7.2.31). □

7.2.7. In this paragraph we prove Theorem 7.2. As mentioned before, to prove this theorem it suffices to verify (7.2.16) and (7.2.17).

Under the assumption of the theorem the process $\mathcal{X}(t)$ has a versions which is a $\mathbb{C}^{0,\infty}$-diffeomorphism of $\mathbb{R}^{\mathrm{d}}$ to $\mathbb{R}^{\mathrm{d}}$ (see Sect. 5.3.2); in the future we consider this version and denote by $\boldsymbol{D}\mathcal{X}$ the corresponding derivative matrix.

With (7.2.25) in mind, define

$$\mathcal{Q} = (\boldsymbol{D}\mathcal{X})^{-1}\hat{\sigma},$$

where $\hat{\sigma}$ is the matrix consisting of the first d_0 columns of $\boldsymbol{\Sigma}$; cf. Sect. 7.3. Then

$$\mathcal{D}\mathcal{X}(t) = \boldsymbol{D}\mathcal{X}(t) \int_{[0,t]} \Big(\boldsymbol{D}\mathcal{X}(s)\Big)^{-1} \widetilde{\boldsymbol{\Sigma}}\big(\mathcal{Z}(s)\big)\Big(\big(\boldsymbol{D}\mathcal{X}(s)\big)^{-1} \widetilde{\boldsymbol{\Sigma}}\big(\mathcal{Z}(s)\big)\Big)^* ds. \tag{7.2.33}$$

To prove (7.2.16) and (7.2.17) we apply Corollary 7.1 with $r = \mathrm{d}$, $\eta = \mathcal{X}$, $\mathcal{E} = \rho$. The only condition to verify is that $\mathcal{D}\mathcal{X}$ satisfies Hypothesis (L), that is, the inverse matrix $(\mathcal{D}\mathcal{X})^{-1}$ exists and its norm, as a random variable, is integrable to every power p.

By Theorem 1.11, the matrix $\boldsymbol{D}\mathcal{X}$ satisfies Hypothesis (L). For $t \in [0, T]$, define the matrix

$$\mathfrak{C}_t = \int_{[0,t]} \Big(\boldsymbol{D}\mathcal{X}(s)\Big)^{-1} \hat{\sigma}\big(\mathcal{Z}(s)\big)\Big(\big(\boldsymbol{D}\mathcal{X}(s)\big)^{-1} \widetilde{\boldsymbol{\Sigma}}\big(\mathcal{Z}(s)\big)\Big)^* ds.$$

It remains to verify that the generalized Hörmander condition (H_1) implies that this matrix satisfies Hypothesis (L) too; together with (7.2.33), this will complete the proof of Theorem 7.2.

Theorem 7.4 *If* (H_1) *holds, then*

$$\sup_{t\in[\varepsilon,T]} \|\mathfrak{C}_t^{-1}\| \in \mathbb{L}_p(\Omega; \mathbb{R}^1)$$

for all $p > 0$ *and* $\varepsilon \in (0, T]$. □

Proof The result is known: [151, Theorem 8.31]. In what follows, we explain how condition (H_1) ensures that the matrix $\mathfrak{C}_t$ is invertible with probability one.

Consider the matrix-valued process $\boldsymbol{D}\mathcal{Z} = \boldsymbol{D}\mathcal{Z}(t)$. Recall that $\mathcal{Z} = (\mathcal{X}, \mathcal{Y})$. Then

$$\boldsymbol{D}\mathcal{Z} = \begin{pmatrix} \boldsymbol{D}\mathcal{X} & \boldsymbol{D}_y\mathcal{X} \\ 0 & \boldsymbol{D}\mathcal{Y} \end{pmatrix},$$

where 0 is a $\mathrm{d}_1 \times \mathrm{d}$-dimensional zero matrix, and

$$(\boldsymbol{D}\mathcal{Z})^{-1} = \begin{pmatrix} (\boldsymbol{D}\mathcal{X})^{-1} & U \\ 0 & (\boldsymbol{D}\mathcal{Y})^{-1} \end{pmatrix},$$

where

$$U = -(\boldsymbol{D}\mathcal{X})^{-1}\boldsymbol{D}_y\mathcal{X}(\boldsymbol{D}\mathcal{Y})^{-1}.$$

As a result, with $\hat{\sigma}^l$ denoting the column number l of the matrix $\hat{\sigma}$,

$$(\boldsymbol{D}\mathcal{Z}^{-1}V^l(Z(t)) := \begin{pmatrix} (\boldsymbol{D}\mathcal{X})^{-1}\hat{\sigma}^l \\ 0 \end{pmatrix}.$$

By construction, $\mathfrak{C}_t^{ij}\xi^i\xi^j \geq 0,\ t > 0,\ \xi \in \mathbb{R}^{\mathrm{d}}$. Define the following sub-spaces in $\mathbb{R}^{\mathrm{d}}$:

1. U_s, the linear space generated by the vectors $(\boldsymbol{D}\mathcal{X})^{-1}(s)\hat{\sigma}^l\big(\mathcal{Z}(s)\big),\ l = 1, 2, \ldots, \mathrm{d}_0$;
2. $\mathscr{U}_t$, the linear space spanned by $\bigcup_{s\leq t} U_s$;
3. $\mathscr{U}_t^+ := \bigcap_{s>t} \mathscr{U}_s$.

Given a non-random non-zero vector ζ from $\mathbb{R}^{\mathrm{d}}$, the event $\{\zeta \in \mathscr{U}_0^+\}$ is $\bigcap_{t>0} \mathscr{F}_t^0$-measurable and consequently non-random. If the matrix $\mathfrak{C}_t$ is degenerate with a positive probability, then $\dim \mathscr{U}_0^+ < \mathrm{d}$. Indeed, let $\boldsymbol{\xi} : \Omega \to \mathbb{R}^{\mathrm{d}}$ be a non-zero random vector such that, with positive probability,

$$\boldsymbol{\xi}^*\mathfrak{C}_t\boldsymbol{\xi} = 0.$$

Because

$$\boldsymbol{\xi}^*\mathfrak{C}_t\boldsymbol{\xi} = \sum_{l=1}^{d_0} \int_{[0,t]} |\boldsymbol{\xi}^*(\boldsymbol{D}\mathcal{X})^{-1}(s)\hat{\sigma}^l\big(\mathcal{Z}(s)\big)|^2 ds,$$

it follows that, with positive probability, the vector ξ is orthogonal to all the vectors $(\mathcal{X}'(s))^{-1}\tilde{\Sigma}^l(Z(s))$, $s \le t$, confirming that $\dim \mathscr{U}_0^+ < \mathrm{d}$.

As a result, we will show that $\mathfrak{C}_t$ is non-degenerate with probability one by showing that $\dim \mathscr{U}_0^+ = \mathrm{d}$. To this end, we take a vector $\theta^\circ \in \mathbb{R}^{\mathrm{d}}$ orthogonal to $\mathscr{U}_0^+$ and show that $\theta^\circ = 0$.

To begin, we add d_1 zeros to θ° and denote by θ the resulting vector in $\mathbb{R}^{\mathrm{d}+\mathrm{d}_1}$.

To simplify the notation, denote by $\bar{\boldsymbol{D}}\mathcal{Z}(t)$ the matrix $(\boldsymbol{D}\mathcal{Z})^{-1}(t)$. The structure of the matrix $\bar{\boldsymbol{D}}\mathcal{Z}(t)$ implies

$$\theta^* \bar{\boldsymbol{D}}\mathcal{Z}(t)\hat{V}^l\big(\mathcal{Z}(t)\big) = 0 \tag{7.2.34}$$

for $t \le \tau$, $l = 1, 2, \ldots, \mathrm{d}_0$.

On the other hand, by (1.13) and the Itô formula,

$$\begin{aligned} d\bar{\boldsymbol{D}}\mathcal{Z}(t) &= -\bar{\boldsymbol{D}}\mathcal{Z}(t)(\boldsymbol{D}V^0)\big(\mathcal{Z}(t)\big)\,dt - \bar{\boldsymbol{D}}\mathcal{Z}(t)(\boldsymbol{D}\hat{V}^l)\big(\mathcal{Z}(t)\big)d\mathcal{W}^l(t) \\ &\quad + \bar{\boldsymbol{D}}\mathcal{Z}(t)(\boldsymbol{D}\hat{V}^l)\big(\mathcal{Z}(t)\big)(\boldsymbol{D}\hat{V}^l)^*\big(\mathcal{Z}(t)\big)\,dt,\ t \in (0, T], \\ \bar{\boldsymbol{D}}\mathcal{Z}(0) &= \mathbb{I}, \end{aligned}$$

where $\mathbb{I}$ is the identity matrix.

Another application of the Itô formula shows that, for every vector field $f \in \mathbb{C}_b^2(\mathbb{R}^{\mathrm{d}+\mathrm{d}_1}, \mathbb{R}^{\mathrm{d}+\mathrm{d}_1})$ and the vector $\theta \in \mathbb{R}^{\mathrm{d}+\mathrm{d}_1}$ satisfying (7.2.34),

$$\begin{aligned} d\Big(\theta^* \bar{\boldsymbol{D}}\mathcal{Z}(t) f\big(\mathcal{Z}(t)\big)\Big) &= \theta^* \bar{\boldsymbol{D}}\mathcal{Z}(t)[\tilde{V}^0, f]\big(\mathcal{Z}(t)\big)\,dt \\ &\quad + \theta^* \bar{\boldsymbol{D}}\mathcal{Z}(t)[\hat{V}^l, f]\big(Z(t)\big)\,d\mathcal{W}^l(t), \end{aligned} \tag{7.2.35}$$

where

$$\tilde{V}^0(z) = \hat{V}^0(z) + \frac{1}{2}\sum_{l=1}^{\mathrm{d}_0+\mathrm{d}_1} [\hat{V}^l, [\hat{V}^l, f]](z).$$

Taking $f(z) = \hat{V}^m(z)$, $m = 1, 2, \ldots, \mathrm{d}_0$, equalities (7.2.34) and (7.2.35) imply

$$\theta^* \bar{\boldsymbol{D}}\mathcal{Z}(t)[\hat{V}^l, \hat{V}^m]\big(\mathcal{Z}(t)\big) = 0 \tag{7.2.36}$$

for $t \le \tau$, $l = 1, 2, \ldots, \mathrm{d}_0 + \mathrm{d}_1$, and $m = 1, 2, \ldots, \mathrm{d}_0$, and also

$$\theta^* \bar{\boldsymbol{D}}\mathcal{Z}(t)[\hat{V}^l, [\hat{V}^l, \hat{V}^m]]\,\big(\mathcal{Z}(t)\big) = 0 \tag{7.2.37}$$

for $t \le \tau$ and $m = 1, 2, \ldots \mathrm{d}_0$.

From (7.2.34)–(7.2.37) it follows that

$$\theta^* \bar{\boldsymbol{D}}\mathcal{Z}(t)[\hat{V}^0, \hat{V}^m](\mathcal{Z}(t)) = 0 \tag{7.2.38}$$

for $t \le \tau$ and $m = 1, 2, \dots, \mathrm{d}_0$.

In other words, for every $l = 0, 1, \dots, \mathrm{d}_0 + \mathrm{d}_1,\ m = 1, 2, \dots, \mathrm{d}_0$, the vector θ is orthogonal to all the vectors $\bar{\boldsymbol{D}}\mathcal{Z}(t)[\hat{V}^l, \hat{V}^m](\mathcal{Z}(t))$ for $t \le \tau$. By induction, the vector θ is orthogonal to all the vectors of the form

$$\bar{\boldsymbol{D}}\mathcal{Z}(t)[\hat{V}^{l_1}, [\hat{V}^{l_2}, \dots, [\hat{V}^{l_n}, \hat{V}^m]\dots](\mathcal{Z}(t))$$

$l_i = 0, 1, \dots, \mathrm{d}_0 + \mathrm{d}_1,\ m = 1, 2, \dots, \mathrm{d}_0$, for $t \le \tau$. Because $\bar{\boldsymbol{D}}\mathcal{Z}(0) = \mathbb{I}$, we conclude that θ is orthogonal to $\hat{V}^1(z_0), \dots, \hat{V}^{\mathrm{d}_0}(z_0)$ and to all the Lie brackets $[\hat{V}^{l_1}, [\hat{V}^{l_2}, \dots, [\hat{V}^{l_n}, \hat{V}^m]\dots](z_0)$, where $l_i = 0, 1, \dots, \mathrm{d}_0 + \mathrm{d}_1$, and $m = 1, 2, \dots, \mathrm{d}_0$. Together with the anti-symmetry of the Lie brackets and the Jacobi identity, this implies that the vector θ is orthogonal to all the Lie brackets of the form

$$[\hat{V}^{l_1}[\hat{V}^{l_2}, \dots, [\hat{V}^{l_n}, \hat{V}^m], \hat{V}^{l_{n+1}}], \dots, \hat{V}^{l_k}](z_0).$$

The last d_1 coordinates for every Lie bracket of this form are equal to 0. Therefore the d-dimensional vector θ° is orthogonal to the projections of $V^1(z_0), \dots, V^m(z_0)$ and

$$[\hat{V}^{l_1}[\hat{V}^{l_2}, \dots, \hat{V}^{l_n}, \hat{V}^m], \hat{V}^{l_{n+1}}], \dots, \hat{V}^{l_k}](z_0)$$

on $\mathbb{R}^{\mathrm{d}}$. By the same arguments, the dimension of the linear space generated by these vectors is the same as that of the linear space generated by their projections on $\mathbb{R}^{\mathrm{d}}$, which means $\theta^\circ = 0$. □

7.3 The Filtering Transition Density and the Fundamental Solution of the Filtering Equation in Hypoelliptic and Superparabolic Cases

7.3.1. In Chap. 6 we studied the filtering measure for the conditional distribution of a diffusion process $X = X(t)$ given another diffusion process $Y = Y(t)$. We showed that the filtering measure has a density satisfying the (forward) filtering equation. A key assumption throughout Chap. 6 was the existence and $\mathbb{L}_p(\Omega, \mathbb{P}; \mathbb{H}^1)$-regularity of the conditional density of $X(0)$ given $Y(0)$, and the main objective of this section is to remove this assumption.

In other words, in terms of the theory of Markov processes, our main object of study is the density of the conditional transition measure, as opposed to the density of the conditional distribution in Chap. 6. In terms of the theory of partial differential equations, we now study the fundamental solution of the filtering equation, as opposed to the solution of an initial value problem.

Warning 7.4 *Besides the existence of the initial filtering density, all other assumptions and notation from Chap. 6 are in force throughout this section; additional assumptions and notation will also be introduced.* □

As in Chap. 6 we consider the $(\mathrm{d}+\mathrm{d}_1)$-dimensional diffusion process $Z = Z(t),\ t \in [0,T]$, defined, in the matrix-vector form, by the system of Itô equations

$$Z(t) = z_0 + \int_{[0,t]} {}_{\mathrm{Z}}b\big(s, Z(s)\big)\, ds + \int_{[0,t]} {}_{\mathrm{Z}}\boldsymbol{\Sigma}\big(s, Z(s)\big)\, d\mathcal{W}(s),$$

with non-random initial condition z_0. Similar to Chap. 6, we split Z into the first d unobserved components X and the remaining d_1 observed components Y. The same splitting applies to the initial condition z_0, the drift coefficient ${}_{\mathrm{Z}}b$, and the Brownian motion $\mathcal{W}$. We also assume that the diffusion matrix ${}_{\mathrm{Z}}\boldsymbol{\Sigma}$ has the canonical form (6.1.2), that is,

$${}_{\mathrm{Z}}\boldsymbol{\Sigma} = \begin{pmatrix} {}_{1}\sigma & \sigma \\ 0 & {}_{0}\sigma \end{pmatrix}.$$

Recall that matrix ${}_{0}\sigma$ is invertible and the function $h = {}_{0}\sigma^{-1}\, {}_{\mathrm{Y}}b$ is bounded.

We are interested in the analytical properties of the filtering transition measure

$$\mathbb{P}\big(X(t) \in \Gamma \big| X(0) = x_0, \mathscr{Y}_t^0\big),$$

in particular, the existence and regularity of its density $\pi = \pi(x_0, t, x)$. We will also address the same questions for the interpolation and extrapolation measures.

7.3.2. In this paragraph we establish the existence of the fundamental solution of the forward linear filtering equation in the hypoelliptic case. To be more precise, the matrix $a = {}_{1}\sigma\, {}_{1}\sigma^* + \sigma\sigma^*$ can be singular, but other assumptions are introduced, namely, a version of Hörmander's condition (cf. Sect. 7.2) and extra smoothness of the coefficients ${}_{\mathrm{Z}}b$ and ${}_{\mathrm{Z}}\boldsymbol{\Sigma}$.

Below is the main notation to be used:

$${}_{\mathrm{X}}B = {}_{\mathrm{X}}b - \sigma h,\quad V^0 = \begin{pmatrix} {}_{\mathrm{X}}B \in \mathbb{R}^{\mathrm{d}} \\ \mathbf{0} \in \mathbb{R}^{\mathrm{d}_1} \\ 1 \in \mathbb{R}^1 \end{pmatrix} \in \mathbb{R}^{\mathrm{d}+\mathrm{d}_1+1},$$

$$\hat{V}^{il}(z) = \begin{cases} {}_{\mathrm{Z}}\boldsymbol{\Sigma}^{il}(z) & \text{for} \quad i = 1, 2, \ldots, \mathrm{d}+\mathrm{d}_1, \\ 0 & \text{for} \quad i = \mathrm{d}+\mathrm{d}_1+1, \end{cases} \quad l = 1, \ldots, \mathrm{d}+\mathrm{d}_1,$$

$$\hat{V}^{i0}(z) := V^{i0}(z) - \frac{1}{2}\sum_{l=1}^{\mathrm{d}+\mathrm{d}_1+1} \hat{V}^{il}_j(z)\hat{V}^{jl}(z),\ i = 1, 2, \dots, \mathrm{d} + \mathrm{d}_1 + 1,$$

$$V^{il}(z) = \hat{V}^{il}(z),\ i = 1, 2, \dots, \mathrm{d} + \mathrm{d}_1 + 1,\ \ l = 1, 2, \dots, \mathrm{d}.$$

In this section we assume that

(D_∞) $\hat{V}^{il} \in \mathbb{C}_b^\infty(\mathbb{R}^{\mathrm{d}+\mathrm{d}_1+1})$, $i = 1, 2, \dots, \mathrm{d} + \mathrm{d}_1 + 1$, $l = 1, 2, \dots, \mathrm{d} + \mathrm{d}_1$.

($\mathrm{H_F}$) **(Hörmander's condition for the filtering problem)**. *For every* $z_0 \in \mathbb{R}^{\mathrm{d}+\mathrm{d}_1}$, *the dimension of the linear space generated by the vector field* $V^1, \dots, V^{\mathrm{d}}$ *and the Lie brackets of the vector fields* $\hat{V}^0, \dots, \hat{V}^{\mathrm{d}+\mathrm{d}_1+1}$ *among which there is at least one vector field from the collection* V^i, $i = 1, 2, \dots, \mathrm{d}$, *is equal to* d.

Just as in Chap. 6 we define the random variables

$$\rho(t, s) = \exp\left\{\int_{[s,t]} h^l\big(r, Z(r)\big)\, dw^l(r) + \frac{1}{2}\int_{[s,t]} h^l h^l\big(r, Z(r)\big) dr\right\},$$

the measure

$$\tilde{\mathbb{P}}(\Gamma) = \int_\Gamma \frac{d\mathbb{P}(\omega)}{\rho(T, 0)},\quad \Gamma \in \mathscr{F},$$

and the operators

$$\mathcal{L}^* f(t, x) := \Big(a^{ij}\big(t, x, Y(t)\big) f(t, x)\Big)_{ij} - \Big({}_{\mathrm{x}}b^i\big(t, x, Y(t)\big) f(t, x)\Big)_i,$$

$$\mathcal{M}^{l,*} f(t, x) := -\Big(\sigma^{il}\big(t, x, Y(t)\big) f(t, x)\Big)_i + h^l\big(t, x, Y(t)\big) f(t, x);$$

as usual, $f_i = \partial f/\partial x^i$, $i = 1, \dots, \mathrm{d}$. Recall that, according to the Girsanov theorem, there exists an $\mathbb{R}^{\mathrm{d}_1}$-valued standard Wiener process $\widetilde{W} = \widetilde{W}(t)$ on the stochastic basis $\widetilde{\mathbb{F}} := (\Omega, \mathscr{F}, \{\mathscr{F}_t\}_{t\in[0,T]}, \tilde{\mathbb{P}})$ such that the process $Z = (X, Y)$, considered on $\widetilde{\mathbb{F}}$, satisfies

$$X(t) = x_0 + \int_{[0,t]} \big({}_{\mathrm{x}}b - \sigma h\big)\big(s, Z(s)\big)\, ds + \int_{[0,t]} {}_1\sigma\big(s, Z(s)\big)\, d\overline{W}(s) + \int_{[0,t]} \sigma\big(s, Z(s)\big)\, d\widetilde{W}(s),$$

$$Y(t) = y_0 + \int_{[0,t]} {}_0\sigma\big(s, Y(s)\big)\, d\widetilde{W}(s),\ t \in [0, T].$$

By Lemma 6.1.1, the σ-algebra $\mathscr{Y}_t^0$ generated by $Y(s)$, $s \le t$, coincides with the σ-algebra $\mathscr{F}_t^0$, the completion with respect $\mathbb{P}$ of the σ-algebra generated by the Wiener process $\tilde{W}(s)$ for $s \le t$. Denote by $\tilde{\mathbb{F}}(\mathscr{Y})$ the probability space $(\Omega; \mathscr{F}; \mathscr{Y}_t^0; \mathbb{P})$ and by $\mathscr{P}(\mathscr{Y})$ the σ-algebra of predictable sets on $\mathbb{F}(\mathscr{Y})$.

On the probability space $\tilde{\mathbb{F}}(\mathscr{Y})$ consider the (forward linear) filtering equation

$$\begin{aligned} du(t,x) &= \mathcal{L}^* u(t,x)\,dt + \mathcal{M}^{l*}\big(t,x,Y(t)\big)u(t,x)\,d\tilde{w}^l(t), \\ (t,x,\omega) &\in (0,T] \times \mathbb{R}^{\mathrm{d}} \times \Omega. \end{aligned} \tag{7.3.1}$$

Definition 7.5 The function $u = u(x_0,t,x,\omega) : \mathbb{R}^{\mathrm{d}} \times (0,T] \times \mathbb{R}^{\mathrm{d}} \times \Omega \to \mathbb{R}^1$ is called the **fundamental solution** of Eq. (7.3.1) if

- $u(x_0,\cdot,\cdot,\cdot)$ is $\overline{\mathscr{B}(]0,T] \times \mathbb{R}^{\mathrm{d}}) \otimes \mathscr{F}}$-measurable for every x_0;
- $u(x_0,\cdot,x,\cdot)$ is $\mathscr{P}(\mathscr{Y})$-measurable for every $x_0, x \in \mathbb{R}^{\mathrm{d}}$;
- $u(x_0,t,\cdot,\omega)$ belongs to $\mathbb{C}^2(\mathbb{R}^{\mathrm{d}})$ for every $t \in]0,T]$ and $x_0 \in \mathbb{R}^{\mathrm{d}}$ on a set of probability *1;*
- for $0 < s < t \le T$,

$$\begin{aligned} u(x_0,t,x) = u(x_0,s,x) &+ \int_{[s,t]} \mathcal{L}^*\big(\tau,x,Y(\tau)\big)u(x_0,\tau,x)\,d\tau \\ &+ \int_{[s,t]} \mathcal{M}^{l*}\big(\tau,x,Y(\tau)\big)u(x_0,\tau,x)\,d\tilde{w}^l(\tau); \end{aligned} \tag{7.3.2}$$

- for every $f \in \mathbb{C}_b^0(\mathbb{R}^{\mathrm{d}})$ and $x_0 \in \mathbb{R}^{\mathrm{d}}$,

$$\lim_{t \downarrow 0} \int_{\mathbb{R}^{\mathrm{d}}} u(x_0,t,x) f(x)\,dx = f(x_0). \tag{7.3.3}$$

□

Theorem 7.5 *Under* (D_∞) *and* ($\mathrm{H_F}$), *Eq. (7.3.1) has a unique fundamental solution and*

(a) *For every* $x_0 \in \mathbb{R}^{\mathrm{d}}$,

$$\mathbb{P}\Big(\omega : u(x_0,\cdot,\omega,\cdot) \in \mathbb{C}^{0,\infty}\big((0,T] \times \mathbb{R}^{\mathrm{d}}\big),\ u(x_0,\cdot,\omega,\cdot) \ge 0\Big) = 1;$$

(b) *For every* $x_0 \in \mathbb{R}^{\mathrm{d}}$, $\varepsilon \in (0,T]$, *and* $p \ge 1$,

$$\mathbb{E} \sup_{t \in [\varepsilon,T]} \sup_{x \in \mathbb{R}^{\mathrm{d}}} |u(x_0,t,x)|^p < \infty. \qquad \square$$

Proof By Theorem 7.1, Eq. (7.3.1), with initial condition μ_0 equal to the point mass at x_0,

$$\mu_0(\omega, \Gamma) = \begin{cases} 1, & x \in \Gamma, \\ 0, & x \notin \Gamma, \end{cases} \tag{7.3.4}$$

has a unique measure-valued solution μ_t and, for every $\psi \in \mathbb{L}_\infty$,

$$\mu_t[\psi] = \tilde{\mathbb{E}}\big[\psi\big(X(t)\big)\tilde{\rho}(t)\big|\mathscr{Y}_t^0\big], \tag{7.3.5}$$

where

$$\tilde{\rho}(t) = \exp\left\{ \int_{[0,t]} h^l\big(s, Z(s)\big)\, d\tilde{w}^l(s) - \frac{1}{2}\int_{[0,t]} h^l h^l\big(s, Z(s)\big)\, ds \right\}.$$

Setting $\mathcal{X}(t) = X(t)$, $\mathcal{Y}(t) = \big(Y(t), t\big)$ and using hypothesis ($\mathrm{H_F}$), we apply Theorem 7.5 to μ_t and conclude that, for $t \in (0, T]$, μ_t has density $u = u(x_0, t, x)$ and

$$u(x_0, \cdot, \cdot) \in \mathbb{C}^{0,\infty}\big((0, T] \times \mathbb{R}^{\mathrm{d}}\big). \tag{7.3.6}$$

Moreover, by Remark 7.2 we have

$$\mathbb{E} \sup_{t \in [\varepsilon, T]} \sup_{x \in \mathbb{R}^{\mathrm{d}}} |u(x_0, t, x)|^p < \infty, \quad \varepsilon \in (0, T], \; p \geq 1. \tag{7.3.7}$$

Equality (7.3.2) now follows from the definition of the measure-valued solution. Equality (7.3.3) follows from continuity of μ_t in t and the equality $\mu_0[f] = f(x_0)$. □

7.3.3. In this paragraph we prove the existence of the fundamental solution of the forward linear filtering equation in the super-parabolic case, so that assumptions (D_∞) and ($\mathrm{H_F}$) are replaced with

(A) *There exists a number* $\delta > 0$ *such that, for all* $(t, x, y) \in (0, T] \times \mathbb{R}^{\mathrm{d}} \times \mathbb{R}^{\mathrm{d}_1}$,

$${}_1\sigma^{ij}(t, x, y)\xi^i\xi^j \geq \delta|\xi|^2, \quad \xi \in \mathbb{R}^{\mathrm{d}}.$$

(D_m) *There exists an* $m \in \mathbb{N}$ *such that, for every* $y \in \mathbb{R}^{\mathrm{d}_1}$, *all the functions* a^{ij}, $\boldsymbol{D}_x a^{ij}$, b^i, σ^{il}, $\boldsymbol{D}_x\sigma^{il}$, *and* h^l *are in* $\mathbb{C}_b^{0,m}\big((0, T] \times \mathbb{R}^{\mathrm{d}}\big)$.

Condition (A) implies uniform non-degeneracy of the matrix a, and is therefore stronger than ($\mathrm{H_F}$). On the other hand, (D_m) is weaker than (D_∞). In particular, (D_m) does not require any smoothness of the coefficients with respect to y.

Let $\mathfrak{n}$ be the least integer greater than $\mathrm{d}/2$ and let $(\cdot,\cdot)_\gamma$ denote the inner product in the Sobolev space $\mathbb{H}^\gamma$.

Definition 7.6 The function $u = u(x_0, t, x, \omega)$ is called a **generalized fundamental solution** of *(7.3.1)* if

- for every $\varepsilon \in (0, T]$ and $x_0 \in \mathbb{R}^{\mathrm{d}}$,

$$u(x_0, \cdot, \cdot, \cdot) \in L_2^{\omega}([\varepsilon, T], \mathscr{P}(\mathscr{Y}); \mathbb{H}^1) \cap \mathbb{L}_1([0, T] \times \Omega; \mathbb{C}_w\mathbb{L}_1);$$

- for all $(x_0, t, \omega) \in \mathbb{R}^{\mathrm{d}} \times [0, T] \times \Omega$ u belongs to the cone of non-negative functions from $\mathbb{L}_1$;
- for all $\psi \in \mathbb{C}_0^{\infty}(\mathbb{R}^{\mathrm{d}})$ and $t \in [0, T]$,

$$\begin{aligned}(u,\psi)_0(x_0,t) = \psi(x_0) + \int_{[0,t]} \Big(&- (a^{ij} u_i(x_0,s), \psi_j)_0 \\ &+ \big((b^i - a^{ij}_j) u_i, \psi\big)_0 \Big)(x_0,s)\, ds + \int_{[0,t]} (\mathcal{M}^l u, \psi)_0(x_0,s)\, d\tilde{w}^l(s);\end{aligned} \tag{7.3.8}$$

- for every function $f \in \mathbb{C}_b^0(\mathbb{R}^{\mathrm{d}})$ and every $x_0 \in \mathbb{R}^{\mathrm{d}}$,

$$\lim_{t \downarrow 0} \int_{\mathbb{R}^{\mathrm{d}}} u(x_0, t, x) f(x)\, dx = f(x_0). \tag{7.3.9}$$

□

Theorem 7.6 *If* (A) *and* (D$_m$) *hold with* $m \geq 2\mathfrak{n}+2$, *then there exists a generalized fundamental solution* $u = u(x_0, t, x, \omega)$ *of the forward linear filtering equation* (7.3.1) *and, for every* $x_0 \in \mathbb{R}^{\mathrm{d}}$,

(a) $u \in \mathbb{L}_2([\varepsilon, T], \mathscr{P}(\mathscr{Y}); \mathbb{H}^{m-2\mathfrak{n}-1}) \bigcap \mathbb{L}_2\big(\Omega; \mathbb{C}([\varepsilon, T]; \mathbb{H}^{m-2\mathfrak{n}-2})\big)$, $\varepsilon \in (0, T]$;
(b) *For every* $t \in [0, T]$ *and* $\psi \in \mathbb{L}_\infty$,

$$\big(u, \psi\big)(x_0, t) = \tilde{\mathbb{E}}[\psi\big(X(t)\big)\rho(t, 0)|\mathscr{Y}_t^0] \quad (\mathbb{P}\text{- a.s.});$$

(c) *If* u_1 *and* u_2 *are two generalized fundamental solutions of the forward linear filtering equation, then*

$$\mathbb{E} \int_{[0,T]} \|u_1(x_0, t) - u_2(x_0, t)\|_1^2\, dt = 0.$$

□

Before proceeding to the proof we give a simple corollary of the theorem.

Corollary 7.2 *If the assumptions of the theorem are fulfilled for some $n \in \mathbb{N}$ such that $m > n + 2(\mathrm{n} + 1) + \mathrm{d}/2$, then the generalized solution $u = u(x_0, t, x)$ of the direct linear filtering equation* (7.3.1) *has a version which is $\mathscr{P}(\mathscr{Y}) \otimes \mathscr{B}(\mathbb{R}^{\mathrm{d}})$-measurable, belongs to $\mathbb{C}^{0,n}\big((0, T] \times \mathbb{R}^{\mathrm{d}}\big)$ for all $\omega \in \Omega$ and $x_0 \in \mathbb{R}^{\mathrm{d}}$, and, when $n \geq 2$, is a classical solution of* (7.3.1) *for $t \in (0, T]$.* □

The proof of this corollary is based on Proposition 4.1 and is similar to the proof of Theorem 4.3; carrying out a detailed argument can be a useful exercise for an interested reader.

Proof of the theorem By Theorem 7.1, Eq. (7.3.1) with the initial condition (7.3.4) has a measure-valued solution μ_t and (7.3.5) holds. To show the existence and regularity of the corresponding density u, several technical constructions are necessary, along the lines of the proof of Lemma 5.3.

Let $\eta = \eta(t)$ be a smooth function such that $\eta(0) = 0$, $\eta(t) > 0$ for $t \in (0, T]$, and, for every $\varepsilon \in (0, T]$, $r = 1, 2, \ldots$,

$$\sup_{t \in [\varepsilon, T]} |\eta^{[r]}(t)|^{-1} < \infty,$$

where

$$\eta^{[r]}(t) = \int\limits_{[0,t]} \int\limits_{[0,t_{r-1}]} \cdots \int\limits_{[0,t_1]} g(s)\, ds dt_1 \ldots dt_{r-1}.$$

By the product rule,

$$\begin{aligned} \mu_t[\psi]\eta(t) = & \int\limits_{[0,t]} \mu_s[\mathcal{L}(s)\psi]\eta(s)\, ds \\ & + \int\limits_{[0,t]} \mu_s[\mathcal{M}^l(s)\psi]\eta(s)\, dw^l(s) + \int\limits_{[0,t]} \mu_s[\psi]\eta'(s)\, ds \end{aligned} \tag{7.3.10}$$

for all $\psi \in \mathbb{C}_0^\infty(\mathbb{R}^{\mathrm{d}})$ and $t \in [0, T]$. Proposition 5.5 implies that there exists a function $v \in \mathbb{L}_2([0, T], \mathscr{P}(\mathscr{Y}); \mathbb{H}^{\mathrm{n}})$ such that, for all $(t, \omega) \in [0, T] \times \Omega$,

$$\mu_t[\psi] = \big(\psi, v(t)\big)_{\mathrm{n}}. \tag{7.3.11}$$

The function v depends on x_0 because μ_t depends on x_0; for notational simplicity, we will not explicitly indicate this dependence on x_0.

Define

$$\bar{u}(t) = v(t)\eta(t).$$

Equalities (7.3.10) and (7.3.10) imply that $\bar{u} \in \mathbb{L}_2([0, T], \mathscr{P}(\mathscr{Y}); \mathbb{H}^{\mathrm{n}})$ and, for every $t \in [0, T]$ and $\psi \in \mathbb{C}_0^\infty(\mathbb{R}^{\mathrm{d}})$,

$$\begin{aligned}\big(\bar{u}(t), \psi\big)_{\mathrm{n}} = \int\limits_{[0,t]} \big(\bar{u}(s), \mathcal{L}\psi\big)_{\mathrm{n}}\, ds + \int\limits_{[0,t]} \big(\bar{u}(s), \mathcal{M}^l\psi\big)_{\mathrm{n}}\, d\tilde{w}^l(s) \\ + \int\limits_{[0,t]} \big(v(s)\eta'(s), \psi\big)_{\mathrm{n}}\, ds.\end{aligned} \tag{7.3.12}$$

Next, define

$$u^{\{1\}}(t) = \Lambda^{-1}\bar{u}(s), \quad \Lambda = \big(\mathbb{I} - \boldsymbol{\Delta}\big)^{1/2}.$$

By Proposition 3.5, $\Lambda^{-1}v\eta' \in \mathbb{L}_2([0, T], \mathscr{P}(\mathscr{Y}); \mathbb{H}^{\mathrm{n}+2})$ and

$$\begin{aligned}\big(u^{\{1\}}, \psi\big)_{\mathrm{n}+1}(t) = \int\limits_{[0,t]} \big(u^{\{1\}}, \mathcal{L}\psi\big)_{\mathrm{n}+1}(s)\, ds + \int\limits_{[0,t]} \big(u^{\{1\}}, \mathcal{M}^l\psi\big)_{\mathrm{n}+1}(s)\, d\tilde{w}^l(s) \\ + \int\limits_{[0,t]} (\Lambda^{-1}v\eta', \psi)_{\mathrm{n}+1}(s)\, ds,\end{aligned} \tag{7.3.13}$$

for every $t \in [0, T]$ and $\psi \in \mathbb{C}_0^\infty(R^d)$.

Given $n \in \mathbb{N} \cup \{0\}$, $\varphi_0 \in \mathbb{L}_2(\Omega, \mathscr{F}_0; \mathbb{H}^n)$, and $f \in \mathbb{L}_2([0, T], \mathscr{P}(\mathscr{Y}); \mathbb{H}^{n-1})$, consider the equation

$$\begin{aligned}(\varphi, \psi)_n(t) = (\varphi_0, \psi)_n + \int\limits_{[0,t]} (\varphi, \mathcal{L}\psi)_n(s)\, ds \\ + \int\limits_{[0,t]} [f(s), \psi]_n ds + \int\limits_{[0,t]} \big(\varphi, \mathcal{M}^l\psi\big)_n(s)\, d\tilde{w}^l(s),\end{aligned} \tag{7.3.14}$$

where $[\cdot, \cdot]_n$ is the CBF of $(\mathbb{H}^{n+1}, \mathbb{H}^n, \mathbb{H}^{n-1})$. A function

$$\varphi \in \mathbb{L}_2^\omega([0, T], \mathscr{P}(\mathscr{Y}); \mathbb{H}^{n+1})$$

satisfying equation (7.3.14) on the interval $[0, T]$ for all $\psi \in \mathbb{C}_0^\infty(\mathbb{R}^{\mathrm{d}})$, $\mathbb{P}$-a.s., will be called a generalized solution of this equation.

Lemma 7.2 *Given $k \in \mathbb{N} \cup \{0\}$ such that $n + k \leq m$, suppose that $\varphi_0 \in \mathbb{L}_2(\Omega, \mathscr{F}_0; \mathbb{H}^{n+k})$ and $f \in \mathbb{L}_2\big([0, T], \mathscr{P}(\mathscr{Y}); \mathbb{H}^{n+k-1}\big)$. Then there exists a unique generalized solution $\varphi = \varphi(t)$ of Eq. (7.3.14) and*

$$\varphi \in \mathbb{L}_2\big([0, T], \mathscr{P}(\mathscr{Y}); \mathbb{H}^{n+k+1}\big) \bigcap \mathbb{L}_2\big(\Omega; \mathbb{C}([0, T]; \mathbb{H}^{n+k})\big). \qquad \square$$

Proof The same arguments as in Proposition 5.5 show that Eq. (7.3.14) is equivalent to

$$u(t) = u_0 + \int_{[0,t]} [A(s)u(s) + f(s)]\,ds + \int_{[0,t]} B(s)u(s)\,d\tilde{w}(s) \tag{7.3.15}$$

in the normal triple $(\mathbb{H}^{n+1}, \mathbb{H}^n, \mathbb{H}^{n-1})$, where

$$[A(s)\varphi, \psi]_n = \big(\varphi, \mathcal{L}(s)\psi\big)_n, \quad \big(B(s)\varphi, \psi\big)_n^l = \big(\varphi, \mathcal{M}^l(s)\psi\big)_n.$$

After that, arguments similar to those used in the proof of Proposition 5.5 (see (5.4.20) and (5.4.24)–(5.4.30)) imply that (7.3.15) satisfies the assumptions of Theorem 3.2 and 3.4 with $\mathbb{V} = \mathbb{H}^{n+k+1}$, $\mathbb{U} = \mathbb{H}^{n+k}$, and $\mathbb{V}' = \mathbb{H}^{n+k-1}$. Application of these theorems to (7.3.15) completes the proof of the lemma. □

The above lemma implies that $u^{\{1\}}$ is the unique generalized solution of (7.3.13) and $u^{\{1\}} \in \mathbb{L}_2\big([0,T], \mathscr{P}(\mathscr{Y}); \mathbb{H}^{\mathfrak{n}+4}\big)$. Hence

$$\Lambda^{-1}v(s)\eta'(s) = \frac{u^{\{1\}}\eta'}{\eta} \in \mathbb{L}_2([0,T], \mathscr{P}(\mathscr{Y}); \mathbb{H}^{\mathfrak{n}+4}).$$

Define

$$u^{\{r\}}(t) := \Lambda^{-1}v(t) \int_{[0,t]} \int_{[0,t_{r-1}]} \cdots \int_{[0,t_1]} \eta(s)\,ds\,dt_1 \ldots dt_{r-1} \equiv \Lambda^{-1}v(t)\eta^{[r]}(t),$$

so that $u^{\{r\}}$ is a generalized solution of (7.3.14) with $n = \mathfrak{n} + 1$, $f(t) = u^{\{r-1\}}(t)$, and $\varphi_0 = 0$. As long as $2r \le m - 3 - \mathfrak{n}$, a repeated application of the lemma yields

$$u^{\{r\}} \in \mathbb{L}_2\big([0,T] \times \Omega, \mathscr{P}(\mathscr{Y}); \mathbb{H}^{\mathfrak{n}+4+2r}\big) \bigcap \mathbb{L}_2\big(\Omega; \mathbb{C}([0,T]; \mathbb{H}^{\mathfrak{n}+4+2r-1})\big),$$

which then implies

$$v \in \mathbb{L}_2\big([\varepsilon,T] \times \Omega, \mathscr{P}(\mathscr{Y}); \mathbb{H}^{\mathfrak{n}+2(r+1)}\big) \bigcap \mathbb{L}_2\big(\Omega; \mathbb{C}\big([\varepsilon,T]; \mathbb{H}^{\mathfrak{n}+2r+1}\big)\big)$$

for every $\varepsilon \in (0,T]$.

Define

$$u(x_0, t) = \Lambda^{\mathfrak{n}}v(t) = \frac{\Lambda^{\mathfrak{n}+1}u^{\{r\}}}{\eta^{[r]}(t)}$$

so that

$$u \in \mathbb{L}_2\big([\varepsilon,T], \mathscr{P}(\mathscr{Y}), \mathbb{H}^{2r+2-\mathfrak{n}}\big) \bigcap \mathbb{L}_2\big(\Omega; \mathbb{C}\big([\varepsilon,T]; \mathbb{H}^{2r+1-\mathfrak{n}}\big)\big),$$

$t \in [0, T]$, $\varepsilon \in (0, T]$. From (7.3.11) it follows that, for $\psi \in \mathbb{C}_0^\infty(\mathbb{R}^{\mathrm{d}})$,

$$\mu_t[\psi] = \big(\psi, u(x_0, t)\big)_0. \tag{7.3.16}$$

Letting $2r = m - 3 - \mathfrak{n}$, we obtain

$$u(x_0, \cdot, \cdot) \in \mathbb{L}_2\big([\varepsilon, T], \mathscr{P}(\mathscr{Y}); \mathbb{H}^{m-2\mathfrak{n}-1}\big) \bigcap \mathbb{L}_2\big(\Omega; \mathbb{C}([\varepsilon, T]; \mathbb{H}^{m-2\mathfrak{n}-2})\big),$$

$\varepsilon \in (0, T]$. By the same arguments as in the proofs of Theorem 5.6 and Corollary 5.6, equality (7.3.16) holds for every $\psi \in \mathbb{L}_\infty$ and the generalized fundamental solution has a version in $\mathbb{L}_1\big([0, T] \times \Omega; \mathbb{C}_w\mathbb{L}_1\big)$ belonging to the cone of non-negative functions in $\mathbb{L}_1$.

By the definition of a measure-valued solution, u satisfies equality (7.3.8) for all $t \in [0, T]$ and $\psi \in \mathbb{C}_0^\infty(\mathbb{R}^{\mathrm{d}})$ on a ω-set of probability 1.

From (7.3.16), using a weak continuity argument, it also follows that u satisfies (7.3.9), confirming that $u = u(x_0, t)$ is a generalized fundamental solution of (7.3.1).

If both u_1 and u_2 are generalized fundamental solutions of this equation, then their difference $\widetilde{u}$ is also a generalized solution of equation (7.3.1) with zero initial condition (in the sense of Definition 4.4). By Theorem 4.5

$$\mathbb{E} \int_{[0,t]} \|\widetilde{u}(s)\|_1^2 ds = 0. \qquad \square$$

7.3.4. Here and in the next two paragraphs we investigate the filtering, interpolation and extrapolation transition probabilities and densities and their analytical properties in the super-parabolic and hypoelliptic cases. Recall that $\mathscr{M}(\mathbb{R}^{\mathrm{d}})$ denotes the space of probability measures on $\mathbb{R}^{\mathrm{d}}$.

As in Sect. 6.1.1, we fix T_0, T_1 so that $0 < T_0 \le T_1 \le T$.

Arguing exactly as in the proof of Theorem 6.3, we obtain from Theorem 7.6 the following result.

Theorem 7.7 *Under the assumptions of Theorem 7.6, there exists a function* $P^t_{\mathscr{Y}}(x_0) : \Omega \to \mathscr{M}(\mathbb{R}^{\mathrm{d}})$ *such that, for every* $x_0 \in \mathbb{R}^{\mathrm{d}}$,

(a) *The measure* $P^t_{\mathscr{Y}}(x_0, \cdot, \omega)$ *is a regular conditional distribution of* $X(t)$ *given* $\mathscr{Y}^0_t$, $t \in (0, T]$;
(b) *For all* $(t, \omega) \in (0, T] \times \Omega'$, *where* $P(\Omega') = 1$, *the measure* $P^t_{\mathscr{Y}}(x_0)$ *is absolutely continuous with respect to the Lebesgue measure on* $\mathbb{R}^{\mathrm{d}}$, *and the density is*

$$\pi(x_0, t, x, \omega) = \frac{u(x_0, t, x, \omega)}{\int_{\mathbb{R}^d} u(x_0, t, x, \omega)\, dx}, \tag{7.3.17}$$

where $u = u(x_0, t, x, \omega)$ is the generalized fundamental solution of (7.3.1);

(c) *For every $\Gamma \in \mathscr{B}(\mathbb{R}^d)$, the function $(t, \omega) \mapsto P^t_{\mathscr{Y}}(x_0, \Gamma, \omega)$ is a $\mathscr{P}(\mathscr{Y})$-measurable, continuous stochastic process.* □

7.3.5. The following theorem can be proved in the same way as Theorem 7.7.

Theorem 7.8 *The conclusions of Theorem* 7.7 *hold under assumptions* (D_∞) *and* (H_F), *this time with u being the fundamental solution of* (7.3.1) *in the sense of Definition* 7.6. □

7.3.6. The functions $\pi = \pi(x_0, t, x, \omega)$ and $u = u(x_0, t, x, \omega)$ will be called the **filtering transition density** and the **unnormalized filtering transition density**, respectively.

The filtering transition density is unique and satisfies the corresponding nonlinear filtering equation:

Theorem 7.9 *Under the assumptions of Theorem* 7.6, *the filtering transition density $\pi = \pi(x_0, t, x, \omega)$ has the following properties for every $x_0 \in \mathbb{R}^d$:*

(a) $\pi \in \mathbb{L}_2([\varepsilon, T] \times \Omega\, \mathscr{P}(\mathscr{Y}); \mathbb{H}^{m-2n-1}) \bigcap \mathbb{L}_2\big(\Omega; \mathbb{C}\big([\varepsilon, T]; \mathbb{H}^{m-2n-2}\big)\big)$;

(b) *It is $\mathscr{P}(\mathscr{Y})$-measurable as a weakly continuous $\mathbb{L}_1$-process;*

(c) *For every $t \in [0, T]$, $\psi \in C_0^\infty(\mathbb{R}^d)$ and ω from the same set of probability one,*

$$
\begin{aligned}
(\pi, \psi)_0(x_0, t) &= \psi(x_0) + \int_{[0,t]} (\pi, \mathcal{L}\psi)_0(x_0, s)\, ds \\
&\quad + \int_{[0,t]} \Big(\big(\pi, \mathcal{M}^l\psi\big)_0(x_0, s) - (\pi, h^l)_0(\pi, \psi)_0(x_0, s)\Big) \\
&\quad \times \Big(d\tilde{w}^l(s) - \big(\pi, h^l\big)_0(x_0, s)\, ds\Big);
\end{aligned} \tag{7.3.18}
$$

(d) *If $\pi^1(x_0, \cdot)$ and $\pi^2(x_0, \cdot)$ belong to*

$$\mathbb{L}_2^\omega([0, T], \mathscr{P}(\mathscr{Y}); \mathbb{L}_1) \cap \mathbb{L}_2^\omega([\varepsilon, T], \mathscr{P}(\mathscr{Y}); \mathbb{H}^1) \cap \mathbb{C}([\varepsilon, T]; \mathscr{P}(\mathscr{Y}); \mathbb{L}_2),$$

$\varepsilon \in (0, T]$, *and satisfy* (7.3.18) *for every $t \in [0, T]$ and $\psi \in \mathbb{C}_0^\infty(\mathbb{R}^d)$, then*

$$\mathbb{E} \int_{[0,T]} \|\pi^1(t) - \pi^2(t)\|_{1,2}\, dt = 0.$$ □

The proof is identical to the proof of Theorem 6.5.

Corollary 7.3 *Under either* (D_∞), (H_F) *or* (A) *and* (D_m) *for all m, there exists a version of π belonging to $\mathbb{C}^{0,\infty}((0, T] \times \mathbb{R}^d)$ for all $(x_0, \omega) \in \mathbb{R}^d \times \Omega$ which is a classical fundamental solution of* (6.2.5).

7.3.7. Here and in the next paragraph we use the notation and assumptions from Sect. 6.4.2. Also, recall that $\mathfrak{n}$ denotes the least integer greater than $\mathrm{d}/2$.

Denote by $v_{[1]}(s,z)$ a continuous version of the r-generalized solution of the backward filtering equation (6.4.1), (6.4.2), and recall that $\mathscr{M}(\mathbb{R}^{\mathrm{d}})$ denotes the space of probability measures on $\mathbb{R}^{\mathrm{d}}$.

Theorem 7.10 *Under either* (D_∞), $(\mathrm{H_F})$ *or* (A), (D_m) *with* $m \geq 2\mathfrak{n}+2$, *there exists a function* $P_{\mathscr{Y}}^{T_1,T_0}(x_0) : \Omega \to \mathscr{M}(\mathbb{R}^{\mathrm{d}})$ *such that*

(a) $P_{\mathscr{Y}}^{T_1,T_0}$ *is a regular conditional distribution of* $X(T_0)$ *relative to* $\mathscr{Y}_{T_1}^{0}$;
(b) $P_{\mathscr{Y}}^{T_1,T_0}(x_0)$ *is absolutely continuous with respect to the Lebesgue measure on* $\mathbb{R}^{\mathrm{d}}$ *and the Radon–Nikodym derivative is defined by the equality*

$$\pi^{T_1,T_0}(x_0,x) = \frac{v_1(T_0,x,Y(t_0))u(x_0,T_0,x)}{\int\limits_{\mathbb{R}^{\mathrm{d}}} v_1(T_0,x,Y(T_0))u(x_0,T_0,x)\,dx}. \tag{7.3.19}$$

□

Remark 7.4 Under conditions (A) and (D_m) equality (7.3.19) is valid in $\mathbb{L}_1$ and u is the generalized solution of (7.3.1), whereas, under (D_∞) and $(\mathrm{H_F})$, equality (7.3.19) holds point-wise in $x \in \mathbb{R}^{\mathrm{d}}$, u is the fundamental solution of Eq. (7.3.1), and v_1 is the classical solution of problem (6.4.1), (6.4.2).

The function $\pi^{T_1,T_0}(x_0,x)$ is called the **interpolation transition density**. Its analytical properties are determined by those of $v_{[1]}$ and u.

7.3.8. Formula

$$\mathbb{E}[f(X(T_1))|\mathscr{Y}_{T_0}^{0}] = \int\limits_{\mathbb{R}^{\mathrm{d}}} v(T_0,x,Y(T_0))\pi(T_0,x)\,dx,$$

established in Sect. 6.4.4, also holds when z_0 is non-random if we use the filtering transition density $\pi(x_0,T_0,x)$ instead of $\pi(T_0,x)$. The result is

Theorem 7.11 *If the process* $Z = Z(t)$ *has transition density* $p(t,z';s,z')$, *then, under either* (D_∞), $(\mathrm{H_F})$ *or* (A), (D_m) *with* $m \geq 2\mathfrak{n}+2$, *there exists a function* $P_{\mathscr{Y}}^{T_1,T_0}(x_0) : \Omega \to \mathscr{M}(\mathbb{R}^{\mathrm{d}})$, *called the* ***extrapolation transition density***, *such that*

(a) $P_{\mathscr{Y}}^{T_1,T_0}$ *is a regular conditional distribution of* $X(T_1)$ *relative to* $\mathscr{Y}_{T_0}^{0}$.

(b) *The measure* $P_{\mathscr{Y}}^{T_1,T_0}$ *is absolutely continuous with respect to the Lebesgue measure on* $\mathbb{R}^d$ *with density*

$$\pi^{T_1,T_0}(x_0,x) = \int_{\mathbb{R}^{d_1}} \int_{\mathbb{R}^d} p\big(T_1,x,y;T_0,x',Y(T_0)\big) \times \pi(x_0,T_0,x')\,dx'dy. \tag{7.3.20}$$

□

Remark 7.5 Formula (7.3.20) should be interpreted in the spirit of Remark 7.4.

□

Chapter 8
Chaos Expansion for Linear Stochastic Evolution Systems

8.1 Introduction

Separation of variables is widely used to study evolution equations. For deterministic equations, there are two variables to separate: time and space; the result is often an orthogonal expansion of the solution in the eigenfunctions of the operator in the equation.

For stochastic equations with deterministic input, the randomness comes only from the Brownian motion driving the equation and becomes a natural variable to separate from the time and space variables. The result is a chaos expansion of the solution, and, similar to the classical Fourier method, leads to new analytical and numerical results about the solution.

Section 8.2 introduces the main definitions and technical tools related to chaos expansion. Section 8.3 illustrates how the chaos method can provide additional information about the solution for some of the equations studied in the previous chapters of the book. Section 8.4 shows how chaos expansion leads to new numerical methods for solving the linear filtering equation. Section 8.5 utilizes the methods of chaos expansion to study the stochastic passive scalar equation. Unlike other equations studied in this book, there are now infinitely many Brownian motions driving the equation, but this technical complication is not an obstacle when it comes to the chaos approach.

8.2 The Propagator

8.2.1. The objective of this section is to consider the solution of a stochastic evolution equation as a random element in the space of square-integrable functionals of the driving Brownian motion, to expand the solution in a suitable Fourier series in this space, and to derive the equations satisfied by the coefficients.

B. L. Rozovsky, S. V. Lototsky, *Stochastic Evolution Systems*, Probability Theory and Stochastic Modelling 89, https://doi.org/10.1007/978-3-319-94893-5_8

Introduce the following objects:

- $W = \big(w^1(t), \ldots, w^r(t)\big)$, an r-dimensional Wiener process on a complete probability space $(\Omega, \mathcal{F}, \mathbb{P})$;
- $(\mathbb{X}, \mathbb{H}, \mathbb{X}')$, a normal triple of Hilbert spaces with canonical bilinear functional $[\cdot, \cdot]$;
- $(\cdot, \cdot)$ and $\|\cdot\|$, the inner product and the norm in $\mathbb{H}$;
- A and B^l, $l = 1, \ldots, \mathrm{d}_1$, linear operators such that A is linear and bounded from $\mathbb{X}$ to $\mathbb{H}$ and each B^l is a linear bounded operator from $\mathbb{X}$ to $\mathbb{H}$;
- $\tilde{\mathscr{F}}^{t_1}_{t_2}$, the sigma-algebra generated by $W(t) - W(s)$, $t_1 \leq s \leq t \leq t_2$.

We will follow the summation convention over repeated indices.

Warning 8.1 *In what follows, both*

$$\int\limits_{[a,b]} \quad \text{and} \quad \int_a^b$$

will be used to denote definite integrals. □

8.2.2. To begin, recall the construction of an orthonormal basis in the Hilbert space $\mathbb{L}_2\big(\Omega, \tilde{\mathscr{F}}^{T_0}_{t^*}, \mathbb{P}\big)$, $t^* \in [T_0, T]$.

Let $\boldsymbol{\alpha}$ be an r-dimensional multi-index, i.e. a collection $\boldsymbol{\alpha} = (\alpha^l_k)_{1\leq l\leq r,\, k\geq 1}$ of non-negative integers such that only finitely many of the α^l_k are different from zero. The set of all such multi-indices will be denoted by $\mathscr{J}$.

For $\boldsymbol{\alpha} \in \mathscr{J}$ define

- $\boldsymbol{\alpha}! := \prod_{k,l} \alpha^l_k!$;
- $|\boldsymbol{\alpha}| := \sum_{l,k} \alpha^k_l$ (length of $\boldsymbol{\alpha}$);
- $d(\boldsymbol{\alpha}) := \max\{k \geq 1 : \alpha^l_k > 0 \text{ for some } 1 \leq l \leq r\}$ (order of $\boldsymbol{\alpha}$).

Every multi-index $\boldsymbol{\alpha}$ with $|\boldsymbol{\alpha}| = k$ can be identified with the set

$$K_{\boldsymbol{\alpha}} = \{(i^{\boldsymbol{\alpha}}_1, q^{\boldsymbol{\alpha}}_1), \ldots, (i^{\boldsymbol{\alpha}}_k, q^{\boldsymbol{\alpha}}_k)\}$$

so that $i^{\boldsymbol{\alpha}}_1 \leq i^{\boldsymbol{\alpha}}_2 \leq \ldots \leq i^{\boldsymbol{\alpha}}_k$ and if $i^{\boldsymbol{\alpha}}_j = i^{\boldsymbol{\alpha}}_{j+1}$, then $q^{\boldsymbol{\alpha}}_j \leq q^{\boldsymbol{\alpha}}_{j+1}$. The first pair $(i^{\boldsymbol{\alpha}}_1, q^{\boldsymbol{\alpha}}_1)$ in $K_{\boldsymbol{\alpha}}$ is the position numbers of the first non-zero element of $\boldsymbol{\alpha}$. The second pair is the same as the first if the first non-zero element of $\boldsymbol{\alpha}$ is greater than one; otherwise, the second pair is the position numbers of the second non-zero element of $\boldsymbol{\alpha}$ and so on. As a result, if $\alpha^q_j > 0$, then exactly α^q_j pairs in $K_{\boldsymbol{\alpha}}$ are (j, q). The set $K_{\boldsymbol{\alpha}}$ will be referred to as **the characteristic set** of the multi-index $\boldsymbol{\alpha}$. For example, if $r = 2$ and

$$\boldsymbol{\alpha} = \begin{pmatrix} 0\,1\,0\,2\,3\,0\,0\,\ldots \\ 1\,2\,0\,0\,0\,1\,0\,\ldots \end{pmatrix},$$

then the non-zero elements are $\alpha_1^2 = \alpha_2^1 = \alpha_1^6 = 1$, $\alpha_2^2 = \alpha_4^1 = 2$, $\alpha_5^1 = 3$, and the characteristic set is

$$K_\alpha = \{(1,2), (2,1), (2,2), (2,2), (4,1), (4,1), (5,1), (5,1), (5,1), (6,2)\}.$$

In future, when there is no danger of confusion, the superscript $\boldsymbol{\alpha}$ in i and q will be omitted. For example, we write (i_j, q_j) used instead of (i_j^α, q_j^α).

For a fixed $t^* \in (T_0, T)$ choose a complete orthonormal system $\{m_k\} = \{m_k(s)\}_{k\geq 1}$ in $\mathbb{L}_2([T_0, t^*])$ and define

$$\xi_{k,l} = \int_{[T_0,t^*]} m_k(s)\, dw^l(s) \tag{8.2.1}$$

so that $\xi_{k,l}$ are independent Gaussian random variables with zero mean and unit variance.

If

$$H(x) := (-1)^n e^{x^2/2} \frac{d^n}{dx^n} e^{-x^2/2} \tag{8.2.2}$$

is the nth Hermite polynomial, then the collection

$$\left\{ \xi_\alpha(W_{T_0,t^*}) := \prod_{k,l} \left(\frac{H_{\alpha_k^l}(\xi_{k,l})}{\sqrt{\alpha_k^l!}} \right), \quad \alpha \in \mathscr{J} \right\} \tag{8.2.3}$$

is an orthonormal system in $\mathbb{L}_2(\Omega, \tilde{\mathscr{F}}_{t^*}^{T_0}, \mathbb{P})$.

8.2.3. A theorem of Cameron and Martin [14] shows that $\{\xi_\alpha(W_{T_0,t^*})\}_{\alpha\in\mathscr{J}}$ is actually a basis in that space.

Theorem 8.1 *If* $\eta \in \mathbb{L}_2(\Omega, \tilde{\mathscr{F}}_{t^*}^{T_0}, \mathbb{P})$, *then*

$$\eta = \sum_{\alpha\in\mathscr{J}} \mathbb{E}[\eta\xi_\alpha(W_{T_0,t^*})]\xi_\alpha(W_{T_0,t^*}) \tag{8.2.4}$$

and

$$\mathbb{E}|\eta|^2 = \sum_{\alpha\in\mathscr{J}} \left|\mathbb{E}\eta\xi_\alpha(W_{T_0,t^*})\right|^2. \tag{8.2.5}$$

This theorem is proved in [14] and [46]. □

8.2.4. Consider the following stochastic evolution equation:

$$u(t) = u_0 + \int_{[T_0,t]} Au(s)\,ds + \int_{[T_0,t]} B^l u(s)\,dw^l(s), \tag{8.2.6}$$

or in the differential form

$$\begin{aligned} du(t) &= Au(t)\,dt + B^l u(t)\,dw^l(t),\ T_0 < t \le T;\\ u(T_0) &= g, \end{aligned} \tag{8.2.7}$$

where $g \in \mathbb{L}_2(\Omega; \mathbb{H})$ is independent of $\tilde{\mathscr{F}}_T^{T_0}$. Assume that Eq. (8.2.6) is coercive. Then there is a unique solution of (8.2.6); this solution is denoted by $u(t; T_0; g)$.

By the Pettis theorem, if $s \in [T_0, t^*]$, $t^* \le T$, and $v \in \mathbb{X}$, then the random variable $\big(u(s; T_0; g), v\big)$ belongs to $\mathbb{L}_2(\Omega, \tilde{\mathscr{F}}_{t^*}^{T_0}, \mathbb{P})$. Therefore, after conditioning on g, it follows from Theorem 8.1 that

$$\mathbb{E}\big(u(s; T_0; g), v\big) = \sum_{\alpha\in\mathscr{J}} \mathbb{E}\big[\big(u(s; T_0; g), v\big)\xi_\alpha(W_{T_0,t^*})\big]\xi_\alpha(W_{T_0,t^*}) \tag{8.2.8}$$

and

$$\mathbb{E}\big|\big(u(s; T_0; g), v\big)\big|^2 = \sum_{\alpha\in\mathscr{J}} \big|\mathbb{E}\big[\big(u(s; T_0; g), v\big)\xi_\alpha(W_{T_0,t^*})\big]\big|^2. \tag{8.2.9}$$

8.2.5. The properties of the solution of (8.2.6) imply that, for every $\boldsymbol{\alpha} \in \mathscr{J}$, the expectation $\mathbb{E}[u(t; T_0; g)\xi_\alpha(W_{T_0,t^*})]$, as a function of t, is a well defined element of $\mathbb{L}_2\big((T_0, t^*); \mathbb{X}\big) \cap \mathbb{C}\big((T_0, t^*); \mathbb{H}\big)$; this element will be denoted by $\dfrac{1}{\sqrt{\alpha!}}\varphi_\alpha(t; T_0; g)$ (the normalizing factor $\dfrac{1}{\sqrt{\alpha!}}$ is introduced for technical reasons). It is shown in the following theorem that the functions $\varphi_\alpha(t; T_0; g)$, $\boldsymbol{\alpha} \in \mathscr{J}$, satisfy a recursive system of deterministic evolution equations.

Theorem 8.2 *Suppose g is independent of $\tilde{\mathscr{F}}_T^{T_0}$ and Eq. (8.2.6) is coercive. If $t^* \in (T_0, T]$ is fixed, then, for every $s \in [T_0, t^*]$, the solution $u(s; T_0; g)$, viewed as an element of $\mathbb{H}$, can be written as*

$$u(s; T_0; g) = \sum_{\alpha\in\mathscr{J}} \frac{1}{\sqrt{\alpha!}}\varphi_\alpha(s; T_0; g)\xi_\alpha(W_{T_0,t^*}), \tag{8.2.10}$$

so that the series converges in $\mathbb{L}_2(\Omega; \mathbb{H})$ and the following Parseval's equality holds:

$$\mathbb{E}\|u(s; T_0; g)\|^2 = \sum_{\alpha\in\mathscr{J}} \frac{1}{\alpha!}\mathbb{E}\|\varphi_\alpha(s; T_0; g)\|^2. \tag{8.2.11}$$

The coefficients of the expansion satisfy the recursive system of deterministic equations

$$\begin{aligned}\frac{\partial \varphi_{\boldsymbol{\alpha}}(s;\,T_0;\,g)}{\partial s} &= A\varphi_{\boldsymbol{\alpha}}(s;\,T_0;\,g) + \sum_{k,l} \alpha_k^l m_k(s) B^l \varphi_{\boldsymbol{\alpha}^-(k,l)}(s;\,T_0;\,g),\\ T_0 < s &\le t^*;\ \varphi_{\boldsymbol{\alpha}}(T_0;\,T_0;\,g) = g1_{\{|\boldsymbol{\alpha}|=0\}},\end{aligned} \tag{8.2.12}$$

where $\alpha = (\alpha_k^l)_{1\le l\le r,\,k\ge 1} \in \mathcal{J}$ *and* $\boldsymbol{\alpha}^-(i,j)$ *stands for the multi-index* $\tilde{\alpha} = (\tilde{\alpha}_k^l)_{1\le l\le r,\,k\ge 1}$ *with*

$$\tilde{\alpha}_k^l = \begin{cases} \alpha_k^l & \text{if } k \ne i \text{ or } l \ne j \text{ or both}\\ \max(0, \alpha_i^j - 1) & \text{if } k = i \text{ and } l = j.\end{cases} \tag{8.2.13}$$

□

Proof Let $\{h_k\}_{k\ge1}$ be an orthonormal basis in $\mathbb{H}$. Then (8.2.9) and the Fubini theorem imply, after conditioning on g,

$$\begin{aligned}\mathbb{E}\|u(s;\,T_0;\,g)\|^2 &= \sum_{\boldsymbol{\alpha}\in\mathcal{J}} \frac{1}{\boldsymbol{\alpha}!} \sum_{k\ge1} |\big(\varphi_{\boldsymbol{\alpha}}(s;\,T_0;\,g), h_k\big)|^2\\ &= \sum_{\boldsymbol{\alpha}\in\mathcal{J}} \frac{1}{\boldsymbol{\alpha}!}\mathbb{E}\|\varphi_{\boldsymbol{\alpha}}(s;\,T_0;\,g)\|^2,\end{aligned}$$

which proves (8.2.11). By linearity, for all $v \in \mathbb{H}$,

$$\sum_{\boldsymbol{\alpha}\in\mathcal{J}} \frac{1}{\sqrt{\boldsymbol{\alpha}!}}\big(\varphi_{\boldsymbol{\alpha}}(s;\,T_0;\,g), v\big) = \Big(\sum_{\boldsymbol{\alpha}\in\mathcal{J}} \frac{1}{\sqrt{\boldsymbol{\alpha}!}}\varphi_{\boldsymbol{\alpha}}(s;\,T_0;\,g), v\Big),$$

and (8.2.10) follows from (8.2.8).

To prove that the coefficients satisfy (8.2.12), define

$$P_t(z) = \exp\Big(\int_{[T_0,t]} \sum_{l=1}^r m_z^l(s)\,dw^l(s) - \frac{1}{2}\int_{[T_0,t]} \sum_{l=1}^r |m_z^l(s)|^2 ds\Big),\ T_0 \le t \le t^*,$$

where $m_z^l = \sum_{k\ge1} m_k(s) z_k^l$ and $\{z_k^l\}$, $l = 1,\ldots,\mathrm{d}_1$, $k = 1,2,\ldots$, is a sequence of real numbers such that $\sum_{k,l} |z_k^l|^2 < \infty$. Also, to simplify the notation, we assume, with no loss of generality, that g is non-random.

It follows from the definition of $P_s(z)$ that

$$dP_s(z) = m_z^l(s) P_s(z)\,dw^l(s),\ T_0 \le s \le t;\ P_{T_0}(z) = 1.$$

Then direct computations show that

$$\xi_{\boldsymbol{\alpha}}(W_{T_0,t^*}) = \frac{1}{\sqrt{\boldsymbol{\alpha}!}}\frac{\partial^{\boldsymbol{\alpha}}}{\partial z^{\boldsymbol{\alpha}}}P_{t^*}(z)\Big|_{z=0},$$

where

$$\frac{\partial^{\boldsymbol{\alpha}}}{\partial z^{\boldsymbol{\alpha}}} = \prod_{k,l}\frac{\partial^{\alpha_k^l}}{(\partial z_k^l)^{\alpha_k^l}},$$

and also, that

$$\mathbb{E}[\eta\xi_{\boldsymbol{\alpha}}(W_{T_0,t^*})] = \frac{\partial^{\boldsymbol{\alpha}}}{\partial z^{\boldsymbol{\alpha}}}\mathbb{E}[\eta P_{t^*}(z)]\Big|_{z=0}$$

for every $\eta \in \mathbb{L}_2(\Omega, \tilde{\mathscr{F}}_{t^*}^{T_0}, \mathbb{P})$. Consequently,

$$\begin{aligned}
\big(\varphi_{\boldsymbol{\alpha}}(s; T_0; g), v\big) &= \frac{\partial^{\alpha}}{\partial z^{\alpha}}\mathbb{E}\big[\big(u(s; T_0; g), v\big)P_{t^*}(z)\big]\Big|_{z=0} \\
&= \frac{\partial^{\boldsymbol{\alpha}}}{\partial z^{\boldsymbol{\alpha}}}\mathbb{E}\big[\big(u(s; T_0; g), v\big)P_s(z)\big]\Big|_{z=0},
\end{aligned}$$

where the second equality follows from the martingale property of the process $s \mapsto P_s(z)$ on $\big(\Omega, \{\tilde{\mathscr{F}}_t^{T_0}\}_{T_0\le t\le t^*}, \mathbb{P}\big)$.

Then (8.2.6) and the Itô formula imply that

$$\begin{aligned}
\big(u(s; T_0; g), v\big)P_s(z) = g + \int\limits_{[T_0,s]} &\Big([Au(\tau; T_0; g), v] \\
&+ \big(B^l u(\tau; T_0; g), v\big)\Big)m_z^l(\tau)P_\tau(z)\,d\tau \\
+ \int\limits_{[T_0,s]} &\Big(\big(B^l u(\tau; T_0; g), v\big) + \big(u(\tau; T_0; g), v\big)m_z^l(s)\Big)P_s(z)\,dw^l(\tau).
\end{aligned}$$

Taking the expectation on both sides of the last equality and setting

$$\varphi(s, z; T_0; g) := \mathbb{E}u(s; T_0; g)P_s(z)$$

results in

$$\begin{aligned}
\big(\varphi(s, z; T_0; g), v\big) = (g, v) + \int\limits_{[T_0,s]} &\Big([A\varphi(\tau, z; T_0; g), v] \\
&+ m_z^l(\tau)\big(B^l\varphi(\tau, z; T_0; g), v\big)\Big)\,d\tau.
\end{aligned}$$

Then (8.2.12) follows after applying the operator $\frac{1}{\sqrt{\alpha!}}\frac{\partial^\alpha}{\partial z^\alpha}$ to both sides of the last equality and setting $z = 0$. □

8.2.6. The proof of Theorem 8.2 shows that system (8.2.12) can be written even when (8.2.6) is not coercive and even when no solution of (8.2.6) is known to exist. The following definition formalizes this observation.

Definition 8.1 The system of Eq. (8.2.12) is called the **propagator** corresponding to (8.2.6). If (8.2.12) happens to have a solution, this solution is called the **chaos solution** of (8.2.6). □

Proposition 8.1 *Equation* (8.2.6) *has a unique chaos solution if, for every* $v_0 \in \mathbb{H}$ *and* $f \in \mathbb{L}_2\big([0,T];\mathbb{X}'\big)$, *the deterministic equation*

$$v(t) = v_0 + \int_{[0,t]} Av(s)\,ds + \int_{[0,t]} f(s)\,ds$$

has a unique solution in the normal triple $(\mathbb{X}, \mathbb{H}, \mathbb{X}')$. □

Proof This is an immediate consequence of the definition and the structure of system (8.2.12). □

8.2.7. We now derive an alternative representation of the chaos solution of (8.2.12). To this end, we assume that the operator A generates a strongly continuous semigroup $\Phi = \Phi_t,\ t \geq 0$, in $\mathbb{H}$ (which is the case under the coercivity assumption) and introduce several additional definitions and constructions:

- $\mathbb{S}^{(k)}$ is the permutation group of the set $\{1, \ldots, k\}$;
- $E_\alpha(s^{(k)}; l^{(k)}) := \sum_{\sigma\in\mathbb{S}^{(k)}} m_{i_1}(s^{\sigma(1)})1_{\{l_{\sigma(1)}=q_1\}}\cdots m_{i_k}(s^{\sigma(k)})1_{\{l_{\sigma(k)}=q_k\}}$, where $\boldsymbol{\alpha} \in \mathcal{J}$ with $|\alpha| = k$ and the characteristic set $\{(i_1, q_1), \ldots, (i_k, q_k)\}$;
- $s^{(k)}$ is the ordered set $(s^1, \ldots, s^k)$; $ds^{(k)} := ds^1 \ldots ds^k$;
- $l^{(k)}$ is the ordered set $(l_1, \ldots, l_k)$;
- $F(t; s^{(k)}; l^{(k)}; g) := \Phi_{t-s^k} B^{l_k} \Phi_{s^k-s^{k-1}} \ldots B^{l_1}\Phi_{s^1-T_0} g,\ k \geq 1,\ g \in \mathbb{H}$;
- $\int_{T_0}^{(k,t)} (\cdots)\,ds^{(k)} := \int_{T_0}^{t}\int_{T_0}^{s^k} \ldots \int_{T_0}^{s^2} (\cdots)\,ds^1 \ldots ds^k$;
- $\sum_{l^{(k)}} := \sum_{l_1,\ldots,l_k=1}^{r}$.

Theorem 8.3 *Assume that* A *generates a strongly continuous semigroup* Φ *in* $\mathbb{H}$ *and the operators* B^l, $l = 1, \ldots, \mathrm{d}_1$, *are linear and bounded from* $\mathbb{X}$ *to* $\mathbb{H}$. *If* $t^* \in (T_0, T]$ *is fixed and* $\boldsymbol{\alpha} \in \mathcal{J}$ *is a multi-index with* $|\boldsymbol{\alpha}| = k$ *and the characteristic set* $\{(i_1, q_1), \ldots, (i_k, q_k)\}$, *then, for* $t \in [T_0, t^*]$, *the corresponding* *solution* $\varphi_\alpha(t; T_0; g)$ *of* (8.2.12) *is given by*

$$\varphi_{\boldsymbol{\alpha}}(t; T_0; g) = \sum_{\sigma \in \mathrm{S}^{(k)}} \sum_{l^{(k)}} \int_{T_0}^{(k,t)} F^k(t; s^{(k)}; l^{(k)}; g) m_{i_{\sigma(k)}}(s^k) 1_{\{l_k = q_{\sigma(k)}\}}$$

$$\cdots m_{i_{\sigma(1)}}(s^1) 1_{\{l_1 = q_{\sigma(1)}\}}\, ds^{(k)},\ k > 1; \tag{8.2.14}$$

$$\varphi_{\boldsymbol{\alpha}}(t; T_0; g) = \int_{T_0}^{t} \Phi_{t-s^1} B^{q_1} \Phi_{s^1 - T_0} g\, m_{i_1}(s^1)\, ds^1,\ k = 1;$$

$$\varphi_{\boldsymbol{\alpha}}(t; T_0; g) = \Phi_{t-T_0} g,\ k = 0,$$

and

$$\sum_{|\boldsymbol{\alpha}|=k} \frac{\|\varphi_\alpha(t; T_0; g)\|^2}{\boldsymbol{\alpha}!} = \sum_{l^{(k)}} \int_{T_0}^{(k,t)} \|F(t; s^{(k)}; l^{(k)}; g)\|^2\, ds^{(k)}. \tag{8.2.15}$$

□

Proof For the sake of simplicity, the arguments T_0 and g will be omitted wherever possible. Representation (8.2.14) automatically holds for $|\boldsymbol{\alpha}| = 0$. Then the general case $|\alpha| \geq 1$ follows by induction using the variation of parameters formula.

To prove (8.2.15), first of all note that

$$\sum_{\sigma \in \mathrm{S}^{(k)}} m_{i_{\sigma(k)}}(s^k) 1_{\{l_k = q_{\sigma(k)}\}} \cdots m_{i_{\sigma(1)}}(s^1) 1_{\{l_1 = q_{\sigma(1)}\}}$$

$$= \sum_{\sigma \in \mathrm{S}^{(k)}} m_{i_k}(s^{\sigma(k)}) 1_{\{l_{\sigma(k)} = q_k\}} \cdots m_{i_1}(s^{\sigma(1)}) 1_{\{l_{\sigma(1)} = q_1\}}.$$

Indeed, every term on the left corresponding to a given $\sigma_0 \in \mathrm{S}^{(k)}$ coincides with the term on the right corresponding to $\sigma_0^{-1} \in \mathrm{S}^{(k)}$.

Then (8.2.14) can be written as

$$\varphi_\alpha(t) = \sum_{l^{(k)}} \int_{T_0}^{(k,t)} F(t; s^{(k)}; l^{(k)}) E_{\boldsymbol{\alpha}}(s^{(k)}; l^{(k)})\, ds^{(k)}.$$

Using the notation

$$G(t; s^{(k)}; l^{(k)}) := \sum_{\sigma \in \mathrm{S}^{(k)}} \Phi_{t - s^{\sigma(k)}} B^{l_{\sigma(k)}} \ldots \Phi_{s^{\sigma(2)} - s^{\sigma(1)}} B^{l_{\sigma(1)}} \Phi_{s^{\sigma(1)} - T_0} g$$

$$1_{s^{\sigma(1)} < \ldots < s^{\sigma(k)} < t},$$

it can be rewritten as

$$\varphi_{\alpha}(t)=\frac{1}{k!}\sum_{l^{(k)}}\int_{[T_0,t^*]^k} G(t;s^{(k)};l^{(k)})E_{\alpha}(s^{(k)};l^{(k)})\,ds^{(k)}. \tag{8.2.16}$$

Since for every $t\in(T_0,t^*)$, $G(t;s^{(k)};l^{(k)})$ is a symmetric function from $\mathbb{L}_2\big(((T_0,t^*)\times\{1,\dots,\mathrm{d}_1\})^k;\mathbb{H}\big)$ with $\mathbb{H}$ separable, and the collection

$$\left\{\frac{E_{\alpha}}{\sqrt{\alpha!k!}},\ |\alpha|=k\right\}$$

is a CONS for the symmetric part of the space, it is possible to write

$$G(t;s^{(k)};l^{(k)})=\sum_{|\beta|=k}\frac{c_{\beta}(t)E_{\beta}(s^{(k)};l^{(k)})}{\sqrt{\beta!k!}}$$

with some $c_{\beta}(t)\in\mathbb{H}$. This and (8.2.16) imply $\|\varphi_{\alpha}(t)\|^2/\alpha!=\|c_{\alpha}\|^2/k!$ and so

$$\begin{aligned}\sum_{|\alpha|=k}\frac{\|\varphi_{\alpha}(t)\|^2}{\alpha!}&=\frac{1}{k!}\sum_{|\alpha|=k}\|c_{\alpha}(t)\|^2=\frac{1}{k!}\int_{[T_0,t^*]^k}\|G(t;s^{(k)};l^{(k)})\|^2\,ds^{(k)}\\&=\frac{1}{k!}\sum_{l^{(k)}}\int_{[T_0,t^*]^k}\Big\|\sum_{\sigma\in\mathcal{S}^{(k)}}\Phi_{t-s^{\sigma(k)}}B^{l_{\sigma(k)}}\dots\Phi_{s^{\sigma(2)}-s^{\sigma(1)}}\\&\qquad B^{l_{\sigma(1)}}\Phi_{s^{\sigma(1)}-T_0}g1_{s^{\sigma(2)}<\dots<s^{\sigma(k)}<t}\Big\|^2ds^{(k)}\\&=\sum_{l^{(k)}}\int_{T_0}^{(k,t)}\|F(t;s^{(k)};l^{(k)})\|^2\,ds^{(k)},\end{aligned}$$

which proves (8.2.15). □

Corollary 8.1 *Assume that* $\|\Phi_t\|_{\mathbb{H}\to\mathbb{H}}\le e^{C_0t}$ *and each* B^l *is bounded from* $\mathbb{H}$ *to* $\mathbb{H}$ *with* $\|B^lv\|^2\le c_l\|v\|^2$. *Define* $C_B=\sum_{l=1}^{r}c_l$. *Then*

$$\sum_{|\alpha|=k}\frac{\|\varphi_{\alpha}(t;T_0;g)\|^2}{\alpha!}\le e^{C_0(t-T_0)}\frac{C_B^k(t-T_0)^k}{k!}\|g\|^2. \tag{8.2.17}$$

□

Proof By assumptions,

$$\|F(t;s^k;l^k;g)\|^2\le e^{C_0(t-T_0)}\Big(\prod_j c_{l_j}\Big)\|g\|^2,$$

and then (8.2.15) implies

$$\sum_{|\alpha|=k} \frac{\|\varphi_\alpha(t;T_0;g)\|^2}{\alpha!} = \sum_{l^k} \int_{T_0}^{(k,t)} \|F(t;s^k;l^k;g)\|^2 ds^k$$

$$e^{C_0(t-T_0)} \left(\sum_{l^k} \left(\prod_j c_{l_j} \right) \right) \frac{(t-T_0)^k}{k!} \|g\|^2,$$

completing the proof. □

Remark 8.1 Under the assumption of the corollary, (8.2.17) holds even with infinitely many Brownian motions as long as $C_B = \sum_l c_l < \infty$. Also note that neither the theorem nor the corollary require the underlying stochastic equation to be coercive or even dissipative. □

8.3 Additional Regularity by Chaos Expansion

8.3.1. Chapter 3 provides general results about well-posedness of linear stochastic evolution systems under coercivity or dissipativity assumptions. When the system input is non-random, chaos expansion can provide additional information about the solution. The objective of this section is to discuss several examples of this type for the equation

$$du = Au\, dt + B^l u\, dw^l,\ 0 < t < T, \tag{8.3.1}$$

in the normal triple $(\mathbb{X}, \mathbb{H}, \mathbb{X}')$. In view of Remark 1.8.1, we do not specify the number of Brownian motions that drive the equation. Recall that $\|\cdot\|$ is the norm in the space $\mathbb{H}$.

8.3.2. As a first application of the chaos expansion to the study of linear stochastic evolution systems, let us look at degenerate parabolic, or dissipative, equations: when $\delta = 0$ in condition (A) from Sect. 3.2.1. Recall that Theorem 3.3 provides a general result, and the result is rather complicated; in fact, too complicated for a quick reproduction. Using the chaos approach, we can establish a much easier version.

Theorem 8.4 *Consider Eq.* (8.3.1) *with non-random* $u(0) = u_0 \in \mathbb{H}$, *and assume that*

$$[Av, v] + \delta_0 \|v\|_{\mathbb{X}}^2 \le C_1 \|v\|^2,\ \delta_0 > 0, \tag{8.3.2}$$

$$2[Av, v] + \sum_{i \ge 1} \|B^i v\|^2 \le C_0 \|v\|^2,\ C_0 \ge 0. \tag{8.3.3}$$

Then (8.3.1) *has a unique chaos solution and*

$$\mathbb{E}\|u(t)\|^2 \leq \|u_0\|^2(1 + C_0te^{C_1t})(1 + C_0te^{C_0t}),\ t \geq 0. \tag{8.3.4}$$

□

Proof We will derive (8.3.4) from equality (8.2.17); note that (8.3.2) ensures the existence of the semigroup Φ_t.

Start by defining, for $n \geq 1$,

$$F_n(t) = \sum_{|\alpha|=n} \frac{\|u_\alpha(t)\|^2}{\alpha!}.$$

By (8.2.15) and using $ds^{(n)} = ds^1 \cdots ds^n$,

$$F_n(t) = \sum_{i_1,\ldots,i_n \geq 1} \int_0^t \int_0^{s^n} \cdots \int_0^{s^2} \|\Phi_{t-s^n} B^{i_n} \cdots \Phi_{s^2-s^1} B^{i_1} \Phi_{s^1} u_0\|^2\, ds^{(n)},$$

which we differentiate with respect to time, using the fundamental theorem of calculus, equality $d\Phi_t/dt = A\Phi_t$, and that Φ_0 is the identity operator:

$$\begin{aligned}\frac{dF_n(t)}{dt} = &\sum_{i_1,\ldots,i_n \geq 1} \int_0^t \int_0^{s^{n-1}} \cdots \int_0^{s^2} \|B^{i_n} \Phi_{t-s^{n-1}} B^{i_{n-1}} \cdots B^{i_1} \Phi_{s^1} u_0\|^2\, ds^{(n-1)}\\ &+ \sum_{i_1,\ldots,i_n \geq 1} \int_0^t \int_0^{s^n} \cdots \int_0^{s^2} 2\Big[A\Phi_{t-s^n} B^{i_n} \cdots B^{i_1} \Phi_{s^1} u_0,\\ &\qquad \Phi_{t,s^n} B^{i_n} \cdots B^{i_1} \Phi_{s^1} u_0\Big]\, ds^{(n)}.\end{aligned}$$

[technically, this is how the formula looks for $n > 1$, and the reader is encouraged to write the corresponding equality for $n = 1$ using (8.2.15)].

Next, re-write (8.3.3) as

$$2[Au, u] \leq -\sum_i \|B^i u\|^2 + C_0\|u\|^2$$

to conclude that

$$\begin{aligned}&\int_0^{s^n} \cdots \int_0^{s^2} 2\Big[A\Phi_{t-s^n} B^{i_n} \cdots B^{i_1} \Phi_{s^1} u_0, \Phi_{t,s^n} B^{i_n} \cdots B^{i_1} \Phi_{s^1} u_0\Big]\, ds^{(n)}\\ &\leq -\sum_{i_{n+1} \geq 1} \int_0^t \int_0^{s^n} \cdots \int_0^{s^2} \|B^{i_{n+1}} \Phi_{t-s^n} B^{i_n} \cdots B^{i_1} \Phi_{s^1} u_0\|^2\, ds^{(n)}\\ &+ C_0 \int_0^t \int_0^{s^n} \cdots \int_0^{s^2} \|\Phi_{t-s^n} B^{i_n} \cdots B^{i_1} \Phi_{s^1} u_0\|^2\, ds^{(n)}.\end{aligned}$$

As a result, for $n \geq 1$,

$$\frac{dF_n(t)}{dt} \leq C_0 F_n(t)$$

$$\sum_{i_1,\dots,i_n \geq 1} \int_0^t \int_0^{s^{n-1}} \cdots \int_0^{s^2} \|B^{i_n} \Phi_{t-s^{n-1}} B^{i_{n-1}} \cdots B^{i_1} \Phi_{s^1} u_0\|^2 \, ds^{(n-1)}$$

$$- \sum_{i_1,\dots,i_{n+1} \geq 1} \int_0^t \int_0^{s^n} \cdots \int_0^{s^2} \|B^{i_{n+1}} \Phi_{t-s^n} B^{i_n} \cdots B^{i_1} \Phi_{s^1} u_0\|^2 \, ds^{(n)}.$$

For $n = 0$, $F_0(t) = \|\Phi_t u_0\|^2$,

$$\frac{dF_0(t)}{dt} = 2[A\Phi_t u_0, \Phi_t u_0] \leq -\sum_{i \geq 1} \|B^i \Phi_t u_0\|^2 + C_0 \|\Phi_t u_0\|^2.$$

Summation over n from 0 to N produces a telescoping sum on the right:

$$\sum_{k=0}^{N} \frac{dF_n(t)}{dt} \leq C_0 \sum_{n=0}^{N} F_n(t)$$

$$- \sum_{i_1,\dots,i_{N+1} \geq 1} \int_0^t \int_0^{s^N} \cdots \int_0^{s^2} \|B^{i_{N+1}} \Phi_{t-s^N} B^{i_N} \cdots B^{i_1} \Phi_{s^1} u_0\|^2 \, ds^{(N)},$$

that is,

$$\frac{d}{dt} \sum_{n=0}^{N} \sum_{|\alpha|=n} \frac{\|u_\alpha(t)\|^2}{\alpha!} \leq C_0 \sum_{n=0}^{N} \sum_{|\alpha|=n} \frac{\|u_\alpha(t)\|^2}{\alpha!},$$

or, with another auxiliary notation

$$F_{0\uparrow N}(t) = \sum_{n=0}^{N} \sum_{|\alpha|=n} \frac{\|u_\alpha(t)\|^2}{\alpha!},$$

$$F_{0\uparrow N}(t) \leq \|u_0\|^2 + C_0 \int_0^t F_{0\uparrow N}(s)\, ds + C_0 \int_0^t \|\Phi_s u_0\|^2 \, ds.$$

By Gronwall's inequality, keeping in mind that, by (8.3.2),

$$\|\Phi_s u_0\|^2 \leq e^{C_1 t} \|u_0\|^2, \ s \leq t,$$

we get

$$F_{0\uparrow N}(t) = \sum_{n=0}^{N} \sum_{|\alpha|=n} \frac{\|u_\alpha(t)\|^2}{\alpha!} \leq \|u_0\|^2(1 + C_0 te^{C_1 t})(1 + C_0 te^{C_0 t}).$$

Then (8.3.4) follows after passing to the limit $N \to \infty$.

This completes the proof of Theorem 8.4. □

8.3.3. If the equation is **fully degenerate**:

$$2[Au, u] + \sum_{i\geq 1} \|B^i u\|^2 = 0,$$

then the natural question is conservation of energy, that is, whether the equality

$$\mathbb{E}\|u(t)\|^2 = \|u_0\|^2$$

holds for all $t > 0$. For example, consider the stochastic transport equation in the Stratonovich form

$$u(t, x) + \sum_{i,k} \int_{[0,t]} u_k(s, x)\sigma^{ik}(x) \circ dw^i(s) = \varphi(x),$$

which becomes a fully degenerate parabolic equation when written in the Itô form. For this equation conservation of energy might not hold if the functions σ^{ik} are not Lipschitz continuous in x, which is related to possible non-uniqueness of the solution of the corresponding flow equation

$$dX^k(t) = \sum_i \sigma^{i,k}\big(X(t)\big)\, dw^i(t).$$

The proof of Theorem 8.4 suggests the following chaos expansion-based criterion for conservation of energy:

$$\lim_{N\to\infty} \sum_{i_1,\ldots,i_{N+1}\geq 1} \int_0^t \int_0^{s^N} \cdots \int_0^{s^2} \|B^{i_{N+1}} \Phi_{t-s^N} B^{i_N} \cdots B^{i_1} \Phi_{s^1} u_0\|^2\, ds^{(N)} = 0.$$

8.3.4. If Eq. (8.3.1) is coercive, then chaos expansion makes it possible to establish better regularity of the solution than what is guaranteed by the basic existence theorem.

Theorem 8.5 *Consider Eq.* (8.3.1) *with non-random initial condition* $u_0 \in \mathbb{H}$. *Assume there exist* $\delta^* > 0$ *and* $C^* \geq 0$ *such that, for all* $v \in \mathbb{X}$,

$$2[Av, v] + \sum_{i\geq 1} \|B^i v\|^2 + \delta^* \|v\|_{\mathbb{X}}^2 \leq C^* \|v\|^2. \tag{8.3.5}$$

Then there exists a number $\mathrm{q} > 1$ *such that, for every* $t > 0$, *the functions* φ_α *from* (8.2.12) *satisfy*

$$\sum_{\alpha \in \mathcal{J}} \frac{\mathrm{q}^{|\alpha|}}{\alpha!} \|\varphi_\alpha(t; 0; u_0)\|^2 < \infty. \tag{8.3.6}$$

□

Proof Coercivity condition (8.3.5) implies that (8.3.2) holds (with $\delta_0 \geq \delta^*$) and

$$\sum_{i \geq 1} \|B^i v\|^2 \leq C_B \|v\|^2_{\mathbb{X}}$$

for some positive number C_B.

We will now use Theorem 8.4 to establish (8.3.6) with

$$\mathrm{q} = 1 + \frac{\delta^*}{C_B}. \tag{8.3.7}$$

Indeed, (8.3.7) and (8.3.5) imply

$$2[Av, v] + \mathrm{q} \sum_{l \geq 1} \|B^l v\|^2 \leq C^* \|v\|^2. \tag{8.3.8}$$

Consider the equation

$$dU = AU\, dt + \sqrt{\mathrm{q}}\, B^l U\, dw^l. \tag{8.3.9}$$

By direct computation, $\mathrm{q}^{|\alpha|/2} \mathbb{E}\Big(u \xi_\alpha(W_{0,T})\Big) = \mathbb{E}\Big(U \xi_\alpha(W_{0,T})\Big)$. Then (8.3.6) follows from Theorem 8.4 applied to Eq. (8.3.9). □

8.3.5. With minor changes, the conclusions of Theorems 8.4 and 8.5 continue to hold for a more general equation

$$du = (Au + f)\, dt + (B^l u + g^l)\, dw^l,\ u|_{t=0} = u_0,$$

as long as the functions $f \in \mathbb{L}_2\Big(\Omega \times (0,T); \mathbb{X}'\Big)$, $g^l \in \mathbb{L}_2\Big(\Omega \times (0,T); \mathbb{H}\Big)$, $u_0 \in \mathbb{L}_2(\Omega; \mathbb{H})$ are independent of W and, in the case of infinite-dimensional W,

$$\sum_{l \geq 1} \int_{[0,T]} \mathbb{E}\|g^l(t)\|^2\, dt < \infty.$$

Still, unlike Theorems 3.1, chaos expansion cannot easily handle $\tilde{\mathcal{F}}^0_t$-adapted functions f, g and operators A, B^l.

8.4 Chaos Expansion and Filtering of Diffusion Processes

8.4.1. In this section we apply chaos expansion to the linear filtering equation, corresponding to the following processes X and Y:

$$\begin{aligned} X(t) &= x_0 + \int_{[0,t]} b\big(X(s)\big)\,ds + \int_{[0,t]} {}_1\sigma\big(X(s)\big)\,d\overline{W}(s) \\ &\quad + \int_{[0,t]} \sigma\big(X(s)\big)\,dW(s) \\ Y(t) &= \int_{[0,t]} h\big(X(s)\big)\,ds + W(t). \end{aligned} \tag{8.4.1}$$

The two main differences from the general filtering model considered in Chap. 6 are

- The coefficients do not depend on time and are non-random; in particular, there is no dependence on the observation process Y;
- ${}_0\sigma \equiv \mathbb{I}$.

  Then the linear filtering Eq. (6.3.1′) becomes

$$\begin{aligned} du(t,x) &= \mathcal{L}^*u(t,x)\,dt + \mathcal{M}^{l,*}u(t,x)\,dY^l(t), \quad 0 < t \le T, \\ u(0,x) &= \pi_0(x). \end{aligned} \tag{8.4.2}$$

Warning 8.2 *Given the nature of the problems discussed in this section, the explicit form of the partial differential operators $\mathcal{L}$ and $\mathcal{M}$ is not important. As a result, lower subscripts will no longer be used to denote partial derivatives; instead, both upper and lower subscripts will be used to index various arrays.*

□

8.4.2. Here is a summary of the main notation, in addition to that introduced in Chap. 6:

- $0 = t_0 < t_1 < \ldots < t_n = T$, a uniform partition of $[0, T]$ with step $\Delta = t_1 - t_0$;
- $\mathfrak{h} = \{h_k,\ k \ge 1\}$, an orthonormal basis in $\mathbb{L}_2(\mathbb{R}^{\mathrm{d}})$;
- $\mathfrak{m} = \{m_k,\ k \ge 1\}$, an orthonormal basis in $\mathbb{L}_2((0, \Delta))$;
- $(\cdot,\cdot)_0$, the inner product in $\mathbb{L}_2(\mathbb{R}^{\mathrm{d}})$;
- $\mathbb{H}^s$, $s \in \mathbb{R}$, the scale of Sobolev spaces on $\mathbb{R}$, cf. Sect. 3.5.3.

8.4.3. Consider the following system of equations:

$$\begin{aligned}\frac{\partial\varphi_{\boldsymbol{\alpha}}(s,x;g)}{\partial s} &= \mathcal{L}^*\varphi_\alpha(s,x;g)\\ &+\sum_{k,\ell}\sqrt{\alpha_\ell^k}\, m_\ell(s)\mathcal{M}^{k,*}\varphi_{\boldsymbol{\alpha}^-(k,\ell)}(s,x;g),\ \ 0<s\le\Delta,\end{aligned} \tag{8.4.3}$$

with initial condition

$$\varphi_{\boldsymbol{\alpha}}(0,x;g)=\begin{cases} g(x), & \text{if } |\boldsymbol{\alpha}|=0,\\ 0, & \text{otherwise.}\end{cases}$$

Define the numbers

$$q^\ell_{\boldsymbol{\alpha},k}=\big(\varphi_{\boldsymbol{\alpha}}(\Delta,\cdot;h_k),h_\ell\big)_0, \tag{8.4.4}$$

the random variables

$$\xi^i_{k,\ell}=\int\limits_{[t_{i-1},t_i]} m_k(s-t_{i-1})\,dY^\ell(s),\ \ \xi^i_{\boldsymbol{\alpha}}=\prod_{k,\ell}\left(\frac{H_{\alpha_k^\ell}(\xi^i_{k,\ell})}{\sqrt{\alpha_k^\ell!}}\right),\quad \boldsymbol{\alpha}\in\mathcal{J}, \tag{8.4.5}$$

and then, by induction, the random variables

$$\psi_j(0)=(\pi_0,h_j)_0;\ \ \psi_j(i)=\sum_{\boldsymbol{\alpha}\in\mathcal{J}}\sum_{k=1}^{\infty}\psi_k(i-1)q^j_{\boldsymbol{\alpha},k}\xi^i_{\boldsymbol{\alpha}}. \tag{8.4.6}$$

Theorem 8.6 *The unnormalized filtering density has representation*

$$u(t_i,x)=\sum_{k=1}^{\infty}\psi_k(i)h_k(x),\ \ 0\le i\le M, \tag{8.4.7}$$

and, for $f\in\mathbb{L}_2(\mathbb{R}^{\mathrm{d}})$, *the unnormalized optimal filter*

$$u_t[f]:=\int\limits_{\mathbb{R}^{\mathrm{d}}} f(x)u(t,x)\,dx$$

has representation

$$u_{t_i}[f]=\sum_{k=1}^{\infty}\psi_k(i)f_k,\ \text{ where } f_k=\int\limits_{\mathbb{R}^{\mathrm{d}}} f(x)h_k(x)\,dx. \tag{8.4.8}$$

□

Proof Recall from Chap. 6 that u is a square integrable $\widetilde{\mathscr{Y}}_t^0$-adapted solution of (8.4.2), and Y is a d_1-dimensional Wiener process under measure $\widetilde{\mathbb{P}}$. By Theorem 3.1 we then conclude that $u(t,x) = \varphi_\alpha\big(t,x;u(t_{i-1},x)\big)\xi^i_\alpha$, $t \in [t_{i-1}, t_i]$. To establish (8.4.7), it remains to write

$$u(t_{i-1},x) = \sum_{k\geq 1}\big(u(t_{i-1},\cdot),h_k\big)_0 h_k(x)$$

and to use linearity of equations (8.4.2) and (8.4.3). Equality (8.4.8) follows from (8.4.7) after integration over $\mathbb{R}^d$. □

8.4.4. The integrals $\int_{\mathbb{R}^d} f(x)h_k(x)\,dx$ may be defined for all $k \geq 1$ even when $f \notin \mathbb{L}_2(\mathbb{R}^d)$. In that case, representation (8.4.8) of the unnormalized optimal filter can still hold; see Theorems 8.8, 8.9, and 8.10 below.

8.4.5. Next, we use Theorem 8.6 to construct recursive approximations of $u(t_i, x)$ and $u_{t_i}[f]$ for all $j \geq 1$. Let K, n, N be positive integers, and denote by $\mathscr{J}^n_N$ the collection of those multi-indices in $\mathscr{J}$ for which $|\boldsymbol{\alpha}| \leq N$ and $\alpha^k_\ell = 0$ if $\ell > n$. Note that $\mathscr{J}^n_N$ is a finite set. With the numbers $q^\ell_{\alpha,k}$ from (8.4.4), we define $\psi^K_\ell(i,N,n)$ by truncating the sums in (8.4.6):

$$\begin{aligned}
&\psi^K_\ell(0,N,n) = (\pi_0, h_\ell)_0,\ \ell = 1,\ldots,K;\\
&\psi^K_\ell(i,N,n) = \sum_{\alpha\in\mathscr{J}^n_N}\sum_{k=1}^{K}\psi^K_k(i-1,N,n)q^\ell_{\alpha,k}\xi^i_\alpha,
\end{aligned} \tag{8.4.9}$$

and then define the approximations of $u(t_i, x)$ and $u_{t_i}[f]$ by

$$\begin{aligned}
&u^n_{N,K}(t_i,x) = \sum_{j=1}^{K}\psi^K_j(i,N,n)h_j(x),\\
&u^n_{N,K;i}[f] = \sum_{j=1}^{K}\psi^K_j(i,N,n)f_j,\quad 0\leq i\leq M,
\end{aligned} \tag{8.4.10}$$

where we assume that the functions f and h_j make it possible to define the numbers $f_j = \int_{\mathbb{R}^d} f(x)h_j(x)\,dx$ for all $j \geq 1$.

8.4.6. The following is an algorithm for computing the approximations of the unnormalized filtering density and filter using (8.4.10).

1. Preliminary computations (before the observations are available)

1. *Choose suitable basis functions* $\{h_k, k = 1,\ldots,K\}$ *in* $\mathbb{L}_2(\mathbb{R}^d)$ *and* $\{m_i, i = 1,\ldots,n\}$ in $L_2((0,\Delta))$.

2. *For* $\boldsymbol{\alpha} \in \mathscr{J}_N^n$ and $k, \ell = 1, \ldots, K$, *use* (8.4.3) *to compute*

$$q_{\alpha,k}^{\ell} = \big(\varphi_\alpha(\Delta, \cdot, h_k), h_\ell\big)_0, \quad f_k = \int_{\mathbb{R}^d} f(x)h_k(x)\,dx,$$

$$\psi_k^K(0, N, n) = \int_{\mathbb{R}^d} \pi_0(x)h_k(x)\,dx.$$

2. *Real-time computations, ith step (as the observations become available): compute* ξ_α^i *according to (8.4.5) and update the coefficients* ψ:

$$\psi_\ell^K(i, N, n) = \sum_{\alpha\in\mathscr{J}_N^n}\sum_{k=1}^{K} \psi_k^K(i-1, N, n) q_{\alpha,k}^{\ell}\xi_\alpha^i \quad \ell = 1, \ldots, K;$$

then, if necessary, compute

$$u_N^{n,K}(t_i, x) = \sum_{\ell=1}^{K} \psi_\ell^K(i, N, n)h_\ell(x) \tag{8.4.11}$$

and/or

$$u_{N,K;i}^n[f] = \sum_{j=1}^{K} \psi_j^K(i, N, n) f_j, \quad \widehat{f}_{N,K;i}^n = \frac{\phi_{N,K;i}^n[f]}{\phi_{N,K;i}^n[1]}. \tag{8.4.12}$$

We call this algorithm the **Spectral Separating Scheme of the First Kind**.

8.4.7. We now present an alternative algorithm for solving the linear filtering Eq. (8.4.2). In the spectral separating scheme of the first kind, the truncation of the expansion in $\mathbb{L}_2(\mathbb{R}^d)$ is done after the truncation of the Cameron–Martin expansion. Now, we will do the truncation in $\mathbb{L}_2(\mathbb{R}^d)$ first.

Let $\mathfrak{h}$ be an orthonormal basis in $\mathbb{L}_2(\mathbb{R}^d)$ such that every function $h_k = h_k(x)$ belongs to the Sobolev space $\mathbb{H}^1(\mathbb{R}^d)$.

Fix a positive integer K. Define the matrices $A^K = (A_{ij}^K,\ i, j = 1, \ldots, K)$ and $B_\ell^K = (B_{\ell,ij}^K,\ i, j = 1, \ldots, K;\ \ell = 1, \ldots, \mathrm{d}_1)$, by

$$A_{ij}^K = (\mathcal{L}^* h_j, h_i)_0, \quad B_{\ell,ij}^K = (\mathcal{M}^{\ell,*} h_j, h_i)_0,$$

and consider the Galerkin approximation $u^K(t, x)$ of $u(t, x)$:

$$u^K(t, x) = \sum_{i=1}^{K} u_i^K(t)h_i(x), \tag{8.4.13}$$

where the vector $u^K(t) = \{u_i^K(t),\ i = 1, \dots, K\}$ is the solution of the system of stochastic ordinary differential equations

$$du^K(t) = A^K u^K(t)\,dt + \sum_{\ell=1}^{\mathrm{d}_1} B_\ell^K u^K(t)\,dY^\ell(t) \tag{8.4.14}$$

with the initial condition $u_i^K(0) = (\pi_0, h_i)_0$. Because the matrices B_ℓ^K, $\ell = 1, \dots, \mathrm{d}_1$, do not, in general, commute with each other, system (8.4.14) has no closed-form solution and must be solved numerically.

Define random variables ξ_α^i according to (8.4.5). Theorem 3.1 implies the following result.

Theorem 8.7 *For every $i = 1, \dots, M$, the solution of* (8.4.14) *can be written in* $\mathbb{L}_2(\Omega, \widetilde{\mathbb{P}}; \mathbb{R}^K)$ *as*

$$u^K(t_i) = \sum_{\alpha \in \mathcal{J}} \varphi_\alpha^K(\Delta; u^K(t_{i-1}))\xi_\alpha^i, \quad i = 1, \dots, M, \tag{8.4.15}$$

where, for $s \in (0, \Delta]$ and $\zeta \in \mathbb{R}^K$, the functions $\varphi_\alpha^K(s; \zeta)$ are the solutions of

$$\frac{\partial \varphi_{\boldsymbol{\alpha}}^K(s;\zeta)}{\partial s} = A^K \varphi_{\boldsymbol{\alpha}}^K(s;\zeta) + \sum_{k,l} \sqrt{\alpha_\ell^k}\, m_\ell(s) B_k^K \varphi_{\boldsymbol{\alpha}^-(k,\ell)}^K(s;\zeta),\ 0 < s \le \Delta, \tag{8.4.16}$$

with initial conditions

$$\varphi_\alpha^K(0;\zeta) = \begin{cases} \zeta, & \text{if } |\boldsymbol{\alpha}| = 0, \\ 0, & \text{if } |\boldsymbol{\alpha}| > 0, \end{cases}$$

and $\boldsymbol{\alpha}^-(k, \ell)$ is defined in (8.2.13). □

8.4.8. To construct a recursive approximation of u^K, fix positive integers N and n and define the set $\mathcal{J}_N^n$ as the collection of multi-indices $\boldsymbol{\alpha} \in \mathcal{J}$ such that $|\boldsymbol{\alpha}| \le N$ and $\alpha_k^\ell = 0$ if $k > n$. The approximation $u_N^{K,n}(t_i)$ of $u^K(t_i)$ is defined by

$$\begin{aligned} u_N^{K,n}(t_0) &= u^K(0), \\ u_N^{K,n}(t_i) &= \sum_{\alpha \in \mathcal{J}_N^n} \varphi_\alpha^K(\Delta; u_N^{K,n}(t_{i-1}))\xi_\alpha^i, \quad i = 1, \dots, M. \end{aligned} \tag{8.4.17}$$

To establish a representation of $u_N^{K,n}(t_i)$ similar to (8.4.11), note that $u_N^{K,n}(t_i)$ is a vector in $\mathbb{R}^K$. Let $\boldsymbol{U} = \{\boldsymbol{u}^j,\ j = 1, \dots, K\}$ be a basis in $\mathbb{R}^K$. The vector $u_N^{K,n}(t_i)$

can then be written as

$$u_N^{K,n}(t_i) = \sum_{j=1}^{K} u_{N,j}^{K,n}(t_i; \boldsymbol{U})\boldsymbol{u}^j,$$

and, by the recursive definition of $u_N^{K,n}(t_i)$,

$$\begin{aligned} u_N^{K,n}(t_i) &= \sum_{\alpha \in \mathscr{J}_N^n} \varphi_{\alpha}^K(\Delta; u_N^{K,n}(t_{i-1}))\xi_{\alpha}^i \\ &= \sum_{\alpha \in \mathscr{J}_N^n} \sum_{j=1}^{K} \varphi_{\alpha}^K(\Delta; \boldsymbol{u}^j) u_{N,j}^{K,n}(t_{i-1}; \boldsymbol{U})\xi_{\alpha}^i. \end{aligned}$$

Because $\varphi_{\alpha}^K(\Delta, \boldsymbol{u}^i)$ is a vector in $\mathbb{R}^K$, we write

$$\varphi_{\alpha}^K(\Delta, \boldsymbol{u}^j) = \sum_{k=1}^{K} q_{jk}^{K,\alpha}(\boldsymbol{U})\boldsymbol{u}^k, \tag{8.4.18}$$

and conclude that

$$u_{N,j}^{K,n}(t_i; \boldsymbol{U}) = \sum_{\alpha \in \mathscr{J}_N^n} \sum_{k=1}^{K} q_{jk}^{K,\alpha}(\boldsymbol{U}) u_{N,k}^{K,n}(t_{i-1}; \boldsymbol{U})\xi_{\alpha}^i. \tag{8.4.19}$$

As a result, if $f_k = \int_{\mathbb{R}^d} f(x)h_k(x)\,dx$ is defined for all $k \geq 1$, then

$$\begin{aligned} u_N^{K,n}(t_i, x) &= \sum_{j,k=1}^{K} u_{N,j}^{K,n}(t_i; \boldsymbol{U})\boldsymbol{u}_k^j h_k(x), \\ u_{N;i}^{K,n}[f] &= \sum_{j,k=1}^{K} u_{N,j}^{K,n}(t_i; \boldsymbol{U})\boldsymbol{u}_k^j f_k \end{aligned} \tag{8.4.20}$$

are the approximations of the unnormalized filtering density and filter.

8.4.9. The following is an algorithm for computing the approximations of the unnormalized filtering density and filter using (8.4.20).

1. Preliminary computations (before the observations are available):

1. *Choose suitable basis functions* $\{h_k,\ k = 1, \ldots, K\}$ *in* $\mathbb{L}_2(\mathbb{R}^d)$, $\{m_k,\ k = 1, \ldots, n\}$ in $\mathbb{L}_2((0, \Delta))$, *and a* ***standard unit basis*** $\{\boldsymbol{u}^j,\ j = 1, \ldots, K\}$ in $\mathbb{R}^K$, *that is,* $\boldsymbol{u}_j^j = 1$, $\boldsymbol{u}_\ell^j = 0$ *otherwise.*

2. *For* $\boldsymbol{\alpha} \in \mathcal{J}_N^n$ and $j,k = 1,\ldots,K$, *use* (8.4.16) *to compute*

$$q_{jk}^{K,\alpha} = \varphi_{\alpha,j}^{K}(\Delta; \boldsymbol{u}^k) f_k = \int_{\mathbb{R}^d} f(x)h_k(x)\,dx u_{N,k}^{K,n}(t_0) \quad = \int_{\mathbb{R}^d} \pi_0(x)h_k(x)\,dx.$$

2. *Real-time computations, ith step (as the observations become available): compute* ξ_{α}^{i}, $\boldsymbol{\alpha} \in \mathcal{J}_N^n$, *according to* (8.4.5) *and update the coefficients* $u_{N,k}^{K,n}$ *as follows:*

$$Q_{jk}^{K}(\xi^i) = \sum_{\alpha \in \mathcal{J}_N^n} q_{jk}^{K,\alpha}\xi_{\alpha}^{i},$$
$$u_{N,j}^{K,n}(t_i) = \sum_{k=1}^{K} Q_{jk}^{K}(\xi^i)u_{N,k}^{K,n}(t_{i-1}), \quad j = 1,\ldots,K; \tag{8.4.21}$$

then, if necessary, compute

$$u_N^{K,n}(t_i,x) = \sum_{j=1}^{K} u_{N,j}^{K,n}(t_i)h_j(x) \tag{8.4.22}$$

and/or

$$u_{N;i}^{K,n}[f] = \sum_{j=1}^{K} f_j u_{N,j}^{K,n}(t_i), \quad \widehat{f}_{N;i}^{K,n} = \frac{u_{N;i}^{K,n}[f]}{u_{N;i}^{K,n}[1]}. \tag{8.4.23}$$

We call this algorithm the **Spectral Separating Scheme of the Second Kind.** *The difference in the notations for the approximations of* $u(t,x)$, $u_t[f]$, *and* $\widehat{f}$ *corresponding to the two schemes is the location of the index* K: *compare* (8.4.20) *with* (8.4.10).

8.4.10. The main advantage of the spectral separating schemes, as compared to most other non-linear filtering algorithms, is that the time consuming computations, including solving partial differential equations and evaluation of integrals, are performed in advance, while the real-time part is relatively simple even when the dimension d of the state process is large. Here are some other features of the spectral separating schemes:

1. If the coefficients do not depend on time, then the amount of preliminary computations does not depend on the number of on-line time steps;
2. Formulas (8.4.12) and (8.4.23) can be used to compute an approximation to $\widehat{f}_{t_i}$, for example, conditional moments, without the time consuming computations of the unnormalized filtering density and the related integrals;

3. Only the coefficients $\psi_j^K(i,n,N)$ or $u_{N,j}^{K,n}(t_i)$ must be updated at every time step; the filtering density and/or filter can be computed independently of each other as needed, for example, at the final time moment.
4. The real-time part of the algorithms can be easily parallelized.
5. If $n = 1$, then each ξ_α^i depends only on the increments $Y_\ell(t_i) - Y_\ell(t_{i-1})$ of the observation process, and the corresponding algorithms can be used for filtering with discrete time observations [97]. For $n > 1$ and $k > 1$, the integral $\int_{t_{i-1}}^{t_i} m_k(s - t_{i-1})\, dY_\ell(s)$ can be reduced to a usual Riemann integral and then approximated by the trapezoidal rule.
6. The implementation of both algorithms does not depend on whether the model is noise-correlated ($\sigma \not\equiv 0$) or not.

Successful implementation of the algorithms requires effective numerical methods for solving deterministic parabolic equations and evaluating integrals, but no special tools from numerical stochastics. On the other hand, successful testing and tuning of the algorithms will require effective numerical methods for stochastic ODEs to simulate the processes X, Y.

Theoretical analysis of the algorithms is possible with little or no change if the model is not time homogeneous, that is, the functions $b, {}_1\sigma, \sigma, h$ depend on time. This time dependence certainly decreases the computational advantages, as the number of preliminary computations will grow substantially and will depend on the number of on-line time steps.

The Wiener chaos approach is far less effective if the coefficients in (8.4.1) depend on the observation process Y, because in that case the corresponding systems (8.4.3) and (8.4.16) have a much more complicated structure and are no longer solvable by induction. The corresponding analysis is still an open problem.

8.4.11. The quality of the approximation for the spectral separating schemes is controlled by four numbers: K, n, N, and Δ. The amount of the preliminary computations and the storage space are controlled by the size of the array q; the size of this array is $K^2|\mathscr{J}_N^n|$, where K is the number of basis functions in $\mathbb{L}_2(\mathbb{R}^{\mathrm{d}})$, and $|\mathscr{J}_N^n|$, the size of the set $\mathscr{J}_N^n$, is the number of Cameron–Martin basis functions. By construction, it is impossible to improve the quality of approximation without increasing K. While increasing n and N should also lead to better approximation, it is essentially impossible to use large values of n and N because of the prohibitively large size of the set $\mathscr{J}_N^n$. For example, if $\mathrm{d}_1 = 1$, the number of elements in the set $\mathscr{J}_5^{10}$ is 740, and this number *more than doubles* for $\mathrm{d}_1 = 2$. A rough asymptotic of $|\mathscr{J}_N^n|$ is $(nN)^{\mathrm{d}_1}$. Accordingly, the convergence of the approximations must be studied with fixed values of n and N: to improve the quality of approximation, we should decrease the time step Δ and increase the number K of spatial basis functions.

8.4.12. The study of convergence of the spectral separating schemes requires a special choice of the bases $\mathfrak{h}$ and $\mathfrak{m}$, as well as extra regularity of the functions $b, {}_1\sigma, \sigma, h$, and π_0.

We begin by specifying the basis $\mathfrak{h}$ in $\mathbb{L}_2(\mathbb{R}^{\mathrm{d}})$. Denote by Γ the set of ordered d-tuples $\gamma = (\gamma_1, \ldots, \gamma_{\mathrm{d}})$ with $\gamma_j = 0, 1, 2, \ldots$. For $\gamma \in \Gamma$ define

$$\overline{H}_\gamma(x) = \prod_{j=1}^{\mathrm{d}} \overline{H}_{\gamma_j}(x_j),$$

where

$$\overline{H}_n(t) = \frac{(-1)^n}{\sqrt{2^n \pi^{1/2} n!}} e^{t^2/2} \frac{d^n}{dt^n} e^{-t^2}, \ n = 0, 1, 2, \ldots.$$

If Λ is the operator

$$\Lambda = -\boldsymbol{\Delta} + (1 + |x|^2), \tag{8.4.24}$$

where $\boldsymbol{\Delta}$ is the Laplace operator, then, by direct computation,

$$\Lambda \overline{H}_\gamma = \lambda_\gamma h_\gamma, \tag{8.4.25}$$

with $\lambda_\gamma = 2\sum_{j=1}^{\mathrm{d}} \gamma_j + \mathrm{d} + 1$.

Next, we introduce an ordering of the set Γ as follows: define $|\gamma| = \sum_{j=1}^{d} \gamma_j$ and then say that $\gamma < \tau$ if $|\gamma| < |\tau|$ or if $|\gamma| = |\tau|$ and $\gamma < \tau$ under the lexicographic ordering, that is, $\gamma_{i_0} < \tau_{i_0}$, where i_0 is the first index for which $\gamma_i \neq \tau_i$. Finally, we define the basis $\mathfrak{h}$, known as the *Hermite basis*, as the collection $\{\overline{H}_\gamma(x),\ \gamma \in \Gamma\}$ together with the above ordering of Γ. By construction, the elements h_k of $\mathfrak{h}$ satisfy

$$\Lambda h_k = \lambda_k h_k, \tag{8.4.26}$$

where $c_1 k^{1/\mathrm{d}} \leq \lambda_k \leq c_2 k^{1/\mathrm{d}}$ and $0 < c_1 < c_2$ do not depend on k. The construction of the Hermite basis implies that each h_k decays at infinity faster than every power of $|x|$, and therefore the number $f_k = \int_{\mathbb{R}^{\mathrm{d}}} f(x) h_k(x)\, dx$ is defined for every $k \geq 1$ and every measurable function f of polynomial growth.

As far as the basis $\mathfrak{m}$ in $\mathbb{L}_2((0, \Delta))$, we use the Fourier cosine basis

$$m_1(s) = \frac{1}{\sqrt{\Delta}}; \quad m_k(s) = \sqrt{\frac{2}{\Delta}} \cos\left(\frac{\pi(k-1)s}{\Delta}\right), \ k > 1; \ 0 \leq s \leq \Delta. \tag{8.4.27}$$

8.4.13.

Definition 8.2 The filtering model (8.4.1) is called ν-regular for some positive integer ν if the functions ${}_1\sigma$ and σ belong to $\mathbb{C}_b^{2\nu+3}$, the functions b and h belong to $\mathbb{C}_b^{2\nu+2}$, and $\Lambda^\nu \pi_0$ belongs to $\mathbb{H}^1$, with Λ as in (8.4.24). □

8.4.14. We are now ready to study the convergence of the spectral separating schemes. Recall that the Spectral Separating Scheme of the First Kind defines the approximations $u^n_{N,K}(t_i,x)$, $u_{t_i}[f]$ of the unnormalized filtering density and filter according to (8.4.10). The following theorem presents the quality of these approximations and establishes the convergence in the limit $\lim_{\Delta\to 0}\lim_{K\to\infty}$ for the noise uncorrelated model.

Theorem 8.8 *Assume that $N \geq 2$, $\sigma \equiv 0$, and the matrix ${}_1\sigma\,{}_1\sigma^*$ is uniformly positive definite. If the filtering model* (8.4.1) *is ν-regular for some $\nu > \mathrm{d}+1$, then*

$$\max_{0\leq i\leq M} \mathbb{E}\|u(t_i,\cdot)-u^n_{N,K}(t_i,\cdot)\|_0 \leq C_0\left(\frac{(C_{11}\Delta)^{N/2}}{\sqrt{(N+1)!}}+\frac{C_{12}\Delta}{\sqrt{n}}\right) + \frac{C_2}{K^{(\nu-d-1)/d}\Delta}. \tag{8.4.28}$$

The number C_0 depends on T and the parameters of the model, that is, the coefficients and the initial condition in (8.4.2)*; the numbers C_{11}, C_{12} depend only on the parameters of the model; the number C_2 depends on ν, T, and the parameters of the model.*

If, in addition, $(1+|x|^2)^{-w} f \in \mathbb{L}_2(\mathbb{R}^{\mathrm{d}})$ for some $w \geq 0$ so that $\nu > \mathrm{d}+1+w$ and $\Lambda^\nu\big((1+|x|^2)^w\pi_0\big) \in \mathbb{H}^1$, then

$$\max_{0\leq i\leq M} \mathbb{E}|u_{t_i}[f]-u^n_{N,K;i}[f]| \leq C_3\left(\frac{(C_{11}\Delta)^{N/2}}{\sqrt{(N+1)!}}+\frac{C_{12}\Delta}{\sqrt{n}}\right) + \frac{C_4}{K^{(\nu-\mathrm{d}-1)/\mathrm{d}}\Delta}. \tag{8.4.29}$$

The numbers C_3, C_4 depend on ν, T, the function f, and the parameters of the model; the numbers C_{11} and C_{12} are the same as in (8.4.28). □

Proof Consider first the local error $\widetilde{\mathbb{E}}\|u(\Delta,\cdot)-u^n_{N,K}(\Delta,\cdot)\|_0^2$. Define

$$u_N(\Delta,x) = \sum_{\alpha\in\mathcal{J}^n_N} \varphi_\alpha(\Delta,x,\pi_0).$$

By direct computation,

$$\widetilde{\mathbb{E}}\|u(\Delta,\cdot)-u^n_N(\Delta,\cdot)\|_0^2 \leq c_1 e^{c_2\Delta}\left(\frac{(c_3\Delta)^{N+1}}{(N+1)!}\|\pi_0\|_0^2+\frac{\Delta^3}{n}\|\pi_0\|_2^2\right), \tag{8.4.30}$$

where the numbers c_1, c_2, c_3 depend only on the coefficients of (8.4.1); recall that $\|\cdot\|_2$ is the norm in the Sobolev space $\mathbb{H}^2$. The first term on the right-hand side of (8.4.30) comes directly from Corollary 8.1; the second term is a consequence of the equality

$$\int_0^\pi g(t)\cos(nt)\,dt = -\frac{1}{n}\int_0^\pi g'(t)\sin(nt)\,dt$$

for a continuously differentiable function g.

Similarly,

$$\widetilde{\mathbb{E}}\|u_N^n(\Delta,\cdot) - u_{N,K}^n(\Delta,\cdot)\|_0^2 \le c_4 e^{c_5\Delta} K^{-2(\nu-\mathrm{d}-1)/\mathrm{d}} \|\Lambda^\nu \pi_0\|_0^2, \tag{8.4.31}$$

where the numbers c_3, c_4 depend on ν and the parameters of the model. We combine (8.4.30), in which $N \ge 2$, and (8.4.31) to get the overall local error

$$\widetilde{\mathbb{E}}\|u(\Delta,\cdot) - u_{N,K}^n(\Delta,\cdot)\|_0^2 \le \left(c_6\Delta^3 + c_7 K^{-2(\nu-\mathrm{d}-1)/\mathrm{d}}\right) e^{c_8\Delta};$$

the global error is then

$$\widetilde{\mathbb{E}}\|u(t_i,\cdot) - u_{N,K}^n(t_i,\cdot)\|_0^2 \le c_9\Delta^2 + c_{10} K^{-2(\nu-\mathrm{d}-1)/\mathrm{d}} \Delta^{-2}.$$

Inequality (8.4.29) follows from (8.4.28) by the Cauchy–Schwarz inequality.

For more details, see [100]. □

8.4.15. The following properties of the functions m_k were essential in the proof of (8.4.30): if $M_k(t) = \int_0^t m_k(s)\,ds$, then $M_k(\Delta) = 0$, $|M_k(t)| \le \sqrt{\Delta}/n$. Any other basis with these properties can also be used, but for now the Fourier cosine basis (8.4.27) appears to be the only one for which these properties are easily verified. The Haar basis, while simplifying calculations of ξ_α^i, results in a *local* error bound (8.4.30) with a slower rate of decay in Δ [13, Corollary 3.8].

8.4.16. The assumption $\sigma \equiv 0$ was also essential for the proof of (8.4.30); without this assumption, a different error bound holds.

Theorem 8.9 *Assume that the matrix* ${}_1\sigma\,{}_1\sigma^*$ *is uniformly positive definite. If the filtering model* (8.4.1) *is* ν*-regular for some* $\nu > \max(4, \mathrm{d}+1)$, *then*

$$\max_{0\le i\le M} \widetilde{\mathbb{E}}\|u(t_i,\cdot) - u_{N,K}^n(t_i,\cdot)\|_0 \le C_1\left(\frac{1}{(1+\delta)^{N/2}} + \frac{1}{\sqrt{n}}\right)\Delta^{1/2} + \frac{C_2}{K^{(\nu-\mathrm{d}-1)/\mathrm{d}}\Delta}. \tag{8.4.32}$$

The number C_1 *depends on* T *and the parameters of the model, that is, the coefficients and the initial condition in the Eq.* (8.4.1); *the number* $\delta > 0$ *depends only on the parameters of the model;* C_2 *depends on* ν, T, *and the parameters of the model.*

If, in addition, $(1+|x|^2)^{-w}f \in \mathbb{L}_2(\mathbb{R}^{\mathrm{d}})$ *for some* $w \geq 0$ *so that* $\nu > \mathrm{d}+1+w$ *and* $\Lambda^{\nu}\big((1+|x|^2)^{w}\pi_0\big) \in \mathbb{H}^1$, *then*

$$\max_{0\leq i\leq M} \mathbb{E}\big|u_{t_i}[f] - u^n_{N,K;i}[f]\big| \leq C_3\left(\frac{1}{(1+\delta)^{N/2}} + \frac{1}{\sqrt{n}}\right)\Delta^{1/2} + \frac{C_4}{K^{(\nu-\mathrm{d}-1)/\mathrm{d}}\Delta}. \tag{8.4.33}$$

The numbers C_3, C_4 *depend on* ν, T, *the function* f, *and the parameters of the model.* □

Proof Once we establish the local error bound of the type (8.4.30), which in this case turns out to be

$$\widetilde{\mathbb{E}}\|u(\Delta,\cdot) - u^n_N(\Delta,\cdot)\|_0^2 \leq c_1 e^{c_2\Delta}\left(\frac{\Delta^2}{(1+\delta)^N}\|\pi_0\|_2^2 + \frac{\Delta^2}{n}\|\pi_0\|_4^2\right), \tag{8.4.34}$$

for a suitable $\delta > 0$, the proof is completed by the same arguments as in Theorem 8.8.

To establish (8.4.34), we use Corollary 1.6 (keeping in mind a different normalization involving $\boldsymbol{\alpha}!$) and put $\mathbb{H} = \mathbb{H}^{\gamma}$ for a suitable γ. Then

$$\sum_{|\boldsymbol{\alpha}|=n} \|\varphi_{\boldsymbol{\alpha}}(t,\cdot,p_0)\|_{\gamma}^2 = \sum_{k_1,\ldots,k_n=1}^{r} \int_0^t \int_0^{s^n} \cdots \int_0^{s^2} \|\Phi^*_{t-s^n}\mathcal{M}^{k_n,*}\cdots\Phi^*_{s^2-s^1}\mathcal{M}^{k_1,*}\Phi^*_{s^1}\pi_0\|_{\gamma}^2\, ds^1\ldots ds^n, \tag{8.4.35}$$

where $\Phi^* = \Phi^*_t$ is the semi-group of the operator $\mathcal{L}^*$. The assumptions of the current theorem imply that the semi-group Φ^* is bounded above by the heat kernel:

$$\|\Phi^*_t f\|_{\gamma} \leq C_1 \int_{\mathbb{R}^{\mathrm{d}}} e^{-C_2|y|^2 t}|\breve{f}(y)|^2(1+|y|^2)^{\gamma}dy \tag{8.4.36}$$

for some positive numbers C_1, C_2, where $\breve{f}$ is the Fourier transform of f; see [27] for details. Notice also that

$$\|\mathcal{M}^{k,*}f\|_{\gamma} \leq C_3(\|f\|_{\gamma} + \|\nabla f\|_{\gamma}),$$

where ∇f is the gradient of f. Then direct computations show that

$$\int_0^t\int_0^s \|\mathcal{M}^{k,*}\Phi^*_{s-s^1}f(s^1)\|_{\gamma}^2\, ds^1 ds \leq C_4\int_0^t \|f(s)\|_{\gamma}^2\, ds.$$

For $n \geq 2$, we combine the last inequality with Theorem 8.5 to conclude that

$$\sum_{k_1,\ldots,k_n=1}^{r} \int_0^{\Delta}\int_0^{s^n}\cdots\int_0^{s^2} \|\Phi^*_{\Delta-s^n}\mathcal{M}^{k_n,*}\cdots\Phi^*_{s^2-s^1}\mathcal{M}^{k_1,*}\Phi^*_{s^1}\pi_0\|_0^2\, ds^1\ldots ds^n$$
$$\leq C_5 \sum_{k_1,\ldots,k_{n-3}}^{r} \int_0^{\Delta}\int_0^{s^n}\cdots\int_0^{s^2}$$
$$\|\mathcal{M}^{k_{n-3},*}\Phi^*_{s^{n-2}-s^{n-3}}\mathcal{M}^{k_{n-3},*}\cdots\Phi^*_{s^2-s^1}\mathcal{M}^{k_1,*}\Phi^*_{s^1}\pi_0\|_2^2\, ds^1\ldots ds^n$$
$$\leq C_6(1+\delta)^{-n}\Delta^2\|\pi_0\|_2^2 \tag{8.4.37}$$

for some $\delta > 0$. Then local error bound (8.4.34) follows by the same arguments as in the proof of Theorem 8.8. The main reason for the factor Δ^2 rather than Δ^3 in (8.4.37) is that the operators $\mathcal{M}^{k,*}$ do not commute with one another when $\sigma \not\equiv 0$. □

8.4.17. If the matrix a is not uniformly positive definite, then the rate of convergence is an open question.

8.4.18. We now establish the rate of convergence for the Spectral Separating Scheme of the Second Kind. Recall that this algorithm defines the approximations $u_N^{K,n}(t_i, x)$, $u_{N;i}^{K,n}$ of the unnormalized filtering density and filter according to (8.4.20). The following theorem presents the quality of these approximations and establishes the convergence in the limit $\lim_{K\to\infty}\lim_{\Delta\to 0}$.

Theorem 8.10 *If the filtering model (8.4.1) is ν-regular for some $\nu > \mathrm{d}+1$, then*

$$\max_{0\leq i\leq M} \mathbb{E}\left\|u(t_i,\cdot) - u_N^{K,n}(t_i,\cdot)\right\|_0 \leq \frac{C_1}{K^{(\nu-d-1)/d}} + C_2\left(\frac{(C_{21}\Delta)^{N/2}}{\sqrt{(N+1)!}} + \frac{C_{22}\Delta^{1/2}}{\sqrt{n}}\right). \tag{8.4.38}$$

The number C_1 depends on ν, T, and the parameters of the model, that is, the coefficients and the initial condition in the Zakai Eq. (8.4.2)*; the number C_2 depends on T, K and the parameters of the model; the numbers C_{21}, C_{22} depend on K and the parameters of the model.*

If, in addition, $(1+|x|^2)^{-w} f \in \mathbb{L}_2(\mathbb{R}^{\mathrm{d}})$ for some $w \geq 0$ so that $\nu > \mathrm{d}+1+w$ and $\Lambda^{\nu}\big((1+|x|^2)^w\pi_0\big) \in \mathbb{H}^1$, then

$$\max_{0\leq i\leq M} \mathbb{E}\left|u_{t_i}[f] - u_{N;i}^{K,n}[f]\right| \leq \frac{C_3}{K^{(\nu-w-\mathrm{d}-1)/\mathrm{d}}} + C_4\left(\frac{(C_{21}\Delta)^{N/2}}{\sqrt{(N+1)!}} + \frac{C_{22}\Delta^{1/2}}{\sqrt{n}}\right). \tag{8.4.39}$$

The number C_3 depends on ν, T, the function f, and the parameters of the model; the number C_4 depends on K, T, the function f, and the parameters of the model; the numbers C_{21}, C_{22} are the same as in (8.4.38). □

Proof The main steps are the same as in the proof of Theorem 8.9, but with the analysis of truncation in space carried out first. The details are in [95], where the interested reader can also find more detailed information about the numbers C_1, C_2, etc. □

8.4.19. Note that, in the Spectral Separating Scheme of the Second Kind, the approximation in space is carried out first, and the Wiener chaos expansion is applied to a system of ordinary differential equations (8.4.14). As a result, unlike Theorems 8.8 and 8.9, the error bound can be established without assuming non-degeneracy of the matrix ${}_1\sigma\,{}_1\sigma^*$. The rate of convergence in Δ for an approximation of the optimal filter for (8.4.1) is, in general, not better than $\Delta^{1/2}$, and both spectral separating schemes achieve this rate. Indeed, for $N \geq 2$, formulas (8.4.33) and (8.4.39) can be written as

$$\max_{0 \leq i \leq M} \mathbb{E}\left|u_{t_i}[f] - u^n_{N,K;i}[f]\right| \leq C_3\Delta^{1/2} + \frac{C_4}{K^{(\nu-\mathrm{w}-\mathrm{d}-1)/\mathrm{d}}\Delta} \tag{8.4.40}$$

and

$$\max_{0 \leq i \leq M} \mathbb{E}\left|u_{t_i}[f] - u^{K,n}_{N;i}[f]\right| \leq \frac{C_3}{K^{(\nu-\mathrm{w}-\mathrm{d}-1)/\mathrm{d}}} + C_4\Delta^{1/2}, \tag{8.4.41}$$

respectively. Note that the error due to truncation in space is $K^{-(\nu-\mathrm{w}-\mathrm{d}-1)/\mathrm{d}}$ in both cases, but, since computation of $u^n_{N,K;i}[f]$ in (8.4.40) involves truncation in space on every time step, this error is multiplied by the number of time steps, and the number of time steps is proportional to $1/\Delta$. The rate of convergence in time is still $\Delta^{1/2}$, since we first take the limit $K \to \infty$.

8.5 An Infinite-Dimensional Example

8.5.1. The following viscous transport equation is used to describe the time evolution of a scalar quantity θ in a given velocity field $\mathbf{v}$:

$$\dot{\theta}(t,x) = \nu\boldsymbol{\Delta}\theta(t,x) - \mathbf{v}(t,x)\cdot\nabla\theta(t,x) + f(t,x);\ x \in \mathbb{R}^{\mathrm{d}},\ \mathrm{d} > 1. \tag{8.5.1}$$

The scalar θ is called passive because it does not affect the velocity field $\mathbf{v}$.

We assume that $\mathbf{v} = \mathbf{v}(t,x) \in \mathbb{R}^{\mathrm{d}}$ is an isotropic Gaussian vector field with zero mean and covariance

$$\mathbb{E}(v^i(t,x)v^j(s,y)) = \delta(t-s)C^{ij}(x-y),$$

where $C = (C^{ij}(x), i, j = 1, \ldots, d)$ is a matrix-valued function so that $C(0)$ is a scalar matrix; with no loss of generality we will assume that $C(0) = I$, the identity matrix.

8.5.2. It is known from [89, Section 10.1] that, for an isotropic Gaussian vector field, the Fourier transform $\hat{C} = \hat{C}(z)$ of the function $C = C(x)$ is

$$\hat{C}(y) = \frac{A_0}{(1+|y|^2)^{(\mathrm{d}+\gamma)/2}}\left(a\frac{yy^*}{|y|^2} + \frac{b}{\mathrm{d}-1}\left(\mathbb{I} - \frac{yy^*}{|y|^2}\right)\right), \tag{8.5.2}$$

where y^* is the row vector $(y_1, \ldots, y_d)$, y is the corresponding column vector, $|y|^2 = y^*y$; and $\gamma > 0$, $a \geq 0$, $b \geq 0$, $A_0 > 0$ are real numbers. Similar to [89], we assume that $0 < \gamma < 2$. This range of values of γ corresponds to a turbulent velocity field $\mathbf{v}$, also known as the generalized Kraichnan model [29]; the original Kraichnan model [63] corresponds to $a = 0$. For small x, the asymptotics of $C^{ij}(x)$ is $(\delta_{ij} - c^{ij}|x|^\gamma)$ [89, Section 10.2].

By direct computation (cf. [4]), the vector field $\mathbf{v} = (v^1, \ldots, v^{\mathrm{d}})$ can be written as

$$v^i(t, x) = \sigma^{ik}(x)\dot{w}^k(t), \tag{8.5.3}$$

where $\{\sigma^k,\ k \geq 1\}$ is an orthonormal basis in the space $\mathbb{H}_C$, the reproducing kernel Hilbert space corresponding to the kernel function C. It is known from [89] that $\mathbb{H}_C$ is all or part of the Sobolev space $\mathbb{H}^{(\mathrm{d}+\gamma)/2}(\mathbb{R}^{\mathrm{d}}; \mathbb{R}^{\mathrm{d}})$.

If $a > 0$ and $b > 0$, then the matrix $\hat{C}$ is invertible and

$$\mathbb{H}_C = \left\{ f \in \mathbb{R}^{\mathrm{d}} : \int_{\mathbb{R}^{\mathrm{d}}} \hat{f}^*(y)\hat{C}^{-1}(y)\hat{f}(y)dy < \infty \right\} = \mathbb{H}^{(\mathrm{d}+\gamma)/2}(\mathbb{R}^{\mathrm{d}}; \mathbb{R}^{\mathrm{d}}),$$

because $\|\hat{C}(y)\| \sim (1+|y|^2)^{-(\mathrm{d}+\gamma)/2}$, $|y| \to \infty$.

If $a > 0$ and $b = 0$, then

$$\mathbb{H}_C = \left\{ f \in \mathbb{R}^{\mathrm{d}} : \int_{\mathbb{R}^{\mathrm{d}}} |\hat{f}(y)|^2(1+|y|^2)^{(\mathrm{d}+\gamma)/2}dy < \infty;\ yy^*\hat{f}(y) = |y|^2\hat{f}(y) \right\},$$

the subset of gradient fields in $\mathbb{H}^{(\mathrm{d}+\gamma)/2}(\mathbb{R}^{\mathrm{d}}; \mathbb{R}^{\mathrm{d}})$, that is, vector fields f for which $\hat{f}(y) = y\hat{F}(y)$ for some scalar $F \in \mathbb{H}^{(\mathrm{d}+\gamma+2)/2}(\mathbb{R}^{\mathrm{d}})$.

If $a = 0$ and $b > 0$, then

$$\mathbb{H}_C = \left\{ f \in \mathbb{R}^{\mathrm{d}} : \int_{\mathbb{R}^{\mathrm{d}}} |\hat{f}(y)|^2(1+|y|^2)^{(\mathrm{d}+\gamma)/2}dy < \infty;\ y^*\hat{f}(y) = 0 \right\},$$

the subset of divergence-free fields in $\mathbb{H}^{(\mathrm{d}+\gamma)/2}(\mathbb{R}^{\mathrm{d}}; \mathbb{R}^{\mathrm{d}})$.

By the embedding theorems, each σ^{ik} is a bounded continuous function on $\mathbb{R}^{\mathrm{d}}$; in fact, every σ^{ik} is Hölder continuous of order $\gamma/2$. In addition, being an element of the corresponding space $\mathbb{H}_C$, each σ_k is a gradient field if $b = 0$ and is divergence free if $a = 0$.

Equation (8.5.1) becomes

$$d\theta(t,x) = \big(\nu \boldsymbol{\Delta}\theta(t,x) + f(t,x)\big)dt - \sum^{k} \sigma_k(x) \cdot \nabla\theta(t,x)dw_k(t). \qquad (8.5.4)$$

8.5.3. We summarize the above constructions in the following **assumptions**:

S1 There is a fixed stochastic basis $\mathbb{F} = (\Omega, \mathcal{F}, \{\mathcal{F}_t\}_{t\geq 0}, \mathbb{P})$ with the usual assumptions and $(w_k(t), k \geq 1, t \geq 0)$ is a collection of independent standard Wiener processes on $\mathbb{F}$.

S2 For each k, the vector field σ^k is an element of the Sobolev space $\mathbb{H}_2^{(\mathrm{d}+\gamma)/2}(\mathbb{R}^\mathrm{d}; \mathbb{R}^\mathrm{d}), 0 < \gamma < 2, \mathrm{d} \geq 2$.

S3 For all x, y in $\mathbb{R}^\mathrm{d}$, $\sum_k \sigma^{ik}(x)\sigma^{jk}(y) = C^{ij}(x - y)$ so that the matrix-valued function $C = C(x)$ satisfies (8.5.2) and $C(0) = I$.

S4 The input data θ_0, f are deterministic and satisfy

$$\theta_0 \in \mathbb{L}_2(\mathbb{R}^\mathrm{d}),\ f \in \mathbb{L}_2((0,T); \mathbb{H}_2^{-1}(\mathbb{R}^\mathrm{d})).$$

Let $\mathbb{Y}$ be a Banach space. If $q > 1$, then

$$\mathbb{L}_{2,q}(\mathbb{Y}) := \Big\{v = \sum_{\alpha\in\mathcal{J}} v_\alpha \xi_\alpha : \sum_{\alpha\in\mathcal{J}} q^{|\alpha|}\|v\|^2_{\mathbb{Y}} < \infty\Big\};$$

if $q \in (0, 1)$, then $\mathbb{L}_{2,q}(\mathbb{Y})$ is the closure of $\mathbb{L}_2(\mathbb{Y})$ with respect to the norm

$$\|v\|_{\mathbb{L}_{2,q}(\mathbb{Y})} = \left(\sum_{\alpha\in\mathcal{J}} q^{|\alpha|}\|v\|^2_{\mathbb{Y}}\right)^{1/2}.$$

Theorem 8.11 *Assume that* $\nu > 0$ *and let* $q < \sqrt{2\nu}$, $k \geq 1$.
Under assumptions **S1–S4**, *there exists a unique chaos solution of* (8.5.4) *and*

$$\|\theta\|^2_{\mathbb{L}_{2,q}(\mathbb{L}_2((0,T);\mathbb{H}_2^1(\mathbb{R}^\mathrm{d})))} + \|\theta\|^2_{\mathbb{L}_{2,q}(\mathbb{C}((0,T);\mathbb{L}_2(\mathbb{R}^\mathrm{d})))}$$
$$\leq C(\nu, q, T)\left(\|\theta_0\|^2_{\mathbb{L}_2(\mathbb{R}^\mathrm{d})} + \|f\|^2_{\mathbb{L}_2((0,T);\mathbb{H}_2^{-1}(\mathbb{R}^\mathrm{d}))}\right).$$

□

Proof This follows by Theorems 8.4 and 8.5. In particular, if $\sqrt{2\nu} > 1$, then $q > 1$ is an admissible choice of the weights. If $\sqrt{2\nu} \leq 1$, then Eq. (8.5.4) does not have a square-integrable solution. If the weight is chosen so that $q = \sqrt{2\nu}$, then Eq. (8.5.1) can still be analyzed in the normal triple $(\mathbb{H}_2^1(\mathbb{R}^d), \mathbb{L}_2(\mathbb{R}^d), \mathbb{H}_2^{-1}(\mathbb{R}^d))$. □

8.5.4. If $\nu = 0$, Eq. (8.5.4) must be interpreted in the sense of Stratonovich:

$$du(t,x) = f(t,x)\,dt - \sigma_k(x)\cdot\nabla\theta(t,x)\circ dw_k(t). \tag{8.5.5}$$

To simplify the presentation, we assume that $f = 0$. If (8.5.2) holds with $a = 0$, then each σ_k is divergence free and (8.5.5) has an equivalent Itô form

$$d\theta(t,x) = \frac{1}{2}\boldsymbol{\Delta}\theta(t,x)\,dt - \sigma^{ik}(x)D_i\theta(t,x)\,dw_k(t). \tag{8.5.6}$$

Equation (8.5.6) is a model of non-viscous turbulent transport [161]. The propagator for (8.5.6) is

$$\begin{aligned}\frac{\partial}{\partial t}\theta_\alpha(t,x) &= \frac{1}{2}\boldsymbol{\Delta}\theta_\alpha(t,x)\\ &- \sum_{i,k}\sqrt{\alpha_i^k}\,\sigma^{jk}D_j\theta_{\alpha^-(i,k)}(t,x)m_i(t),\ 0 < t \le T,\end{aligned} \tag{8.5.7}$$

with initial condition $\theta_\alpha(0,x) = \theta_0(x)1_{(|\alpha|=0)}$.

The following result about solvability of (8.5.6) is an adaptation of the ideas and methods from Chap. 5 to the chaos solution.

Theorem 8.12 *In addition to* **S1–S4**, *assume that each σ_k is divergence free. Then there exists a unique chaos solution $\theta = \theta(t,x)$ of (8.5.6). This solution has the following properties:*

(i) *For every $\varphi \in \mathbb{C}_0^\infty(\mathbb{R}^d)$ and all $t \in [0,T]$, the equality*

$$(\theta,\varphi)(t) = (\theta_0,\varphi) + \frac{1}{2}\int_0^t (\theta,\boldsymbol{\Delta}\varphi)(s)\,ds + \int_0^t (\theta,\sigma^{ik}D_i\varphi)\,dw_k(s) \tag{8.5.8}$$

holds in $\mathbb{L}_2(\Omega)$, where $(\cdot,\cdot)$ is the inner product in $\mathbb{L}_2(\mathbb{R}^d)$.

(ii) *If $X = X_{t,x}$ is a weak solution of*

$$X_{t,x} = x + \int_0^t \sigma_k\left(X_{s,x}\right)dw_k(s), \tag{8.5.9}$$

then, for each $t \in [0,T]$,

$$\theta(t,x) = \mathbb{E}\left(\theta_0\left(X_{t,x}\right)|\mathcal{F}_t^W\right). \tag{8.5.10}$$

(iii) *For* $1 \le p < \infty$ *and* $r \in \mathbb{R}$*, define* $\mathbb{L}_{p,(r)}(\mathbb{R}^d)$ *as the Banach space of measurable functions with norm*

$$\|f\|^p_{\mathbb{L}_{p,(r)}(\mathbb{R}^d)} = \int_{\mathbb{R}^d} |f(x)|^p (1+|x|^2)^{pr/2} dx$$

is finite. Then there exists a number K *depending only on* p, r *so that, for each* $t > 0$,

$$\mathbb{E}\|\theta\|^p_{\mathbb{L}_{p,(r)}(\mathbb{R}^d)}(t) \le e^{Kt}\|\theta_0\|^p_{\mathbb{L}_{p,(r)}(\mathbb{R}^d)}. \tag{8.5.11}$$

In particular, if $r = 0$*, then* $K = 0$. □

8.5.5. In this paragraph we use the passive scalar equation to illustrate an application of chaos expansion to the computation of statistical moments of the solution.

Let $\theta = \theta(t, x)$ be the chaos solution of (8.5.6). Properties of the basis functions ξ_α imply that, for all s, t and almost all x, y,

$$\mathbb{E}\theta\,(t, x) = \theta_\alpha\,(t, x)\,1_{|\alpha|=0}$$

and

$$\mathbb{E}\theta\,(t, x)\,\theta\,(s, y) = \sum_{\alpha \in \mathscr{J}} \theta_\alpha\,(t, x)\,\theta_\alpha\,(s, y)\,.$$

If the initial condition θ_0 belongs to $\mathbb{L}_2(\mathbb{R}^d) \cap \mathbb{L}_p(\mathbb{R}^d)$ for $p \ge 3$, then, by (8.5.11), higher-order moments of θ exist. To obtain the expressions of the higher-order moments in terms of the coefficients θ_α, we need some auxiliary constructions.

For $\boldsymbol{\alpha}, \boldsymbol{\beta} \in \mathscr{J}$, define $\boldsymbol{\alpha} + \boldsymbol{\beta}$ as the multi-index with components $\alpha_i^k + \beta_i^k$. Similarly, we define the multi-indices $|\boldsymbol{\alpha} - \boldsymbol{\beta}|$ and $\boldsymbol{\alpha} \wedge \boldsymbol{\beta} = \min(\boldsymbol{\alpha}, \boldsymbol{\beta})$. We write $\boldsymbol{\beta} \le \boldsymbol{\alpha}$ if and only if $\beta_i^k \le \alpha_i^k$ for all $i, k \ge 1$. If $\boldsymbol{\beta} \le \boldsymbol{\alpha}$, we define

$$\binom{\boldsymbol{\alpha}}{\boldsymbol{\beta}} := \prod_{i,k} \frac{\alpha_i^k!}{\beta_i^k!(\alpha_i^k - \beta_i^k)!}.$$

Definition 8.3 We say that a triple of multi-indices $(\boldsymbol{\alpha}, \boldsymbol{\beta}, \boldsymbol{\gamma})$ is complete and write $(\boldsymbol{\alpha}, \boldsymbol{\beta}, \boldsymbol{\gamma}) \in \triangle$ if all the entries of the multi-index $\boldsymbol{\alpha} + \boldsymbol{\beta} + \boldsymbol{\gamma}$ are even numbers and $|\boldsymbol{\alpha} - \boldsymbol{\beta}| \le \boldsymbol{\gamma} \le \boldsymbol{\alpha} + \boldsymbol{\beta}$. For fixed $\boldsymbol{\alpha}, \boldsymbol{\beta} \in \mathscr{J}$, we write

$$\triangle\,(\boldsymbol{\alpha}) := \left\{\boldsymbol{\gamma}, \boldsymbol{\mu} \in \mathscr{J} : (\boldsymbol{\alpha}, \boldsymbol{\gamma}, \boldsymbol{\mu}) \in \triangle\right\}$$

and

$$\triangle(\boldsymbol{\alpha}, \boldsymbol{\beta}) := \{\boldsymbol{\gamma} \in \mathscr{J} : (\boldsymbol{\alpha}, \boldsymbol{\beta}, \boldsymbol{\gamma}) \in \triangle\}.$$

□

For $(\alpha, \beta, \gamma) \in \Delta$, we define

$$\Psi(\alpha, \beta, \gamma) := \sqrt{\alpha!\beta!\gamma!}\left(\left(\frac{\alpha-\beta+\gamma}{2}\right)!\left(\frac{\beta-\alpha+\gamma}{2}\right)!\left(\frac{\alpha+\beta-\gamma}{2}\right)!\right)^{-1}. \tag{8.5.12}$$

Note that the triple (α, β, γ) is complete if and only if any permutation of the triple (α, β, γ) is complete. Similarly, the value of $\Psi(\alpha, \beta, \gamma)$ is invariant under permutation of the arguments.

We also define

$$C(\gamma, \beta, \mu) := \left[\binom{\gamma+\beta-2\mu}{\gamma-\mu}\binom{\gamma}{\mu}\binom{\beta}{\mu}\right]^{1/2}, \ \mu \leq \gamma \wedge \beta. \tag{8.5.13}$$

Direct computations show that if f is a function on $\mathcal{J}$, then for $\gamma, \beta \in \mathcal{J}$,

$$\sum_{\mu \leq \gamma \wedge \beta} C(\gamma, \beta, \mu) f(\gamma+\beta-2\mu) = \sum_{\mu \in (\gamma, \beta)} f(\mu)\Phi(\gamma, \beta, \mu). \tag{8.5.14}$$

The next theorem presents the formulas for the third and fourth moments of the solution of Eq. (8.5.6) in terms of the coefficients θ_α.

Theorem 8.13 *In addition to* **S1–S4**, *assume that each σ^k is divergence free and the initial condition θ_0 belongs to $\mathbb{L}_2(\mathbb{R}^d) \cap \mathbb{L}_4(\mathbb{R}^d)$. Then*

$$\mathbb{E}\theta(t,x)\theta(t',x')\theta(s,y) = \sum_{(\alpha,\beta,\gamma)\in\Delta} \Psi(\alpha,\beta,\gamma)\theta_\alpha(t,x)\theta_\beta(t',x')\theta_\gamma(s,y) \tag{8.5.15}$$

and

$$\mathbb{E}\theta(t,x)\theta(t',x')\theta(s,y)\theta(s',y') \tag{8.5.16}$$
$$= \sum_{\rho\in\Delta(\alpha,\beta)\cap\Delta(\gamma,\kappa)} \Psi(\alpha,\beta,\rho)\Psi(\rho,\gamma,\kappa)\theta_\alpha(t,x)\theta_\beta(t',x')\theta_\gamma(s,y)\theta_\kappa(s',y').$$

□

Proof It is known [110] that

$$\xi_\gamma\xi_\beta = \sum_{\mu\leq\gamma\wedge\beta} C(\gamma,\beta,\mu)\xi_{\gamma+\beta-2\mu}. \tag{8.5.17}$$

Let us consider the triple product $\xi_\alpha \xi_\beta \xi_\gamma$. By (8.5.17),

$$\mathbb{E}\xi_\alpha \xi_\beta \xi_\gamma = \mathbb{E} \sum_{\mu \in \triangle(\alpha,\beta)} \xi_\gamma \xi_\mu \Psi(\alpha, \beta, \mu) = \begin{cases} \Psi(\alpha, \beta, \gamma), & (\alpha, \beta, \gamma) \in \triangle; \\ 0, & \text{otherwise.} \end{cases} \tag{8.5.18}$$

Equality (8.5.15) now follows. To compute the fourth moment, note that

$$\begin{aligned} \xi_\alpha \xi_\beta \xi_\gamma &= \sum_{\mu \le \alpha \wedge \beta} C(\alpha, \beta, \mu)\, \xi_{\alpha+\beta-2\mu} \xi_\gamma \\ &= \sum_{\mu \le \alpha \wedge \beta} C(\alpha, \beta, \mu) \sum_{\rho \le (\alpha+\beta-2\mu) \wedge \gamma} C(\alpha + \beta - 2\mu, \gamma, \rho)\, \xi_{\alpha+\beta+\gamma-2\mu-2\rho}. \end{aligned} \tag{8.5.19}$$

Repeated applications of (8.5.14) yield

$$\begin{aligned} \xi_\alpha \xi_\beta \xi_\gamma &= \sum_{\mu \le \alpha \wedge \beta} C(\alpha, \beta, \mu) \sum_{\rho \in \triangle(\alpha+\beta-2\mu, \gamma)} \xi_\rho \Psi(\alpha + \beta - 2\mu, \gamma, \rho) \\ &= \sum_{\mu \in \triangle(\alpha,\beta)} \sum_{\rho \in \triangle(\mu,\gamma)} \Psi(\alpha, \beta, \mu)\, \Psi(\mu, \gamma, \rho)\, \xi_\rho. \end{aligned}$$

Thus,

$$\begin{aligned} \mathbb{E}\xi_\alpha \xi_\beta \xi_\gamma \xi_\kappa &= \sum_{\mu \in \triangle(\alpha,\beta)} \sum_{\rho \in \triangle(\mu,\gamma)} \Psi(\alpha, \beta, \mu)\, \Psi(\mu, \gamma, \rho)\, 1_{\{\mu=\kappa\}} \\ &= \sum_{\rho \in \triangle(\alpha,\beta) \cap \triangle(\gamma,\kappa)} \Psi(\alpha, \beta, \rho)\, \Psi(\rho, \gamma, \kappa). \end{aligned}$$

Equality (8.5.16) now follows. □

In the same way, one can get formulas for fifth- and higher-order moments.

Remark 8.2 Expressions (8.5.15) and (8.5.16) do not depend on the structure of Eq. (8.5.6) and can be used to compute the third and fourth moments of any random field with a known chaos expansion. The interested reader should keep in mind that the formulas for the moments of orders higher than two should be interpreted with care. In fact, they represent the pseudo-moments; see [114] for details. □

8.5.6. To conclude the section, we take another look at Eq. (8.5.4). By reducing the smoothness assumptions on σ^k, it is possible to consider velocity fields $\mathbf{v}$ that are more turbulent than in the Kraichnan model, for example,

$$v^i(t,x) = \sum_{k\geq 1} \sigma^{ik}(x)\dot{w}^k(t), \tag{8.5.20}$$

where $\{\sigma^k,\ k \geq 1\}$ is an orthonormal basis in $\mathbb{L}_2(\mathbb{R}^d;\mathbb{R}^d)$. With $\mathbf{v}$ as in (8.5.20), the passive scalar Eq. (8.5.4) becomes

$$\dot{\theta}(t,x) = \nu\boldsymbol{\Delta}\theta(t,x) + f(t,x) - \nabla\theta(t,x)\cdot\dot{W}(t,x), \tag{8.5.21}$$

where $\dot{W} = \dot{W}(t,x)$ is a d-dimensional space-time white noise and the Itô stochastic differential is used.

Let $\mathbb{Y}$ be a Banach space. For $\boldsymbol{\alpha} \in \mathcal{J}$ and a sequence $\mathfrak{q} = \{q_l,\ l \geq 1\}$ with $q_l > 0$,

$$\mathfrak{q}^{\alpha} := \prod_{l,k} q_l^{\alpha_k^l}.$$

If $\mathfrak{q} = \{q_l,\ l \geq 1\}$ is a sequence with $q_l > 1$, then

$$\mathbb{L}_{2,\mathfrak{q}}(\mathbb{Y}) := \Big\{v = \sum_{\alpha\in\mathcal{J}} v_\alpha \xi_\alpha : \sum_{\alpha\in\mathcal{J}} \mathfrak{q}^\alpha \|v\|^2_{\mathbb{Y}} < \infty\Big\};$$

if $q_l \in (0,1)$, then $\mathbb{L}_{2,\mathfrak{q}}(\mathbb{Y})$ is the closure of $\mathbb{L}_2(\mathbb{Y})$ with respect to the norm

$$\|v\|_{\mathbb{L}_{2,\mathfrak{q}}(\mathbb{Y})} = \left(\sum_{\alpha\in\mathcal{J}} \mathfrak{q}^\alpha \|v\|^2_{\mathbb{Y}}\right)^{1/2}.$$

The following result is an extension of Theorem 8.6.

Theorem 8.14 *Suppose that $\nu > 0$ is a real number, each σ^{ik} is a bounded measurable function, and the input data are deterministic and satisfy $u_0 \in \mathbb{L}_2(\mathbb{R}^d)$, $f \in \mathbb{L}_2\big((0,T);\mathbb{H}_2^{-1}(\mathbb{R}^d)\big)$.*

Fix $\varepsilon > 0$ and let $\mathfrak{q} = \{q_k,\ k \geq 1\}$ be a sequence such that, for all $x, y \in \mathbb{R}^d$,

$$2\nu|y|^2 - \sum_{k\geq 1} q_k^2 \sigma^{ik}(x)\sigma^{jk}(x) y_i y_j \geq \varepsilon |y|^2.$$

Then, for every $T > 0$, there exists a unique chaos solution θ of equation

$$d\theta(t,x) = \big(\nu\boldsymbol{\Delta}\theta(t,x) + f(t,x)\big)\,dt - \sigma^k(x)\cdot\nabla\theta(t,x)\,dw^k(t) \tag{8.5.22}$$

and

$$\|\theta\|^2_{\mathbb{L}_{2,q}(\mathbb{L}_2((0,T);\mathbb{H}^1_2(\mathbb{R}^d)))} + \|\theta\|^2_{\mathbb{L}_{2,q}(\mathbb{C}((0,T);\mathbb{L}_2(\mathbb{R}^d)))}$$
$$\leq C(\nu, q, T)\left(\|\theta_0\|^2_{\mathbb{L}_2(\mathbb{R}^d)} + \|f\|^2_{\mathbb{L}_2((0,T);\mathbb{H}^{-1}_2(\mathbb{R}^d))}\right).$$

□

Remark 8.3 If $\max_i \sup_x |\sigma^{ik}(x)| \leq C_k$, $k \geq 1$, then a possible choice of q is

$$q_k = (\delta\nu)^{1/2}/(\mathrm{d}2^k C_k),\ 0 < \delta < 2.$$

If $\max_{i,j} \sup_x |\sigma^{ik}(x)\sigma^{jk}(x)| \leq C_\sigma$, then a possible choice of q is

$$q_k = \varepsilon\big(2\nu/(C_\sigma \mathrm{d})\big)^{1/2},\ 0 < \varepsilon < 1.$$

□

Notes

Chapter 1

The proof of Theorem 1.17 (Itô–Ventcel's formula) follows Rozovskiĭ [130]. Lemma 1.3 is due to Ventcel [153]. The Itô–Ventcel formula was rediscovered and generalized by Bismut [9] and Kunita [78]. A standard reference on stochastic calculus is Karatzas and Shreve [58].

Chapter 2

Stochastic integrals with respect to square integrable martingales taking values in a Hilbert space were first systematically investigated in Kunita [77]. Afterwards the stochastic integration theory based on martingales taking values in infinite dimensional spaces was developed mainly by Métivier and his students (see [106, 107] and the references there). There is considerable overlap between the results presented in Sects. 2.2–2.4 with those of [107]. However the construction of a stochastic integral in this chapter differs from those developed in [77] and [107]. It is based on the idea outlined in an article by Krylov, Rozovskiĭ [72].

The notion of a normal triple given in Sect. 2.5 is a version of Gelfand's triple. Theorem 2.13 (the statement and the proof) is almost identical to Theorem 2.10 from [72]. In this connection see also Gyöngy, Krylov [40] and Grigelionis, Mikulevicius [39].

Chapter 3

The existence and uniqueness theorem for LSES (3.1.1) in the coercive case follows Pardoux [121, 124]. The results of Sect. 3.3 appears to be new. Similar results were

B. L. Rozovsky, S. V. Lototsky, *Stochastic Evolution Systems*, Probability Theory and Stochastic Modelling 89, https://doi.org/10.1007/978-3-319-94893-5

announced in Rozovskiĭ [132]. Theorem 3.4 is new. The exposition of the results concerning the approximation of LSES (3.1.1), the Markov property of its solution (Sect. 3.4), and the solvability of the first (Dirichlet) boundary problem (Sect. 3.5) follows Krylov, Rozovskiĭ [72], where corresponding results were developed for non-linear systems. We note that all the results of Sects. 3.2, 3.4, and 3.5 could be carried over to the case of monotone coercive non-linear systems (see Krylov, Rozovskiĭ [72], Pardoux [121]).

Chapter 4

Sections 4.2 and 4.3 are based on Krylov, Rozovskiĭ [69, 74]. The super-parabolic condition for a second-order stochastic partial differential equation (SPDE) was introduced independently by Pardoux [121] and Krylov, Rozovskiĭ [68]. The results of Sect. 4.4 are due to the first author.

The proofs of the existence theorems for parabolic and super-parabolic Itô equations given in this chapter are based on the Galerkin method. This method is still applicable in much more general situations. In some particular cases, e.g. if the operator L is not random, results could be obtained by semi-group methods. For details, see [16, 19, 104, 131, 141, 142].

Chapter 5

The averaging over characteristic (AOC) formulas (5.2.2) and (5.2.3), and Corollary 5.2 (maximum principle) are due to Krylov, Rozovskiĭ [73, 74]. Formula (5.2.3) in the case of a uniformly non-degenerate matrix $(A^{ij}) := (2a^{ij} - \sum \sigma^{il}\sigma^{jl})$ was proved by Pardoux [124]. This non-degeneracy assumption appears to be rather restrictive. For example, for both Liouville's equations for diffusion processes, $(A^{ij}) \equiv 0$. Subsequently, formulas similar to (5.2.2) and (5.2.3) for classical solutions of the corresponding problems were obtained by Kunita [82].

Theorem 5.2 for $r = 0$ was proved by Krylov [66]. A statement very close to that in Lemma 5.1 can be found in Hida [45].

The forward and backward Liouville's equations, and in particular, the forward equation of inverse diffusion and the backward diffusion equation were derived by Krylov, Rozovskiĭ [71, 73, 74], and Rozovskiĭ [134, 135]. Independently, the backward diffusion equation was obtained under different assumptions and by a different method in Kunita [81].

Theorem 5.4 was first published in Krylov, Rozovskiĭ [75]. An equivalent to the forward equation of inverse diffusion and a formula similar to (5.3.6) were derived in Kunita [78] under additional assumptions that the coefficients are non-random and possess some extra derivatives.

The derivation of the backward diffusion equation is taken from Krylov, Rozovskii [73]. Another development can be found in Malliavin [102]. See also the book of Ikeda and Watanabe [51].

That the mapping $X(t, \cdot) : x \to X(t, x)$ is a diffeomorphism has been known to many authors [8, 9, 51, 73, 75, 78, 83, etc.].

Theorem 5.5 is due to the first author. The idea to reduce a second-order parabolic Itô equation to a second-order parabolic deterministic equation (although to one with random coefficients) goes back to Ventcel [153]. Subsequently this idea was systematically used by Rozovskiĭ [129, 130] in the study of the filtering equations. Later an analogous idea was used by Kunita [80].

The averaging over characteristic formula (5.4.3) and its corollaries are due to the first author. Note that the methods used in the proof had been used earlier in Krylov, Rozovskiĭ [70], and Rozovskiĭ [133] in the study of absolute continuity of the filtering measure with respect to the Lebesgue measure.

Theorem 5.7 is well known (see e.g. Lipster, Shiryayev [93], Rozovskiĭ, Shiryayev [138], and also Krylov, Rozovskiĭ [70]).

Chapter 6

Different versions of the Bayes formula were traditionally used in the development of filtering theory. The references are e.g. Kallianpur [56], Lipster, Shiryayev [93]. Note that in these books the reader can find the general theory of filtering, interpolation and extrapolation for semimartingales. Lemma 6.1 is taken from Loéve [94]. Theorem 6.1 is new, but certainly has predecessors (see e.g. [56], [93] cited above).

Section 6.3 is based on Rozovskiĭ [133]. Proposition 6.2 is in Lipster, Shiryayev [93]. Similar problems for discontinuous processes were considered by Grigelionis, Mikulevicius [39].

A forward linear filtering equation (for non-normalized filtering density) was first derived by Zakai [166] in a particular case. The equivalence of the forward linear filtering equation and non-linear filtering equations in quite a general situation was proved in Rozovskiĭ, Shiryayev [138] (see also Krylov, Rozovskiĭ [70] and Lipster, Shiryayev [93]).

Theorems 6.6–6.8 are due to the first author. The first results about the backward filtering equation were obtained in Kushner [86] and Pardoux [124].

Results related to those of the present chapter were published earlier by many authors. For example, filtering in bounded domains was considered in Margulis [104], Pardoux [123].

Chapter 7

In 1967 Hörmander published his famous results on the hypoellipticity of second-order degenerate parabolic equations. Since then, these results have been elaborated on by many authors. Malliavin [102, 103] provided the first probabilistic proof of Hörmander's theorem. Later Bismut [10, 11] developed a somewhat different probabilistic approach to the problem, which was more or less equivalent to that of Malliavin.

In this chapter we establish the hypoelliptic property of Itô's second-order parabolic equations using the basis of Bismut's version of Malliavin calculus.

Theorem 7.1 overlaps partly with the result of Kunita [79]. The prototype of Theorem 7.3 was developed for filtering equations in Bismut, Michel [11] and Kusuoka, Stroock [87]. The idea to prove the hypoellipticity property of deterministic second-order parabolic-elliptic equations via the application of Proposition 7.1 belongs to Malliavin. A formula for stochastic integration by parts in a form close to ours was first derived in Haussmann [44], where it was used in the investigation of the structure of square integrable martingales. Bismut [10] showed that it is an indispensable tool in the probabilistic study of hypoellipticity. The general scheme of the proof of Theorem 7.3 runs along the lines of Veretennikov [154]. A complete proof of Theorem 7.4 is in Stroock [151].

The proof of the existence and uniqueness of a generalized fundamental solution of the forward linear filtering equation in the super-parabolic case mainly follows Rozovskiĭ, Shimizu [137].

The results of Sect. 7.3 concerning the existence of conditional transition densities and their analytical properties are similar to those of Sects. 6.4.2–6.4.5.

In the hypoelliptic case similar results were obtained in Bismut, Michel [11], Kusuoka, Stroock [87], and Michel [111].

Further references are Chaleyat-Maurel, Michel [15], Ichihara, Kunita [50], Kunita [80], Nualart [118], Shikegawa [140], Stroock [148–151], Veretennikov [155], and Zakai [167].

Chapter 8

Shortly after Cameron and Martin [14] introduced the orthonormal basis in the space of square integrable functionals of the Brownian motion, K. Itô [54] established an equivalent form of the chaos expansion using multiple integrals. Kunita [78]–[80] used this form to study the linear filtering equation, whereas Ocone [119] and Wong [164] used it to analyze a more general (not necessarily diffusion) filtering problem; see also Budhiraja and Kallianpur [13]. Krylov and Veretennikov [76] developed a systematic chaos-based approach to the study of stochastic ordinary differential equations.

The Spectral Separating Scheme of the First Kind was first suggested in [113] and analyzed in [100] for the filtering model with no correlation between the state and observations noise; the general analysis is in [96]. The Spectral Separating Scheme of the Second Kind was introduced and analyzed in the Ph.D. dissertations of the second author (SL), under the supervision of the first author (BR).

Stochastic passive scalar and related equations have also been studied using the white noise approach in the spaces of Hida distributions [22, 126]. A summary of the related results can be found in [48, Section 4.3]. Theorem 8.12 is proved in [99] and, in a slightly weaker form, in [98]. Theorem 8.14 is proved in [99].

References

1. Arnold, V.I.: Ordinary Differential Equations. MIT Press, Cambridge (1973)
2. Arnold, L.: Dynamics of synergetic systems. In: Proceedings of the International Symposium on Synergetics, Bielefeld, 1979, pp. 107–118. Springer, Berlin (1980)
3. Balakrishnan, A.V.: Introduction to Optimization Theory in a Hilbert Space. Springer, New York (1971)
4. Baxendale, P., Hariss, T.E.: Isotropic stochastic flows. Ann. Probab. **14**(4), 1155–1179 (1986)
5. Bellman, R.: Stability Theory of Differential Equations. McGraw-Hill, New York (1953)
6. Bensoussan, A.: Filtrage optimale des systemes lineaires. Dunod, Paris (1971)
7. Bensoussan, A., Lions J.-L.: Applications des inéquations variationelles an contrôle stochastique. Dunod, Paris (1978)
8. Bismut, J.-M.: Mécanique aléatoire. Lecture Notes in Mathematics, vol. 866. Springer, Berlin (1981)
9. Bismut, J.-M.: A generalized formula of Itô and some other properties of stochastic flows. Z. Wahrsch. **55**, 331–350 (1981)
10. Bismut, J.-M.: Martingales, the Malliavin calculus and hypoellipticity under general Hörmander's condition. Z. Wahrsch. **56**, 469–505 (1981)
11. Bismut, J.-M., Michel, D.: Diffusions conditionnelles. J. Funct. Anal. **44**, 174–211 (1981); II. Ibid **45**, 274–292 (1982)
12. Blagoveshchenskii, Y.N., Freidlin, M.I.: Some properties of diffusion processes depending on a parameter. Sov. Math. Dokl. **138**, 633–636 (1961)
13. Budhiraja, A., Kallianpur, G.: Approximation to solutions of Zakai equations using multiple Wiener and Stratonovich expansions. Stochastics **56**(3–4), 271–315 (1996)
14. Cameron, R.H., Martin, W.T.: The orthogonal development of nonlinear functionals in series of Fourier–Hermite functionals. Ann. Math. **48**(2), 385–392 (1947)
15. Chaleyat-Maurel, M., Michel, D.: Hypoellipticity theorems and conditional laws. Z. Wahrsch. **65**, 573–597 (1984)
16. Curtain, R.F., Pritchard, A.J.: Infinite Dimensional Linear Systems Theory. Lecture Notes in Control and Information Sciences, vol. 8. Springer, Berlin (1978)
17. Da Prato, G., Zabczyk, J.: Stochastic Equations in Infinite Dimensions. Cambridge University Press, Cambridge (1992) [2nd edn. (2016)]
18. Dalećkii, Y.L.: Infinite-dimensional elliptic operators and parabolic equations connected with them. Russ. Math. Surv. **22**(4), 1–53 (1967)
19. Dalećkii, Y.L., Fomin, S.V.: Measures and Differential Equations in Infinite-Dimensional Spaces. Nauka, Moscow (1983) [in Russian]

B. L. Rozovsky, S. V. Lototsky, *Stochastic Evolution Systems*, Probability Theory and Stochastic Modelling 89, https://doi.org/10.1007/978-3-319-94893-5

20. Davis, M.B.A.: Pathwise nonlinear filtering. In: Hazewinkel, M., Willems, J.C. (eds.) Stochastic Systems: The Mathematics of Filtering and Identification. Proceedings of the NATO Advanced Study Institute, Les Aros, 1980, pp. 505–528. D. Reidel Publ. Co., Dordrecht (1981)
21. Dawson, D.A.: Stochastic evolution equations and related measure processes. J. Multivar. Anal. **5**(1), 1–52 (1975)
22. Deck, T., Potthoff, J.: On a class of stochastic partial differential equations related to turbulent transport. Probab. Theory Relat. Fields **111**, 101–122 (1998)
23. Dellacherie, C.: Capacités et processus stochastiques. Springer, Berlin (1972)
24. Dellacheire, C., Meyer, P.A.: Probabilités et potentials. Theorie des martingales. Herman, Paris (1980)
25. Doob, J.L.: Stochastic Processes. Wiley, New York; Chapman and Hall, London (1953)
26. Dunford, N., Schwarz, J.T.: Linear Operators. Part I: General Theory. Interscience Publishers, New York (1958)
27. Eidelman, S.D.: Parabolic Systems. Wolters-Noordhoff, Groningen (1969)
28. Ershov, M.P.: Sequential estimation of diffusion processes. Theory Probab. Appl. **15**(4), 705–717 (1970)
29. Gawedzki, K., Vergassola, M.: Phase transition in the passive scalar advection. Physica **D138**, 63–90 (2000)
30. Gelafand, I.M., Vilenkin, N.Y.: Generalized Functions. Vol. 4: Applications of Harmonic Analysis. Academic Press, New York (1969)
31. Gikhman, I.I.: On some differential equations with random functions. Ukrain. Matem. Z. **2**(3), 45–69 (1950) [in Russian]
32. Gikhman, I.I.: On the theory of differential equations of random processes. Ukrain. Matem. Z. **2**(4), 37–63 (1950); II. ibid **3**(3), 317–339 (1951) [in Russian]
33. Gikhman, I.I.: The boundary value problem for a stochastic equation of parabolic type. Ukrain. Matem. Z. **3**(5), 483–489 (1979) [in Russian]
34. Gikhman, I.I.: Qualitative Methods of Investigations of Non-linear Equations and Non-linear Oscillation, pp. 25–59. Institut Matematiki AN USSR, Kiev (1981) [in Russian]
35. Gihman, I.I., Skorokhod, A.V.: Stochastic Differential Equations. Springer, Berlin (1972)
36. Gihman, I.I., Skorokhod, A.V.: The Theory of Stochastic Processes. Springer, Berlin (1974)
37. Gihman, I.I., Skorokhod, A.V.: Stochastic Differential Equations and Its Applications. Naukova Dumka, Kiev (1982) [in Russian]
38. Gikhman, I.I., Mestechkina, T.M.: The Cauchy problem for stochastic first-order partial differential equations. Theory Random Process. **11**, 25–28 (1983)
39. Grigelionis, B., Mikulevicius, R.: Lecture Notes in Control and Information Sciences, vol. 49, pp. 49–88. Springer, Berlin (1983)
40. Gyöngy, I., Krylov, N.V.: On stochastics equations with respect to semimartingales. II. Itô formula in Banach spaces. Stochastics **6**(3–4), 153–173 (1981/82)
41. Fleming, W.H.: Distributed parameter stochastic systems in population biology. Lecture Notes in Economics and Mathematical Systems, vol. 107, pp. 179–191 (1975)
42. Fleming, W.H., Rishel, R.W.: Deterministic and Stochastic Optimal Control. Springer, Berlin (1975)
43. Krein, S.G. (ed.): Functional Analysis. Wolters-Noordhoff, Groningen (1972)
44. Haussmann, U.: On the integral representation of Ito processes. Stochastics **3**, 17–27 (1979)
45. Hida, T.: Stationary Stochastic Processes. Princeton University Press, Princeton (1970)
46. Hida, T., Brownian Motion. Springer, Berlin (1979)
47. Hida, T., Streit, L.: On quantum theory in terms of white noise. Nagoya Math. J. **68**(12), 21–34 (1977)
48. Holden, H., Øksendal, B., Ubøe, J., Zhang, T.: Stochastic Partial Differential Equations. Birkhauser, Boston (1996)
49. Hörmander, L.: Hypoelliptic second order differential equations. Acta Math. **119**, 147–171 (1967)

50. Ichihara, K., Kunita, H.: A classification of the second order degenerate elliptic operators and its probabilistic characterization. Z. Wahrsch. Verw. Gebiete **30**, 235–254 (1974)
51. Ikeda, N., Watanabe, S.: Stochastic Differential Equations and Diffusion Processes. North-Holland, Amsterdam; Kodansha, Tokyo (1981)
52. Itô, K.: On stochastic integral equation. Proc. Jpn. Acad. **22**, 32–35 (1946)
53. Itô, K.: On Stochastic Differential Equations. American Mathematical Society, New York (1951)
54. Itô, K.: Multiple Wiener integral. J. Math. Soc. Jpn. **3**, 157–169 (1951)
55. Itô, K., McKean, H.P.: Diffusion Processes and Their Sample Paths. Springer, Berlin (1965)
56. Kallianpur, G.: Stochastic Filtering Theory. Springer, New York (1980)
57. Kalman, R.E., Bucy, R.S.: New results in linear filtering and prediction theory. Trans. ASME D **83**, 95–108 (1961)
58. Karatzas, I., Shreve, S.: Brownian Motion and Stochastic Calculus, 2nd edn. Springer, New York (1991)
59. Klyackin, V.I.: Statistical Description of Dynamical Systems with Fluctuating Parameters. Nauka, Moscow (1975) [in Russian]
60. Klyackin, V.I.: Stochastic Equations and Waves in Random Heterogeneous Medium. Nauka, Moscow (1980) [in Russian]
61. Kolmogorov, A.N.: Ueber die analytischen Methoden in der WahrscheinlichkeitSrechnung. Math. Ann. **104**, 415–458 (1931)
62. Kolmogorov, A.N.: Interpolation and extrapolation of stationary stochastic series. Izv. Akad. Nauk SSSR. Ser. Mat. **5**(1), 3–14 (1941) [in Russian]
63. Kraicnnan, R.H.: Small-scale structure of a scalar field convected by turbulence. Phys. Fluids **11**, 945–963 (1968)
64. Krein, S.G., Petunin, Y.U., Semenov, E.M.: Interpolation of Linear Operators. American Mathematical Society, Providence, RI (1982)
65. Krylov, N.V.: Controlled Diffusion Processes. Springer, Berlin (1980)
66. Krylov, N.V.: Some new results in the theory of controlled diffusion processes. Math. USSR Sb. **37**(1), 133–149 (1980)
67. Krylov, N.V.: An analytic approach to SPDEs. In: Rozovskii, B.L., Carmona, R. (eds.) Stochastic Partial Differential Equations. Six Prospectives. Mathematical Surveys and Monographs, pp. 185–242. American Mathematical Society, Providence (1999)
68. Krylov, N.V., Rozovskii, B.L.: On Cauchy problem for superparabolic stochastic differential equations. In: Proceedings of III Soviet-Japan Symposium on Probability Theory and Mathematical Statistics, Tashkent, pp. 171–173 (1974)
69. Krylov, N.V., Rozovskii, B.L.: On the Cauchy problem for linear stochastic partial differential equations. Math. USSR Izv. **11**(6), 1267–1284 (1977)
70. Krylov, N.V., Rozovskii, B.L.: On conditional distributions of diffusion processes. Math. USSR Izv. **12**(2), 336–356 (1978)
71. Krylov, N.V., Rozovskii, B.L.: On complete integrals of Ito's equations. Usp. Mat. Nauk **35**(4), 147 (1980) [in Russian]
72. Krylov, N.V., Rozovskii, B.L.: Stochastic evolution equations. J. Sov. Math. **16**, 1233–1276 (1981)
73. Krylov, N.V., Rozovskii, B.L.: On the first integrals and Liouville equations for diffusion processes. Lecture Notes in Control and Information Sciences. Stochastic Differential Systems. Proceedings of the 3rd IFIP-WG 7/1 Working Conference, pp. 36, 117–125. Visegard, Hungary (1980). Springer, Berlin (1981)
74. Krylov, N.V., Rozovskii, B.L.: Characteristics of degenerating second-order parabolic Itô equations. J. Sov. Math. **32**, 336–348 (1982)
75. Krylov, N.V., Rozovski, B.L.: Stochastic partial differential equations and diffusion processes. Russ. Math. Surv. **37**(6), 75–95 (1982)
76. Krylov, N.V., Veretennikov, A.Y.: On explicit formulas for solutions of stochastic equations. Math. USSR Sb. **29**(2), 239–256 (1976)

77. Kunita, H.: Stochastic integrals based on martingales taking values in Hilbert space. Nagoya Math. J. **38**(1), 41–52 (1970)
78. Kunita, H.: On the decomposition of solutions of stochastic differential equations. In: Durham Conference on Stochastic Integrals. Lecture Notes in Mathematics, vol. 851, pp. 213–255. Springer, Berlin (1981)
79. Kunita, H.: Cauchy problem for stochastic partial differential equations arising in nonlinear filtering theory. Syst. Control Lett. **1**(1), 37–41 (1981)
80. Kunita, H.: Densities of a measure valued process governed by a stochastic partial differential equation. Syst. Control Lett. **1**(2), 100–104 (1981)
81. Kunita, H.: On backward stochastic differential equations. Stochastics **6**, 293–313 (1982)
82. Kunita, H.: Stochastic partial differential equations connected with non-linear filtering. In: Nonlinear Filtering and Stochastic Control (Cortona, 1981). Lecture Notes in Mathematics, vol. 972, pp. 100–168. Springer, Berlin (1982)
83. Kunita, H.: Stochastic partial differential equations connected with non-linear filtering. In: Nonlinear Filtering and Stochastic Control (Cortona, 1981). Lecture Notes in Mathematics, vol. 1097, pp. 149–303. Springer, Berlin (1984)
84. Kuo, H.-H.: Gaussian Measures in Banach Spaces. Lecture Notes in Mathematics. Springer, Berlin (1975)
85. Kuratovski, K.: Topology, vol. I. Academic, New York (1966)
86. Kushner, H.J.: Probability Methods for Approximations in Stochastic Control and for Elliptic Equations. Academic Press, New York (1977)
87. Kusuoka, S., Strook, D.W.: The partial Malliavin calculus and its applications to nonlinear filtering. Stochastics **12**, 83–142 (1984)
88. Ladyzhenskaya, O.A., Solonnikov, V.A., Uraltseva, N.N.: Linear and Quasilinear Equations of Parabolic Type. American Mathematical Society, Providence, RI (1968)
89. Le Jan, Y., Raiimond, O.: Integration of Brownian vector fields. Ann. Probab. **30**(2), 826–873 (2002)
90. Lévy, P.: Processus stochastiques et mouvement Brownien. Gauthier-Villars, Paris (1965)
91. Lions, J.-L., Magenes, E.: Problemes aux limites non homogénes et applications, vol. 1. Dunod, Paris (1968)
92. Liptser, R.S., Siryayev, A.N.: Theory of Martingales. Kluwer, Dordrecht (1989)
93. Liptser, R.S., Siryayev, A.N.: Statistics of Random Processes, I, II, 2nd edn. Springer, New York (2001)
94. Loéve, M.: Probability Theory. D. Van Nostrand Company Inc., Princeton (1960)
95. Lototsky, S.V.: Nonlinear filtering of diffusion processes in correlated noise: analysis by separation of variables. Appl. Math. Optim. **47**(2), 167–194 (2003)
96. Lototsky, S.V.: Wiener chaos and nonlinear filtering. Appl. Math. Optim. **54**(3), 265–291 (2006)
97. Lototsky, S.V., Rozovskii, B.L.: Recursive nonlinear filter for a continuous-discrete time model: separation of parameters and observations. IEEE Trans. Autom. Control **43**(8), 1154–1158 (1998)
98. Lototsky, S.V., Rozovskii, B.L.: Passive scalar equation in a turbulent incompressible Gaussian velocity field. Russ. Math. Surv. **59**(2), 297–312 (2004)
99. Lototsky, S.V., Rozovskii, B.L.: Wiener chaos solutions of linear stochastic evolution equations. Ann. Probab. **34**(2), 638–662 (2006)
100. Lototsky, S.V., Mikulevicius, R., Rozovskii, B.L.: Nonlinear filtering revisited: a spectral approach. SIAM J. Control Optim. **35**(2), 435–461 (1997)
101. Mahno, S.Y.: Limit theorems for stochastic equations with partial derivatives. In: International Symposium on Stochastic Differential Equations: Abstracts of Communications, Vilnius, pp. 73–77 (1978)
102. Malliavin, P.: Stochastic calculus of variation and hypoelliptic operators. In: Itô, K. (ed.) Proceedings of International Symposium on Stochastic Differential Equations Kyoto, 1976, pp. 195–265. Kimokuniya, Tokyo (1978)

103. Malliavin, P.: C^k-hypoellipticity with degeneracy. In: Friedman, A., Pinsky, M. (eds.) Stochastic Analysis, pp. 199–214. Academic, New York (1978)
104. Margulis, L.G.: Markovian Random Processes and Applications, pp. 1, 50–63. Saratov State University, Saratov (1980) [in Russian]
105. Margulis, L.G., Rozovskii, B.L.: Fundamental solutions of stochastic partial differential equations and filtration of diffusion processes. Usp. Mat. Nauk **33**(2), 197 (1978) [in Russian]
106. Métivier, M.: Semimartingales, a Course on Stochastic Processes. Walter de Gruyter, Berlin (1982)
107. Métivier, M., Pellaumoil S.: Stochastic Integration. Academic, New York (1980)
108. Meyer, P.A.: Probability and Potentials. Blaisdel, Waltham, MA (1966)
109. Meyer, P.A.: Notes sur les integrales stochastiques I. Integrales Hilbertiennes. In: Séminaire de Prob. XI. Lecture Notes in Mathematics, vol. 581, pp. 446–463. Springer, Berlin (1977)
110. Meyer, P.A.: Quantum Probability for Probabilitists. Lecture Notes in Mathematics, p. 1538. Springer, Berlin (1993)
111. Michel, D.: Régularité des lois conditionnelles en théorie du filtrage non-linéaire et calcul des variations stochastique. J. Funct. Anal. **41**(1), 8–36 (1981)
112. Mikulevicius, R.: Stochastic Navier–Stokes equations for turbulent flows. SIAM J. Math. Anal. **35**(5), 1250–1310 (2004)
113. Mikulevicius, R., Rozovskii, B.L.: Separation of observations and parameters in nonlinear filtering. In: Proceedings of the 32nd IEEE Conference on Decision and Control, pp. 1564–1559 (1993)
114. Mikulevicius, R., Rozovskii, B.L.: Linear parabolic stochastic PDE's and Wiener chaos. SIAM J. Math. Anal. **29**(2), 452–480 (1998)
115. Mikulevicius, R., Rozovskii, B.L.: Global L_2-solutions of stochastic Navier–Stokes equations. Ann. Probab. **33**(1), 137–176 (2005)
116. Monin, A.S., Yaglom, A.M.: Statistical Fluid Mechanics: Mechanics of Turbulence. MIT Press, Cambridge (1971)
117. Nikolskii, S.M.: Approximation of Functions of Several Variables and Embedding Theorems. Nauka, Moscow; Springer, Berlin (1975)
118. Nualart, D.: Mallavian Calculus and Related Topics. Springer, New York (1995)
119. Ocone, D.: Multiple integral expansions for nonlinear filtering. Stochastics **10**(1), 1–30 (1983)
120. Oleinik, O.A., Radkevich, E.V.: Second Order Equations with Nonnegative Characteristic Form. American Mathematical Society, Providence, RI (1978)
121. Pardoux, E.: Equations aux dérivees partielles stochastiques non linéaries monotones. Etude de solutions fortes de type Itô : Thes. P. (1975)
122. Pardoux, E.: Integrales stochastiques hilbertiennes, Univ. Paris-Dauphine, Cahiers de math. de la decis. no. 7617 (1976)
123. Pardoux, E.: Filtrage de diffusions avec conditions frontiéres: Caractérisation de la densitf conditionnelle. In: Dacunha-Castelle, D., van Cutsem, B. (eds.) Journées de Statistique des Processus Stochastiques. Lecture Notes in Mathematics, vol. 636, pp. 163–188. Springer, Berlin (1978)
124. Pardoux, E.: Stochastic partial differential equations and filtering of diffusion processes. Stochastics **3**, 127–167 (1979)
125. Pardoux, E.: Equations du filtrage non-linéaire de la prédiction et du lissage. Stochastics **6**, 193–231 (1982)
126. Potthoff, J., Våge, G., Watanabe, H.: Generalized solutions of linear parabolic stochastic partial differential equations. Appl. Math. Optim. **38**, 95–107 (1998)
127. Riesz, F., Sz.-Nagy, B.: Lecons d'analyse fonctionelle. Akadémiai Kiado, Budapest (1972)
128. Rozanov, I.A.: Markov Random Fields. Springer, New York (1982)
129. Rozovskii, B.L.: On stochastic equations arising in filtering of Markov processes. Ph.D. Thes., Moscow State Univ., (1972) [in Russian]
130. Rozovskii, B.L.: The Ito-Wentzell formula. Mosc. Univ. Math. Bull. **28**(1), 22–26 (1973)
131. Rozovskii, B.L.: On stochastic partial differential equations. Math. USSR Sb. **25**, 295–322 (1975)

132. Rozovskii, B.L.: On Itô equations in Hilbert spaces. In: Proceedings of the Second International Vilnius Conference in Probability and Mathematical Statistics, Vilnius, vol. 3, pp. 196–197 (1977)
133. Rozovskii, B.L.: On conditional distributions of degenerate diffusion processes. Theory Probab. Appl. **25**(1), 147–151 (1980)
134. Rozovskii, B.L.: Liouville equations for a diffusion Markov process. In: XIV All Union School in Probability and Statistics Bakuriani, 1980: Proceedings, pp. 26–28. Mecnieraba, Tbilisi (1980) [in Russian]
135. Rozovskii, B.L.: Backward diffusion. In: Proceedings of the Third International Vilnius Conference in Probability and Mathematical Statistics, Vilnius, vol. 3, pp. 291–292 (1981)
136. Rozovskii, B.L.: Stochastic Evolution Systems. Kluwer, Dordrecht (1990)
137. Rozovskii, B.L., Shimizu, A.: Smoothness of solutions of stochastic evolution equations and the existence of a filtering transition density. Nagoya Math. J. **84**, 195–208 (1981)
138. Rozovskii, B.L., Shiryaev, A.N.: Reduced form of non-linear filtering. In: Suppl. to Prepr. of IFAC Symp. on Stochastic Control, Budapest, pp. 59–61 (1974)
139. Rytov, S.M., Kravtsov, Y.A., Tatarskij, V.I.: Introduction to Statistical Radiophysics Part II, Random Fields. Nauka, Moscow (1978) [in Russian]
140. Shigekawa, I.: Derivatives of Wiener functionals and absolute continuity of induced measures. J. Math. Kyoto Univ. **20**, 263–289 (1980)
141. Shimizu, A.: Construction of a solution of a certain evolution equation. Nagoya Math. J. **66**(10), 23–36 (1977)
142. Shimizu, A.: Construction of a solution of a certain evolution equation. II. Nagoya Math. J. **71**(1) 181–198 (1978)
143. Shubin, M.A.: Pseudodifferential Operators and Spectral Theory. Springer, Berlin (1987)
144. Simon, B.: The $P(\varphi)_2$ Euclidean (Quantum) Field Theory. Princeton University Press, Princeton NJ (1974)
145. Skorohod, A.V.: Operator stochastic differential equations and stochastic semigroups. Russ. Math. Surv. **37**(6), 177–204 (1982)
146. Skorohod, A.V.: Random Linear Operators. D. Reidel Pub. Co., Dordrecht (1984)
147. Sobolev, S.L.: Applications of Functional Analysis in Mathematical Physics. American Mathematical Society, Providence, RI (1963)
148. Stroock, D.W.: The Malliavin calculus and its applications. In: Stochastic Integrals. Lecture Notes in Mathematics. Springer, Berlin, vol. 851, pp. 394–432 (1981)
149. Stroock, D.W.: The Malliavin calculus and its applications to second order parabolic differential equations. Math. Syst. Theory **14**, 25–65 (1981)
150. Stroock, D.W.: The Malliavin calculus, a functional analytic approach. J. Funct. Anal. **44**, 212–257 (1981)
151. Stroock, D.W.: Some applications of stochastic calculus to partial differential equations. Lecture Notes in Mathematics, vol. 976, pp. 268–382. Springer, Berlin (1983)
152. Stroock, D.W., Varadhan, S.R.S.: Multidimensional Diffusion Processes. Springer, New York (1979)
153. Ventcel, A.D.: On equations in the theory of conditional Markov processes. Theory Probab. Appl. **10**(2), 357–361 (1965)
154. Verentennikov, A.Y.: A probabilistic approach to hypoellipticity. Russ. Math. Surv. **38**, 127–140 (1983)
155. Veretennikov, A.Y.: Probabilistic problems in the theory of hypoellipticity. Math. USSR Izv. **25**(3), 455–473 (1985)
156. Veršik, A.M., Ladyzenskaja, O.A.: The evolution of measures as determined by Navier-Stokes equations, and the solvability of the Cauchy problem for the statistical Hopf equation. Sov. Math. Dokl. **17**(1), 23–25 (1976)
157. Viot, M.: Solutions faibles d'équations aux dérivées partielles stochastiques non linéaires: Thes. doct. sci. Univ. Pierre et Marie Curie. P. (1976)
158. Vishik, M.I., Fursikov, A.V.: Mathematical Problems of Statistical Hydromechanics. Kluwer, Dordrecht (1988)

159. Višik, M.I., Komeč, A.I.: The Navier-Stokes stochastic system and corresponding Kolmogorov equations. Sov. Math. Dokl. **23**(2), 444–447 (1981)
160. Višik, M.I., Komeč, A.I.: Generalized solutions of the inverse Kolmogorov equation that corresponds to the stochastic Navier-Stokes system. In: Proceedings of Petrovski Seminar, vol. 8, pp. 86–119. Moscow State Univ. Pub., Moscow (1982) [in Russian]
161. Weinan, E., Vanden Eijden, E.: Generalized flows, intrinsic stochasticity, and turbulent transport. Proc. Natl. Acad. Sci. **97**(15), 8200–8205 (2000)
162. Wiener, N.: Differential space. J. Math. Phys. **2**, 131–174 (1923)
163. Wiener, N.: Extrapolation, Interpolation and Smoothing of Stationary Time Series. Wiley, New York (1949)
164. Wong, E.: Explicit solutions to a class of nonlinear filtering problems. Stochastics **5**(4), 311–321 (1981)
165. Yosida, K.: Functional Analysis. Springer, Berlin (1965)
166. Zakai, M.: On the optimal filtering of diffusion processes. Z. Wahrsch. **11**, 230–243 (1969)
167. Zakai, M.: The Malliavin calculus. Acta Appl. Math. **3**, 175–207 (1985)

Index

B. L. Rozovsky, S. V. Lototsky, *Stochastic Evolution Systems*, Probability Theory
and Stochastic Modelling 89, https://doi.org/10.1007/978-3-319-94893-5